THE CROMBERGERS OF SEVILLE

The History of a
Printing and Merchant Dynasty

THE CROMBERGERS OF SEVILLE

THE HISTORY OF A PRINTING AND MERCHANT DYNASTY

CLIVE GRIFFIN

Fellow and Tutor in Spanish,
Trinity College, Oxford

CLARENDON PRESS · OXFORD
1988

Oxford University Press, Walton Street, Oxford OX2 6DP
Oxford New York Toronto
Delhi Bombay Calcutta Madras Karachi
Petaling Jaya Singapore Hong Kong Tokyo
Nairobi Dar es Salaam Cape Town
Melbourne Auckland
and associated companies in
Berlin Ibadan

Oxford is a trade mark of Oxford University Press

Published in the United States
by Oxford University Press, New York

British Library Cataloguing in Publication Data
Griffin, Clive
The Crombergers of Seville: the history
of a printing and merchant dynasty.
1. Cromberger (Family) 2. Printing—
Spain—Seville—History
I. Title
686.2'092'2 Z232. C9/
ISBN 0–19–815831–9

Library of Congress Cataloging in Publication Data
Griffin , Clive.
The Crombergers of Seville.
Bibliography: p.
Includes index.
1. Cromberger family. 2. Printing—Spain—Seville—
History—16th century. 3. Seville (Spain)—Commerce—
History—16th century. 4. Seville (Spain)—Imprints.
5. Bibliography—Early printed books—16th century.
6. Printers—Spain—Biography. 7. Merchants—Spain—
Biography. I. Title.
Z232.C93G74 1987 686.2'0946'86 86–33251
ISBN 0–19–815831–9

Set by Litho Link Limited
Printed in Great Britain
at the University Printing House, Oxford
by David Stanford
Printer to the University

PREFACE

THIS book is the first full-length study of a major printing-office in sixteenth-century Spain. Although we have detailed accounts of the activities of Aldus Manutius at Venice, Christopher Plantin in the Low Countries, and the Estiennes in France, there has been little published about individual Spanish presses operating during the period when those other major figures were printing their books and when Spain was itself a world power. The Crombergers' office is particularly important because it dominated printing in Seville at a time when the city was the largest centre of population in Castile, the hub of economic expansion through the opening up of the Indies, the focus of spiritual reform, and the centre of Castilian book-production. The Crombergers' output of books during the first half of the century was both large and influential, many important literary and spiritual works first appearing under their imprint. Yet their press has received less attention from scholars than it certainly deserves. Spanish printing and the Crombergers' role in it during the early years of the sixteenth century have not, however, been entirely ignored. The period from 1501 to 1520 was studied by the late F. J. Norton in two splendid books, but no scholar has carried this pioneering work into the 1520s or beyond, and even Norton's account of the Crombergers' early activities is necessarily brief, given the wide scope of his history of Spanish printing.[1]

The present study is thus intended to fill an important gap in our knowledge of the history of printing in Spain. It is based both on documents, most of which are unpublished, and on a close scrutiny of books printed by the Crombergers. Information from these two sources has been combined and set in a wider historical context to present a full account of this press. The approach adopted draws upon both that commonly associated with the French *historiens du livre* (with their emphasis on the role of the book in society, the cultural context of printing, and the book as a commodity) and that taken by descriptive bibliographers working in a more Anglo-Saxon tradition.[2]

[1] F. J. Norton, *Printing in Spain 1501–1520* (Cambridge, 1966); id., *A Descriptive Catalogue of Printing in Spain and Portugal 1501–1520* (Cambridge, 1978). These two books have provided the major inspiration for this study.

[2] For a succinct comparison of the French and Anglo-Saxon approaches see Kenneth E. Carpenter (ed.), *Books and Society in History: Papers of the Association of College and Research Libraries, Rare Books and Manuscripts Preconference, 24–28 June, 1980, Boston, Massachusetts* (New York and London, 1983), pp. vii-ix. For a masterly example of the methods of the *historiens du livre* see Henri-Jean Martin and Roger Chartier (eds.), *Le Livre conquérant: du Moyen Âge au milieu du XVIIe siècle* (Paris, 1983); vol. i of eid. (eds.), *Histoire de l'édition française* (Paris, 1983–).

Despite the small number of studies of early printers in Spain, there is no lack of evidence about the presses, for Spain is rich in notarial archives. One such archive, the Archivo de Protocolos in Seville, proved the most useful source of documents for a study of the Crombergers. Some documents from the Archivo de Protocolos had already been published by two local scholars, Gestoso y Pérez and Hazañas y la Rúa, and their collections were an invaluable starting-point.[3] Neither author, however, had discovered more than a fraction of the documents contained in the Archivo de Protocolos which are concerned with printing; both died before their researches were completed; and neither attempted to write an account of an individual Seville press or a history of printing in the city as a whole. Given the litigious nature of Spaniards in the sixteenth century, the Archivo de Protocolos is a veritable goldmine of information on many aspects of life in Seville: economics, trade, industry, political and racial problems, institutions, the arts, and even contemporary *mores*—the frequent documents entitled 'perdón de virgo' and 'perdón de cuernos' (pardons purchased from the aggrieved parties by seducers of maidens or by cuckolders), for example, present a view of marital and sexual relations which might cause raised eyebrows among those brought up on an undiluted fare of a certain sort of Spanish Golden-Age literature.

The Crombergers appeared before a notary at least three times a week (including Sundays and, on occasion, Christmas Day). They accordingly left a mass of documents: wills, dowry-agreements, contracts for the sale and purchase of land and property, shipping manifests, inventories of goods auctioned or left in the estate of a deceased member of the family, powers of attorney, contracts to print, and even records of gifts from one member of the family to another. Their account books and correspondence unfortunately do not survive, and there are no hard data available about their beliefs or personalities, for the documents held in the Archivo de Protocolos are strictly notarial. The researcher therefore has to read between the lines of these documents if he is to build up a picture of Seville's leading printers. And the Archivo de Protocolos is not without shortcomings of other kinds.

It is composed of the in-house copies of documents drawn up in twenty-four notarial offices ('oficios', or 'escribanías')—one for each of the twenty-four notaries public who were allowed to operate in the city; it is arranged by notary, and within each office, by year. The lamentable state of the building in which the documents are now housed has resulted in the complete records of some offices having been virtually destroyed by damp. Some whole books ('libros') and many individual documents, were removed by early scholars; papers have been transferred willy-nilly from one book to another (the books are frequently unbound); and some books have, for obscure reasons, been hidden in inaccessible

[3] See the works listed in 'Abbreviations' below.

parts of the building by irresponsible researchers. Nevertheless, the archive is at least open to scholars—something which was not always the case. At a very rough estimate, some five million folios survive for the sixteenth century alone. The sheer quantity of the documentation is daunting. There is no index, apart from a useful list of papers concerning the Indies, a catalogue of documents involving Jews or Moors, and the odd indexes made by individual notaries covering a few years of their papers.[4] Pike and Hoffman imply that investigation in the archive is somewhat facilitated by the fact that clients generally employed a single notary or small group of notaries whose offices were located near the client's residence.[5] This is untrue of the Crombergers, at least, who used the services of seventeen notarial offices scattered all over the city. Occasionally one document contains a reference to another drawn up elsewhere which can then be tracked down with relative ease, but such cross-references are rare. The researcher is therefore obliged to examine all the books in all the offices for all the years in which he is interested. While this is less of an obstacle for certain sorts of investigation, it makes research particularly tiresome for somebody interested in an individual or a family. It is therefore scarcely surprising that many scholars have been reluctant to use this archive. Nevertheless, Klaus Wagner's recent study of the minor Seville printer, Martín de Montesdoca, shows how productive a systematic scrutiny of this archive can be, and the present book provides further proof of the importance of such sources—something long known to historians of printing in France.[6]

Other archives containing useful information on the Crombergers in particular or on contemporary printing in general are the Archivo General de Indias and the Archivo de la Catedral, both in Seville, the Archivo Histórico Nacional in Madrid, and the Archivo General de la Nación in Mexico City. The first was used to investigate the family's interests in the Indies. Numerous files ('legajos') were consulted, particularly in the appropriate section, *Contratación*, but also in *Justicia* and *Indiferente General*. The chapter minutes *(Actas Capitulares)* held in the Archivo de la Catedral provided evidence about the printing of liturgical editions. Some information on printing in general was encountered in documents from the Archivo Histórico Nacional, while the Archivo General de la Nación was found

[4] *Catálogo de los fondos americanos del Archivo de Protocolos de Sevilla*, Colección de documentos inéditos para la historia de Hispano-América, 10, 11, 14, etc., 5 vols. (Madrid, Barcelona, Buenos Aires, 1930–7); and Klaus Wagner, *Regesto de documentos del Archivo de Protocolos de Sevilla referentes a judíos y moros* (Seville, 1978).

[5] Ruth Pike, *Aristocrats and Traders: Sevillian Society in the Sixteenth Century* (Ithaca and London, 1972), p. 95; and Paul E. Hoffman, 'The Archivo de Protocolos de Sevilla', *Itinerario: Leyden Centre for the History of European Expansion*, 5 (1981), 39–45 (p. 41).

[6] Klaus Wagner, *Martín de Montesdoca y su prensa: contribución al estudio de la imprenta y de la bibliografía sevillanas del siglo XVI* (Seville, 1982). Cristóbal Pérez Pastor had, of course, used legal archives extensively for his studies of printing in several Castilian cities (see Daniel Eisenberg, ' "Una uña de la gran bestia" ', *Journal of Hispanic Philology*, 8 [1984], 93–8).

to contain a good deal of evidence of the Crombergers' commercial activities in the New World.

As it is hoped that this book will be of interest to readers who are not specialists in things Spanish, all quotations from documents and printed sources cited in the text and footnotes to the text (but not in the Appendices, which are more specialized) have been translated into modern English. The titles of books have, however, been left in the original language. Titles have generally been regularized and modernized except when a particular edition is being referred to; in this case, the title is given in the form used on the title-page of that edition. In the index of editions, Spanish titles are modernized.

This study is divided into two parts, a division which reflects the approach noted above. Part One is a history of the Crombergers and their press in which their activities as printers are set in the context of their wider business interests. No account of a press is complete if only bibliographical evidence is considered; this part is therefore largely based on archival sources. It is hoped that the story of the rise and fall of a family of artisans-cum-merchants working at a critical period of Seville's development will be of interest to historians as well as to students of early printing.

A printer is, however, of particular interest to us because his products were books, and a study of these products can provide evidence for the cultural and spiritual atmosphere of an age. Part Two therefore focuses on the Crombergers' editions. It contains both a survey of the press's output—with its attendant insight into reading-habits in the first half of the sixteenth century, something which is of relevance to students of Spanish letters—and a more detailed study of the printers' material, which should be of interest to bibliographers.

An indispensable part of any study of a printing-house is a reliable list of the editions which came from its presses and a list of the libraries where copies can be consulted. Most of the editions printed by the Crombergers of which copies survive have been examined personally in England, Spain, Portugal, Mexico, and Cuba. When a copy has not been accessible, every effort has been made to examine a microfilm copy, photographs, or photocopies. Of the remaining books which I have not seen personally, some have been described for me by patient librarians or colleagues, while descriptions of the others have on occasion been taken from reliable secondary sources. Detailed descriptions are provided in Appendix One. Appendices Two, Three, and Four are illustrative indexes of the types, woodcuts, ornamental material, and ornamental initials used by the Crombergers. However useful to bibliographers, librarians, and researchers these appendices prove, they are likely to be of only limited interest to most readers. Moreover, the cost of setting up the bibliographical descriptions in type and the reproduction of the accompanying illustrative material would, at best, have made this book inordinately expensive and, at worst, unpublishable. The four appendices have therefore been reproduced from typescript in the form of microfiches which are to be found at the end of the printed section of this study. Although the book can be read without consulting the microfiches, references to individual types, descriptions, and blocks reproduced there have been included in the text, especially in the more technical sections of Chapters Seven and Eight. It is hoped that any inconvenience caused to the scholar by having to use a microfiche-reader will be compensated by the provision of a large amount of detailed material which would otherwise have been suppressed.

At the end of the printed section an index of Cromberger editions is provided for easy consultation. The items are numbered, and these numbers in brackets refer to the detailed descriptions contained in Appendix One. The pages in the microfiche section are numbered in a new sequence, each page number being preceded by an asterisk.

Many printers are mentioned in the text, notes, and appendices. The form of the early printers' names presents some problems. I have employed the forms most commonly encountered while recognizing that this involves some inconsistency (thus 'Aldus Manutius' but 'Josse Bade'; 'Magnus Herbst', but 'Jacobo Cromberger').

My debts to those who have helped me with this book are enormous, but none greater than to two men whom it is a pleasure to thank in print. Professor Peter Russell supervised the doctoral thesis on which the present study is based. I have benefited not only from his wisdom and erudition, his patience and encouragement during periods when other duties prevented me from pursuing my research, and his willingness to set aside his own work and read with minute attention long drafts when they were eventually written, but also from the example of his own research and publications, which have had such a profound effect on several generations of scholars.

Dr Klaus Wagner, who is the leading expert on the history of printing in Seville and whose works on the subject are models of their kind, generously offered me his assistance and friendship during my frequent stays in the city. He guided my first attempts to study in the Archivo de Protocolos and the Biblioteca Colombina, disinterestedly shared with me the fruits of his own long years of researches in Spanish archives, and extended to me the hospitality of his own home in Seville.

Archivists and librarians from almost all the institutions mentioned in this study have been patient and helpful far beyond the call of duty. I am particularly beholden to Rafael Lago Martínez of the Archivo de Protocolos, Seville; John Wainwright and the staff of the Taylor Institution Library, Oxford; David Rogers of the Bodleian; Dennis Rhodes of the British Library (formerly the British

Museum), London; Jeanne Veyrin-Forrer and her colleagues in the Réserve of the Département des Imprimés of the Bibliothèque Nationale, Paris; Isabel Cepeda of the Biblioteca Nacional, Lisbon; Drs Pointner and Schneiders of the Bayerische Staatsbibliothek, Munich; Pierrette Limacher of the Bibliothèque de la Sorbonne, Paris; María Dolores Vives, librarian of D. Bartolomé March Servera's private collection in Madrid; María Brey de Rodríguez-Moñino, Madrid; and Amadeu-J. Soberanas i Lleó of the Biblioteca de Catalunya, Barcelona.

Innumerable scholars, colleagues, and friends have provided me with information and advice. At the risk of ungrateful omissions, I should like to thank in particular Ian Archer, Benjamin Arnold, Peter Bakewell, Theodore Beardsley Jr., Roberto Berenstein, Robin Briggs, Peter Carey, Ivo Castro, Don Cruickshank, Jack Gibbs, José Luis Gotor, Leofranc Holford-Strevens, Oldřich Kašpar, Gordon Kinder, Jeremy Lawrance, Francis Lamport, Michael Maclagan, David Mackenzie, Ian Michael, Jaime Moll, F. J. Norton, Antonio Odriozola, Francisco de la Peña, José Simón Díaz, Ron Truman, and Jesús Yhmoff Cabrera.

Research abroad can be a lonely business. I owe a special debt of gratitude to José Herrera Sánchez, Matilde Mármol Otero, Ángeles Pérez Esteban, Rosario Pérez Esteban, and María del Carmen Esteban Ocaña in Seville; Chris Pratt, Darrell Williams, and Peter Shackleford in Madrid; Jim Amelang in Barcelona; Laura Salinas, Jorge Gómez, and George Foulkes in Mexico City; and Jorge Rodríguez Iglesias in Havana, for their repeated hospitality and good company over many years.

My brother, Nigel, and my wife, Sue, both of whom read and commented judiciously on the manuscript of this study, have been a continual source of inspiration. The former has always been generous with his time, scholarship, and inexhaustible energy. The latter ceaselessly encouraged me to complete this project, tolerated my prolonged absences abroad, and provided me with an atmosphere of absolute calm in which to work. Without her this book would not have been written.

Oxford, July 1985

TO MY FIRST AND BEST TEACHERS
JACK AND SHELAGH GRIFFIN

ACKNOWLEDGEMENTS

PERMISSION has kindly been given by the following publishers, libraries, institutions, and individuals for the reproduction of types and blocks contained in Appendices Two, Three, and Four.

In Austria: Österreichische Nationalbibliothek, Vienna. In Britain: Cambridge University Press for permission to reproduce a plate from F. J. Norton, *Printing in Spain 1501–1520*; Maggs Bros. Ltd.; Methuen & Co. Ltd. for permission to reproduce a plate from Oscar Jennings, *Early Woodcut Initials*; British Library, London; John Rylands Library, Manchester; the Master and Fellows of Balliol College, Oxford; F. J. Norton. In France: Bibliothèque Nationale, Paris; Bibliothèque de la Sorbonne, Paris. In Italy: Archivio Storico Civico e Biblioteca Trivulziana, Milan; Biblioteche Riunite Civica e Negroni, Novara. In Mexico: Fondo de Cultura Económica. In Portugal: Biblioteca Central da Faculdade de Medicina, Coimbra; Biblioteca Geral da Universidade, Coimbra; Biblioteca Pública e Arquivo Distrital, Évora; Biblioteca da Ajuda, Lisbon; Biblioteca Nacional, Lisbon. In Spain: Biblioteca de Catalunya, Barcelona; Biblioteca Universitaria, Granada; Biblioteca Nacional, Madrid; Biblioteca Universitaria, Seville. In the United States: Boston Athenaeum; Houghton Library, Harvard; University Library, Michigan.

I am beholden to the Ilustre Colegio Notarial de Sevilla for allowing me access to its Archivo de Protocolos.

I am grateful to the following bodies for research and travel grants: the Curators of the Taylor Institution, Oxford; the Board of the Faculty of Medieval and Modern Languages and Literature, Oxford; the Ford Foundation; the Inter-Faculty Committee for Latin-American Studies, Oxford; the Humanities Research Centre of the Australian National University, Canberra; and my society, Trinity College, Oxford.

CONTENTS

LIST OF FIGURES AND TABLES

ABBREVIATIONS

AHNM	=	Archivo Histórico Nacional, Madrid
AGNMex.	=	Archivo General de la Nación, Mexico City
ACS	=	Archivo de la Catedral, Seville
AGIS	=	Archivo General de Indias, Seville
APS	=	Archivo de Protocolos, Seville
Gestoso	=	José Gestoso y Pérez, *Noticias inéditas de impresores sevillanos* (Seville, 1924)
Hazañas	=	Joaquín Hazañas y la Rúa, *La imprenta en Sevilla: noticias inéditas de sus impresores desde la introducción del arte tipográfico en esta ciudad hasta el siglo XIX*, 2 vols. (Seville, 1945–9)
mr(s)	=	*maravedí(s)*

TABLE 1.* *Money*

The following denominations used in Spain are referred to:

maravedí (*mr*)	unit of account
silver *real*	equivalent to 34 *mrs*
gold *ducado*	equivalent to 375 *mrs* for most of the period (after 1537 it became a unit of account; its disappearance as a coin led to its replacement by the gold *escudo* which was initially equivalent to 350 *mrs*)
gold *peso*	equivalent to 450 *mrs* when a *peso de minas* or *peso de ley perfecta*. The fineness varied so that a *peso de oro de tipuzque*, for example, was equivalent to only 272 *mrs*, while a *peso corriente* was worth approximately 300 *mrs*. In most documents the fineness of the *pesos* is not specified
silver *marco*	equivalent to *c.* 2,200 *mrs*

* This table is based on information contained in John Lynch, *Spain under the Habsburgs*, 2nd edn., 2 vols. (Oxford, 1981), i. 369; Arthur Scott Aiton, *Antonio de Mendoza: First Viceroy of New Spain* (New York, 1927), p. 114; Clarence Henry Haring, *Trade and Navigation between Spain and the Indies in the Time of the Habsburgs* (Cambridge, Mass., 1918), p. 175; id., *The Spanish Empire in America*, 3rd edn. (New York and London, 1963), pp. 286–9.

TABLE II.* *Prices*

(All figures are in *mrs*)

fanega (dry measure) = 55.5 litres; *libra* = 0.46 kg.; *libra carnicera* = 0.92 kg.; *arroba* = 11.5 kg.; *quintal* = 46 kg. (these weights and measures are applicable to Castile).

	1500–10	1510–20	1520–30	1530–40	1540–50	1550–60
Beans *(fanega)*	115	113	—	237	277	—
Beef *(libra carnicera)*	7	7	—	11	15	15
Tallow candles *(libra)*	8	10	—	14	13	25
Chick-peas *(fanega)*	227	144	—	249	355	478
Dried codfish *(lib. carn.)*	7	3	—	8	12	—
Olive oil *(arroba)*	93	113	—	187	215	223
Salt pork *(lib. carn.)*	11	12	—	23	40	—
Sugar *(arroba)*	310	483	—	698	938	937
Wine *(arroba)*	41	48	—	79	115	115
Chicken (one)	—	—	—	—	20	35
Soap *(arroba)*	—	—	—	—	—	196
Slave (mature black male)	10,000	11,000	11,000	—	—	—
Slave (mature Moorish male)	15,000	14,000	13,000	—	—	—
Horse	—	—	—	—	6,000	—
Mule	—	—	12,000	—	—	—
Paper (sheet)	—	—	—	—	1	—
Type-metal *(quintal)*	—	—	—	—	—	3,112
Egg (one)	—	—	—	—	2	—
Sardines (100)	—	—	—	—	12	—
Bread (loaf)	—	—	—	—	2.5	—
Shoes (pair)	—	—	—	—	57	—
Hose (pair)	—	—	—	—	40	—
Scribe's charges:						
will & codicil	—	—	—	—	680	—
power of attorney	—	—	—	—	102	—
transcript	—	—	—	—	40	40
Annual cost of keeping a student from Seville at Salamanca	—	—	17,000	—	—	—

* The figures above the line are mainly based upon E. J. Hamilton, *American Treasure and the Price Revolution in Spain 1501–1650* (Cambridge, Mass., 1934), Appendix III. His figures have been averaged over ten-year periods. His prices are for Andalusia as a whole, but his sources are largely from Seville. The figures below the line come from my own research in Seville archives as well as from Cristóbal Espejo, 'La carestía de la vida en el siglo XVI', *Revista de archivos, bibliotecas y museos*, 41 (1920), 36–54, 170–204, 329–54; Pike, *Aristocrats and Traders*, p. 149; Alfonso Franco Silva, *La esclavitud en Sevilla y su tierra a fines de la Edad Media* (Seville, 1979); Klaus Wagner, 'Apuntes para el coste de vida en Sevilla: agosto 1544—febrero 1545', *Archivo hispalense*, 2nd ser., 55, No. 170 (1972), 119–30; Klaus Wagner, '¿Qué costaron los estudios universitarios en Salamanca a principios de siglo XVI?', *Archivo hispalense*, 2nd ser., 59, No. 180 (1976), 149–51.

TABLE III.* *Wages*

(All figures are in *mrs*)

	1500–10	1510–20	1520–30	1530–40	1540–50	1550–60
Book-keeper (per year)	—	1,000	1,000	1,000	1,500	—
Cart-loader (per day)	26	33	34	34	43	—
Female cook (per year)	1,224	—	1,632	1,632	2,040	2,040
Labourer (per day)	27	30	34	32	37	39
Maidservant (per year)	1,224	1,224	1,632	1,632	1,958	—
Pressman (per month)	—	1,219	—	—	—	—
'Naipero' (per month)	—	—	1,125	—	—	—
Puller in press (per month)	—	—	938	—	—	—
Apprentice pressman (per month)	—	—	—	—	—	375
Skilled slave (per month)	200–300	—	—	—	—	—
Female slave, as maid (when paid; per week)	78	—	—	—	—	—
Doctor (per visit)	—	—	—	—	136	—
Barber-surgeon:						
bloodletting	—	—	—	—	68	—
tooth extraction	—	—	—	—	34	—
shave	—	—	—	—	34	—
On board ship:						
Scrivener (per month)	—	—	—	—	1,875	3,350
Sailor (per month)	938	—	—	—	938	1,050
Doctor (per month)	—	—	—	—	1,500	3,750
Pilot (per month)	1,500	—	—	—	2,625	2,625
Captain (per year)	30,000	—	—	—	74,813	—
General of Fleet (per year)	49,875	—	—	—	—	825,000

* The figures above the line are based on information contained in E. J. Hamilton, *American Treasure*, Appendix vii, and are for New and Old Castile. The figures below it are from my own researches in Seville archives as well as from C. Haring, *Trade and Navigation*, Appendix ix and p. 317, and Franco Silva, *La esclavitud en Sevilla*. It should be remembered that labourers received a meal as part of their wages, while workers employed on a yearly basis normally received free board and lodging.

PART ONE

The Crombergers and their Press

CHAPTER I

INTRODUCTION: PRINTING IN SPAIN AND SEVILLE

Printing in Fifteenth-century Spain

THE first place of printing in Spain was probably Segovia where, in about 1472, Johannes Parix of Heidelberg printed an undated *Sinodal de Segovia*.[1] It had taken some twenty years for the craft to reach the Peninsula from the Rhine Valley, but within the next two decades all the major centres of population in Spain, with the exception of Moorish Granada, had their own presses. Most of the early printers were Germans or Central Europeans and, in the case of those few sites where native Spaniards had been the pioneers of the craft, such as Seville, Toledo, and a number of smaller towns, native printers had within a few years ceased to work and were replaced either by immigrants or by nobody at all.[2]

[1] Historical and bibliographical studies of printing in Spain in the incunable age are more plentiful than for later periods. For the 15th c. there are, in particular, the works of Konrad Haebler and Francisco Vindel: Haebler, *Bibliografía ibérica del siglo XV: enumeración de todos los libros impresos en España y Portugal hasta el año de 1500*, 2 vols. (The Hague and Leipzig, 1904–17); *The Early Printers of Spain and Portugal*, Bibliographical Society's Monographs, 4 (London, 1897); Vindel, *El arte tipográfico en España durante el siglo XV*, 9 vols. and an appendix (Madrid, 1945–54). For the first books printed in Spain see Antonio Odriozola, 'Los protoincunables (1472–1479) impresos por Juan Parix en Segovia (España) y Toulouse (Francia) con tres figuras', *Gutenberg-Jahrbuch* (1976), 130–7; id. 'Los libros impresos por Juan Parix en Segovia y Toulouse y los atribuibles a Turner y Parix en esta última ciudad (1472–1478) (una investigación sobre protoincunables)', in J. M. Ruiz Asencio *et al.*, *Homenaje a D. Agustín Millares Carlo*, 2 vols. (Las Palmas, 1975), i. 281–308 and 18 plates; and Laurence Witten, 'The Earliest Books Printed in Spain', *Papers of the Bibliographical Society of America*, 53 (1959), 91–113.

[2] The first presses in 15th-c. Spanish towns were: *1472?*: Segovia: Johannes Parix of Heidelberg; *1473?*: Barcelona: Henricus Botel of Embich, Georgius vom Holtz of Hoeltingen, and Johannes Plank of Halle; *1473?*: Valencia: Lambert Palmart of Cologne; *1475*: Zaragoza: Matheus Flander; *1477*: Seville: Alfonso del Puerto, Bartolomé Segura, and Antonio Martínez (Spaniards); *1477*: Tortosa: Pedro Brun of Geneva and Nicolaus Spindeler of Zwickau; *1479*: Lérida: Henricus Botel; before *1481*: Montalbán: Juan de Lucena (Spaniard); *1481*: Salamanca: printer unknown; *1481?*: Valladolid: printer unknown; *1482*: Zamora: Antonio de Centenera (Spaniard); *1482*: Burgos: Fadrique de Basilea (Friedrich Biel?) (German); *1482*: Guadalajara: Salomo ben Moise Levi Alkabiz (Spaniard); *1483*: Gerona; printer unknown; *1483?*: Toledo: Juan Vázquez (Spaniard); *1484*: Tarragona: Nicolaus Spindeler; *1484*: Huete: Álvaro de Castro (Spaniard); *1484*: Murcia: Alfonso Fernández de Córdoba (Spaniard); *1485*: Majorca: Nicolaus Calafat (Majorcan); *1485*: Híjar: Elieser Alantansi (Spaniard); *1489*: Coria: Bartholomeus de Lila (Fleming); *1490?*: Pamplona: Arnao Guillén de Brocar (French); *1495?*: Mondoñedo: printer unknown; *1496*: Granada: Meinardus Ungut (German?) and Johannes Pegnitzer of Nuremberg; *1496*: Monterrey: Johann Gherlinc (German?); *1499*: Montserrat: Johann Luschner (German). This list is based on Leslie Sheppard and George Painter, 'Introduction to the Presses', in *Catalogue of Books Printed in the XVth Century*

Spain was not unusual in this, for the early printers almost everywhere in Europe either were Germans or, like William Caxton, had received their training in Germany.

From the outset printing was a business. The first printers in Valencia, for example, had not gone to that city in answer to a humanist scholar's invitation, but had been summoned by Jakob Vizlandt, a factor of the largest import–export firm of the fifteenth century, the Great Trading Company of Ravensburg.[3] In the west of the Peninsula, Valentim Fernandes of Moravia, the most notable printer of Portuguese incunables, worked as a broker and translator for the numerous German merchants resident in Lisbon.[4] Even the greatest printers of Europe had to forge a close relationship with capitalists, as is clearly demonstrated by Lowry's study of the business connections in Venice between Aldus Manutius, Andrea Torresani, and Pierfrancesco Barbarigo.[5] The substantial capital required to equip a printing-shop, the high price of paper, the slow and uncertain return on their investment, and particularly problems of distribution, meant that printers tended to establish their presses in important centres of commerce.[6] It was therefore only natural that the first printers who settled in Eastern Spain should set up their presses in Barcelona and Valencia, which were important commercially as well as culturally.

The early printers of Catalonia and Valencia did not share the humanist aspirations of an Aldus or a Josse Bade; they were artisans who printed one year in one city, the next in another. Their offices were small and reflected in a more marked fashion the history of early printing almost everywhere: a press would be set up, would produce a mere handful of books, and would then disappear. The hazardous nature of the business in Spain, as elsewhere, is revealed by the custom of syndicates of merchants or booksellers joining company to finance an edition, thus spreading the risk of loss.[7] The capitalists and merchants who backed the presses and knew how to distribute manufactured goods along the established trade routes treated books just as they did other merchandise, as a commodity to be produced and sold in order to realize a profit. Many of the

now in the British Museum [= *BMC*], 12 vols. (London, 1908–71), x, pp. xxxvi-lxxv, and differs in some details from Norton, *Printing in Spain*, p. 158.

[3] Jorge Rubió y Balaguer, 'Integración de los impresores alemanes en la vida social y económica de Cataluña y Valencia en los siglos XV—XVI', *Spanische Forschungen der Görresgesellschaft*, 1st ser., 20 (1962), 103–22 (pp. 104–5).

[4] Sheppard and Painter, 'Introduction to the Presses', p. lxxiv.

[5] Martin Lowry, *The World of Aldus Manutius: Business and Scholarship in Renaissance Venice* (Oxford, 1979), pp. 72–108.

[6] S. H. Steinberg, *Five Hundred Years of Printing*, 3rd edn. rev. James Moran (Harmondsworth, 1974), p. 46. David Rogers, 'Printing in 1478: The Background to the First Press at Oxford', *Journal of the Printing Historical Society*, 13 (1978–9), 67–77 (p. 75).

[7] Lowry, *The World of Aldus Manutius*, pp. 82–3; Klaus Wagner, 'El negocio de las *Siete Partidas*', *La Bibliofilia*, 78 (1976), 67–82.

printers in the East of Spain found themselves at the mercy of hard-headed financiers and suppliers as Rubió y Balaguer's picture of the early printers at Barcelona reveals:

Printers in Barcelona . . . were constantly beset by creditors, behind with their contractural obligations, and ever fearful of ending up in prison. On more than one occasion notarial documents tell us that the printers had been forced to pawn their punches and sell presses, types, and even their own beds.[8]

The scale of printing in Spain was modest even if some Spanish incunables are among the finest ever printed. Printers there could not hope to compete with the presses of France, Italy, and Germany, which mass-produced books for export. For example, at the turn of the century Venice boasted some 150 printing-shops, while in Spain it was exceptional even by 1520 for three or four offices to be operating in a single city.[9] There is clear evidence in Spain of a flourishing trade in imported editions, largely in the hands of foreign merchants and booksellers who dealt especially in what might be termed the 'international book'—editions of the classics, legal text-books, theology, and other academic works. In 1489, for instance, there are records of Venetian printed books being sold in various parts of Spain.[10] Since paper would normally have to be imported for these editions even if they were printed in Spain, for domestic production of paper was always inadequate, transport costs were the same as those incurred in the import of foreign books. As local printers had no international market themselves, they were unable to compete with the more prestigious editions of foreign presses.[11] From the earliest days Spanish printing was largely limited to books intended for national or frequently just local markets, but even these were not entirely supplied by Spanish presses. Liturgical books for Spanish dioceses were frequently printed abroad, while orders for other books might go to a cheaper or more efficient competitor in France or Italy even though they could have been printed in Spain.[12] Thus, for example, two editions of Alfonso the Wise's compilation of laws, the *Siete partidas*, had been printed at Seville in 1491. The first of them (Haebler

[8] Rubió, 'Integración', p. 112.

[9] Steinberg, *Five Hundred Years*, p. 73; Norton, *Printing in Spain*, p. 85. Spain and Portugal produced only about 3 per cent of all incunables in Europe (see Hendrik D. L. Vervliet, 'Printing in Fifteenth Century Antwerp', in Simon van Venlo, *Little Book of the Office or Service of the Mass: Boexken van der Officien ofte Dienst der Missen*, ed. Ludo Simons, 2 vols. [Antwerp, 1982], ii. 43–9 [p. 46], and Martin and Chartier, *Le Livre conquérant*, p. 174).

[10] Lucien Febvre and Henri-Jean Martin, *L'Apparition du livre* (Paris, 1958), p. 344.

[11] On similar reasons influencing Caxton's decision to print almost exclusively in English and on the collapse of contemporary presses in England which attempted to compete with continental printers in the production of the 'international book', see N. F. Blake, *Caxton and His World* (London, 1969), pp. 200–1, 211–12.

[12] Norton, *Printing in Spain*, p. 134. There are even cases of popular works of entertainment destined for the Spanish market being printed in France in the 15th c. (see Martin and Chartier, *Le Livre conquérant*, p. 210).

518) had been commissioned from the press of Meinardus Ungut (Meinardo Ungut) and Stanislaus Polonus (Stanislao Polono) by Guido de Labezaris, a Genoese bookseller resident in Seville, and Juan de Porras, who was later to become a printer in his own right at Salamanca. The second, signed two months later (Haebler 519), was the work of the other major Seville printers, the 'Compañeros alemanes', and had been commissioned by a Novarese bookseller from Toledo, Melchor Gorricio, and a colleague of his from Salamanca, Rodrigo de Escobar. When, at the end of the decade, a company which probably consisted of Labezaris, Porras, and another Genoese, Lázaro de Gazanis, decided to publish a further edition of the *Siete partidas*, the order went to the Venetian press of Luc'Antonio Giunti even though surviving members of the two Seville printing-offices were still in business.[13]

There was an awareness in Spain that the domestic printers had their limitations. When Salamanca University decided in 1506 to publish a monumental Latin edition of El Tostado, it automatically turned again to Venice to find a suitable printer, although a local one was considered up to printing a vernacular collection of the same author's works.[14] The humanist Gonzalo García de Santa María put a good face on things in his prologue to a bilingual edition of the widely-used school-text, the *Catón*, which Paulus Hurus printed at Zaragoza in 1493 or 1494 (Haebler 140). He praised the printer's expertise despite the poor quality of the paper available to him. But it turns out to be faint praise when one reads the revealing clause, 'At least it is certainly the case that *for a book printed in Spain* this edition is of high quality as regards spelling and accuracy' (emphasis added).[15] This all serves to make Arnao Guillén de Brocar's printing of the Polyglot Bible at Alcalá early in the sixteenth century an even more remarkable achievement.

However, the commissioning and importation of books from abroad was not a consequence of the invention of printing. Recent research indicates that there had long been a lively trade in foreign manuscripts in Castile which increased dramatically during the fifteenth century with the rise there of vernacular lay humanism. Italy, and in particular Florence, had supplied about one-third of all manuscript books to be found in Castile at this time, while large numbers of illuminated devotional works for lay readers, especially books of hours, were dispatched to Spain from Flanders from the middle of the fifteenth century onwards and continued to be imported long after printing had been established in the Peninsula. This is evidence of a significant demand for books in Spain before the arrival of the presses, and that this demand outstripped the capacity of the

[13] Klaus Wagner, 'El negocio', pp. 67–72.

[14] Norton, *Printing in Spain*, pp. 134–5.

[15] Gonzalo García de Santa María, *El Catón en latín y en romance (1493/94)*, facsimile, ed. A. Pérez Gómez, Incunables poéticos castellanos, 9 (Cieza, 1964), fol. a_2^b.

commercial *scriptoria* known to have flourished in Castile in the fifteenth century.[16] Painter draws similar conclusions for the final quarter of the fifteenth century from a study of Spanish incunables:

> the circumstances that in Spain folios fell not far short of quartos in numbers of editions, and moreover attained nearly double in leaf-numbers and therefore perhaps nearly quadruple in area of print, may well seem a sign of a healthy book-trade, in which the demand of readers and the capital resources of their suppliers sufficed not only for the catch crops of smaller works but for a substantial harvest of large volumes.[17]

Printing did not, then, as is often assumed, immediately create a demand. It supplied one which already existed. But the established pattern of the importation of certain sorts of books (now printed ones from Venice, Lyons, and Paris rather than manuscript ones from Florence, Bologna, and Bruges) continued as before.[18]

Spanish presses, run or commissioned as they were by businessmen with experience of the book-trade, did not attempt to reduplicate in a cheaper or more attractive form 'international editions' already on the market. Rather, they concentrated from the first on jobbing printing—indulgences, official forms, certificates, and vernicles—commissioned by the local secular or ecclesiastical authorities, as well as on the more ambitious liturgical books printed for the latter.[19] Much of the work they undertook on their own behalf or on commission from booksellers and merchants was similarly intended for local consumption: compilations of laws, school-books, almanacs, and devotional or popular literature in the vernacular.[20] This emphasis on printing for the vernacular market is clearly demonstrated by Painter's analysis of book production in fifteenth-century Spain: 54.3 per cent of surviving Spanish incunables were printed in the vernacular. This figure compares with 21 per cent for Italy, 24 per cent for Germany, and 35 per cent for France during the first fifty years of printing.[21]

However, although a demand existed for books printed in Spain, it would still have been strictly limited. Even in fifteenth-century Venice printers had found to their cost that markets were finite.[22] Far from having to be introduced to satisfy

[16] Jeremy Lawrance, 'Nuño de Guzmán: Life and Works' (D.Phil. thesis, Oxford, 1983), pp. 98–100. Lawrance suggests that a major reason for the importation of these manuscript books was the lack of sufficient exemplars in Spain (or at least in Castile) for the scribes to copy. See also id., 'The Spread of Lay Literacy in Late Medieval Castile', *Bulletin of Hispanic Studies*, 62 (1985), 79–94.

[17] George Painter, 'General Introduction' to *BMC*, x, pp. ix-xxxv (pp. ix-x).

[18] Febvre and Martin, *L'Apparition*, pp. 344–6.

[19] Steinberg, *Five Hundred Years*, p. 94, maintains that printers in Spain depended on ecclesiastical patronage more than anywhere else in Europe.

[20] Rubió, 'Integración', p. 122.

[21] Painter, 'General Introduction', p. xxvii; Harry Carter, *A View of Early Typography up to about 1600* (Oxford, 1969), p. 59. Febvre and Martin, *L'Apparition*, p. 388, claim that only 22 per cent of incunables printed in Europe as a whole were in the vernacular.

[22] Lowry, *The World of Aldus Manutius*, p. 15.

demand, printing was not considered indispensable in Spain even after its arrival there. Indeed, García de Santa María said of the art in 1494 'although it is not necessary, I cannot deny that it is beneficial'.[23] At first it does not seem to have stimulated demand but merely reproduced in a cheaper form books which were in the main already sought after in manuscript.[24] Printing does, however, seem to have reduced the price of books in Spain.[25] This might seem an obvious point to make, but evidence for the cost of manuscripts and printed material is not plentiful and is difficult to assess; we know, for instance, that some sorts of second-hand manuscripts were remarkably cheap in Spain even before the advent of printing.[26] Nevertheless, new luxury manuscripts were still expensive in the fifteenth century and few readers could afford them. On the other hand, although there is evidence that the purchase of some printed books involved a substantial investment—a book printed by Lambert Palmart and bought at Valencia in 1482 cost as much as a lion destined for the royal menagerie there—it is significant that García de Santa María noted in his will of 1519 that the advent of printing had considerably reduced the value of his library.[27] Indeed, in his brief history of printing, Jacobo Cromberger claimed that one of the merits of the craft was that it had reduced the cost of books, although it must be admitted that he laid even more emphasis on the fact that printing had solved the problem of the shortage of reading matter.[28]

Whether the increased availability of books had the effect of creating a new readership and encouraging literacy, at least in the early years, is open to serious doubt if the results of Berger's researches in Valencia are applicable to the rest of Spain. Printing did not markedly increase the body of readers in that city; even well into the sixteenth century the number of owners of books was scarcely affected. The same groups of readers just owned more volumes.[29] This provides

[23] García de Santa María, *El Catón*, fol. a_2^b.

[24] Rogers, 'Printing in 1478', p. 77; Blake, *Caxton*, p. 72.

[25] Vicente Beltrán de Heredia, OP, *Cartulario de la Universidad de Salamanca*, 6 vols. (Salamanca, 1970–2), ii. 43; José María Madurell Marimón, *Documentos para la historia de la imprenta y librería en Barcelona (1474–1533)*, with notes by Jorge Rubió y Balaguer (Barcelona, 1955), p. 29*, claims that books had decreased in value so much by 1485 at Vich that they were no longer accepted as surety against debts.

[26] Lawrance, 'Nuño de Guzmán', pp. 96–8.

[27] Luis Gil Fernández, *Panorama social del humanismo español (1500–1800)* (Madrid, 1981), pp. 666–8.

[28] Alfonso de la Torre, *Vision delectable de la philosophia* (Seville, Jacobo and Juan Cromberger, 1526), fol. 80r. The Seville printer, Alfonso del Puerto, had earlier stressed the increased availability rather than the cheapness of books brought about by printing (see the colophon to his edition of Diego de Valera's *Crónica de España abreviada* [Seville, 1482] reproduced in Vindel, *El arte tipográfico*, v. 51).

[29] Philippe Berger, 'La Lecture à Valence de 1474 à 1504', *Mélanges de la Casa de Velázquez*, 11 (1975), 99–118; id., 'La Lecture à Valence de 1474 à 1560 (évolution des comportements en fonction des milieux sociaux)', in Manuel C. Díaz y Díaz *et al.*, *Livre et lecture en Espagne et en France sous l'Ancien Régime: colloque de la Casa de Velázquez* (Paris, 1981), pp. 97–110.

a further indication that, although the demand for books had increased dramatically in fifteenth-century Spain before the arrival of printing, it was still restricted. Thus in the first decade of the sixteenth century there was a fall in the output of the Spanish presses, one reason for which may well have been over-production in the 1490s. When we consider that only some six hundred editions had been printed in that decade, an idea of the size of the market for books printed in Spain can be gained.[30] An indication that printers were unable to keep their presses fully employed with local commissions is the work they undertook for customers all over the country. At Seville Ungut and Polono printed a Toledo breviary in 1493 (Haebler 90) and a missal for the diocese of Jaén some six years later (Haebler 443[5]), while the 'Compañeros alemanes' printed a Toledo manual in 1494 (Haebler 397) and Alfonso Cámara's *Epitoma sive compilatio de sacramentis* for the Dean and Chapter of Cuenca in 1496 (Haebler 115). Similarly, in 1497 Paulus Hurus printed in Aragon a manual for the Castilian see of Burgos (Haebler 394).

Either through a desire to avoid excessive competition for the limited market available to the Spanish presses, or because the local market affected the choice of titles published, or because printers used whatever exemplars were available locally, there was some degree of specialization in certain centres, and this became more marked in the sixteenth century. Thus the press established at Salamanca in 1481 produced mainly the sort of Latin books demanded by the readership of a university city, while Fadrique de Basilea's press at Burgos was remarkable for its illustrated editions of popular literary works in the vernacular. It is therefore not surprising that Burgos developed in the sixteenth century into a centre for the printing of *pliegos sueltos*, or chap-books.[31]

Printing in Spain in the First Half of the Sixteenth Century

The history of Fadrique de Basilea shows that the division between the incunable age and the sixteenth century is merely a bibliographical convenience. He had practised his craft at Burgos for over thirty years when he disappeared from the records in 1517, thus exemplifying both the continuity between the two centuries and also the remarkable longevity of those fifteenth-century presses which had been able to establish themselves securely in Spanish cities. Joan Rosembach of Heidelberg, who had begun printing in Spain by 1492, continued to work at various sites in Catalonia for some thirty-eight years, while Arnao Guillén de

[30] Norton, *Printing in Spain*, p. 117. Estimated production figures for Spain are: 1472–90, fewer than 300 editions; 1491–1500, at least 600 editions; 1501–10, about 500 editions; 1511–20, about 800 editions.

[31] Sheppard and Painter, 'Introduction to the Presses', pp. lvii–lviii, lxiii; Norton, *Printing in Spain*, p. 61. For similar specialization in 15th-c. France see Martin and Chartier, *Le Livre conquérant*, pp. 177–93.

Brocar printed in Castile and Navarre for thirty-four years until his death in 1524. Some presses which were established in Spain at the turn of the century were similarly long-lived. The German Jorge Coci (Georg Koch?) worked at Zaragoza from 1499 until 1536; Joan Joffre of Briançon ran his presses at Valencia from 1498 until 1530; and books were issued by the Crombergers at Seville from 1504 until at least the mid-1550s.

The continuity between the two centuries is reflected in the titles produced (see Chapter Six), the material employed by the printers (see Chapters Seven and Eight), and, in some areas of Spain, by the organization of the trade. In Barcelona and Valencia printers remained in the sixteenth century as dependent upon the merchants and booksellers who supplied the capital for the purchase of paper and production of editions as they had been in the fifteenth. In Seville, on the other hand, although editions were financed by booksellers at least in the early years of the new century, it appears that some printers were better able to overcome this dependence upon capitalists and issued books on their own behalf. This was certainly the case with Jacobo Cromberger. Throughout the century, however, an important role in the book-trade was still played by foreigners. Just as the famous printer Anton Koberger of Nuremberg had sent his nephew to fifteenth-century Barcelona to open a bookshop for imported editions there, so in the following century Italian firms such as the Giunti and the Portonarii not only spread their tentacles into the increasingly important centre of commerce and printing at Lyons, but also sent members of their families to Spain as agents, some of whom became printers in their own right. It is not uncharacteristic of the trade in Spain that one of the most important book-merchants of Medina del Campo at the end of the century, Benoît Boyer, should have been an immigrant.

There is a view still held by some historians of printing that it is at best boorish and at worst downright perverse to take an interest in mean economic matters in a study of printing, for were printers not the prime movers or, at least, the disseminators of ideological, spiritual, and cultural currents, without whom the great upheavals of sixteenth-century Europe would not have been fuelled? Some students of the early presses emphasize the elevated concerns of the scholar-printer, in whose house the erudite and the noble rubbed shoulders with the craftsman while they debated a doubtful manuscript reading, corrected the proofs of a learned edition, or wittingly fostered a spiritual or political revolution without a care for the lowly concerns of profit-making or the ever-present demands of their creditors. However attractive such a picture, it is scarcely true of the great mass of printers.[32] Recent research into the history of even the great Aldus reveals that

[32] Elizabeth L. Eisenstein, *The Printing Press as an Agent of Change: Communications and Cultural Transformations in Early-Modern Europe*, 2 vols. (Cambridge, 1979), takes this elevated but misleading approach; see David Shaw's critical review in *The Library*, 6th ser., 3 (1981), 261–3. For a clear

he was unable to ignore commercial concerns as he tried to implement his scholarly programme of publishing the classics.[33] The ambitions of most printers, even of skilled and conscientious ones like Jacobo Cromberger, cannot have been very different from those of Thomas Platter of Basle, who bluntly stated in his autobiography: 'when I saw how Hervagius and printers like him ran a profitable business and enjoyed a handsome income despite putting in very little work, I realized that this was the job for me'.[34] This is not to say that all printers were exclusively concerned with profit. There were, particularly outside Spain, some printers of considerable erudition and scholarly ambitions, while in Seville some of Jacobo Cromberger's products—in particular his 1528 edition of Lucan's *Pharsalia*—may have been an indication of aspirations more elevated than was evident in most of his products. But the importance of the economics of printing should never be underestimated.

Possibly even more than elsewhere, printers in Spain had to be businessmen if they were to survive, especially if they had freed themselves from the publishers and had begun to issue editions themselves. Spanish printers of the early sixteenth century were surrounded by evidence of colleagues who had failed to establish their operation on a sound economic foundation and who had collapsed, especially during the first decade of the century. The famous printer of Catalonia, Johann Luschner, had crashed at Barcelona in 1507, while in Castile Hans Gysser's Salmantine press had disappeared two years later, doubtless as the result of a financial crisis, and he had become a mere employee in Porras's office.[35] There is evidence that printers had to engage in other commercial activities unless they were fortunate enough to corner the market in one of the few lucrative areas of jobbing printing. Brocar, for example, made his money not from the magnificent editions for which he is now remembered, but from the privilege which he enjoyed on the best-selling works of the grammarian, Antonio de Nebrija, and by his appointment as joint printer of the indulgences of the Santa Cruzada, for which purpose he established branch offices in Toledo and Valladolid, where the monasteries holding the privileges on these indulgences, San Pedro Mártir and Santa María del Prado, were located.[36] Juan Varela de Salamanca was not only given the same appointment at Toledo, but was a prosperous merchant in Seville as well as being a printer there. His career exemplifies the need felt by printers in Spain to diversify their interests; thus Miguel de Eguía, who took over Brocar's

examination of the predominance of business over artistic or ideological considerations among the early printers, see Rogers, 'Printing in 1478', pp. 67–77.

[33] Lowry, *The World of Aldus Manutius, passim*; Gil Fernández, *Panorama social*, pp. 574–5; Blake, *Caxton*, pp. 45, 83.

[34] Thomas Platter, *Autobiographie*, tr. Marie Helmer, Cahiers des Annales, 22 (Paris, 1964), p. 105.

[35] Norton, *Printing in Spain*, p. 26.

[36] Norton, *Printing in Spain*, p. 33.

presses and his privileges on indulgences and Nebrija's *Artes de gramática*, combined his activities as a printer with considerable involvement in trade in his native Estella.[37] Jacobo Cromberger and his heirs typify this diversification of interests by successful Spanish printers at the time: the family was able to marry its printing, publishing, and mercantile interests in both Spain and the Americas and thus minimize the risk of financial disaster.[38]

Few printers in Spain can, therefore, have had more than the most modest ambitions to emulate the scholarly presses of an Aldus or a Bade. Eguía's strident criticism in humanist vein of Spanish printing is frequently quoted:

> How accursed we are in Spain, where our printing offices ceaselessly pour out common, and sometimes even obscene doggerel, tasteless ditties, and works which are yet more worthless than these.[39]

But this should not deceive us into thinking him a Spanish Aldus whose sole interest was in bringing humanist learning to a benighted Spain. Before publishing an edition he would go to a group of scholars at Alcalá who told him which books were unavailable in Spain and, no doubt, which ones would sell. On one occasion their advice was that he should print Erasmus' works, and they were right: Eguía's Castilian edition of the *Enchiridion* which he issued in 1526 was a best-seller, as is witnessed by the scramble among other Spanish presses to reprint it.[40] But, despite his own apparent interest in Erasmus, Eguía was equally capable of printing a work attacking the Dutchman when there was money to be made from it.[41] Moreover, when he printed an anthology of Greek texts for the students of Alcalá, he realized that, however remote from the despised 'common doggerel' and however necessary for students who were at that time suffering from an acute shortage of Greek material in the university, the edition was unlikely to produce a handsome return on his investment and he therefore obliged the unfortunate editor to finance the whole edition himself.[42]

Eguía, like his predecessor Brocar, was largely printing for a captive market,

[37] José Goñi Gaztambide, 'El impresor Miguel de Eguía procesado por la Inquisición (c. 1495–1546)', *Hispania sacra*, 1 (1948), 35–88 (pp. 42, 58–66). On Brocar's and Eguía's privileges for the printing of indulgences, see Luis Fernández, SJ, 'La real imprenta del Monasterio de Nuestra Señora del Prado (1481–1835)', in Benito Colombás *et al.*, *Studia Hieronymiana: VI Centenario de la Orden de San Jerónimo*, 2 vols. (Madrid, 1973), ii. 137–214 (pp. 149–61).

[38] In England, Caxton is an example of a printer who was also a prosperous merchant (see Blake, *Caxton*, pp. 34–5).

[39] Eguía's dedication, addressed to the Archbishop of Toledo, of an edition of the *Precatio dominica* (Alcalá, 1525), cited by Gil Fernández, *Panorama social*, p. 573.

[40] Marcel Bataillon, *Erasmo y España: estudios sobre la historia espiritual del siglo XVI*, tr. Antonio Alatorre, 2nd Spanish-language edn. (Mexico, 1966), p. lvi.

[41] Eugenio Asensio, 'El erasmismo y las corrientes espirituales afines', *Revista de filología española*, 36 (1952), 31–99 (pp. 78–81).

[42] Gil Fernández, *Panorama social*, pp. 572–5.

for they managed what was in effect a university press at Alcalá. Their output suggests a growing specialization in some Spanish printing centres. This is illustrated by the contrast between the sort of books which Brocar printed at Alcalá and those which he produced in his other establishments at Pamplona, Logroño, Toledo, and Valladolid.[43] Similarly, at Salamanca, Porras followed the tradition set by his fifteenth-century predecessors in that city and specialized in academic texts, while at Toledo there was a tendency to print in the vernacular for the more popular end of the market. Joan Joffre's output at Valencia indicates a different sort of specialization, for he printed numerous books written by local authors and intended for local consumption.[44] This concentration on a certain sort of printing was not unknown in Seville, which boasted the most flourishing book-trade in Castile, and became the centre for popular printing in sixteenth-century Spain, at least until Philip II situated his capital in Madrid.[45] A contract drawn up between Miguel de Eguía and Jacobo Cromberger, the founder of Seville's most prolific press, illustrates this facet of the trade:

> Inasmuch as I, the aforementioned Jacomo Moramberga, have my home, press, and books in this city of Seville, and the said Miguel de Guía has his home, establishment, and printing-office in Alcalá de Henares, and the books which we, the said parties, print in our two establishments are of different sorts, we therefore . . . make the following agreement that the said Miguel de Guía should send me, the aforementioned Jacomo Moramberga, in the said city of Seville, all the products of his press that he wishes so that I, the said Jacomo Moramberga, can sell them in the city of Seville . . . and likewise the said Jacomo Moramberga can send to the said Miguel de Guía's establishment in the town of Alcalá de Henares all the books I wish from my own presses so that the said Miguel de Guía can sell them.[46]

Cromberger and Eguía exchanged books because the output of their presses was different and an agreement would not be to the disadvantage of either. It is a nice irony that Eguía is here seen arranging to sell books from a press which was well known for printing just the sort of editions he had so roundly condemned in the very year he signed this contract.

As had been the case in the previous century, Spanish presses did not supply the demand for books in Spain. Large shipments of foreign editions continued to be imported through Seville and also through the fairs at Medina del Campo, where booksellers and publishers gathered to distribute and purchase their wares.

[43] F. J. Norton, 'Typographical Evidence as an Aid to the Identification and Dating of Unsigned Spanish Books of the Sixteenth Century', *Iberoromania*, 2 (1970), 96–103 (p. 99).

[44] Norton, *Printing in Spain*, pp. 49–55, 81–2.

[45] Norton, *Printing in Spain*, p. 6; Sir Henry Thomas, 'The Output of Spanish Books in the Sixteenth Century', *The Library*, 4th ser., 1 (1920–1), 69–94.

[46] Document dated 20 Oct. 1525 (APS, Oficio 4, Libro 3 of 1525, fols. 142^{v}–143^{v}), partially transcribed by Gestoso, pp. 22–3.

The 'international book' continued to be produced outside Spain, and Spanish printers concentrated much of their efforts on printing vernacular editions for the local market. In Seville, for example, copies of 'international books' were ordered by local booksellers from suppliers in Lyons and the Low Countries.[47] As the century progressed, however, even editions printed exclusively for Spanish readers (or, at least, for readers of Spanish) were imported.[48] This poses the question: why did printed books continue to flood into the country and why did the Spanish presses fail to increase their output to satisfy the demand? One factor is that the book-trade in the Peninsula had, as we have already seen, traditionally relied on imports even before the age of printing; this reliance naturally increased in the sixteenth century, the prestige and scholarship of foreign editions, especially of the classics or academic text-books, being so great that domestic presses could not attempt to rival them. It also appears that, as much of Castile's industry was being weakened by a series of factors including the 'price revolution' and its consequences, books, like other manufactured goods, were produced more cheaply abroad and imports undercut domestic producers. Spanish printers would thus have been restricted to those titles, intended for an exclusively local or national readership, which could not profitably be mass-produced on foreign presses.

Another explanation may be that Spanish presses were just unable to supply the demand for books in the country, and imports filled the vacuum. The general inability of Spanish presses to mass-produce books or to undertake large projects could be explained by lack of materials (and, particularly, of paper), the shortage of skilled workers, restrictive legislation, and the absence of capital investment—possibly both because the returns on other forms of speculation were higher and because wealth was so often spent on conspicuous consumption in Spain.[49] It is worth remembering that when a major printing project like the Alcalá Polyglot Bible was undertaken, an enormous amount of capital had to be injected into Brocar's press by Cisneros; such patrons were not plentiful. Contemporary commentators noted the shortcomings of sixteenth-century printing in Spain in terms of both quality and quantity. In 1523 Francisco de Vergara lamented that

> we live in a country where idleness and greed have caused both the mechanical and liberal arts to stagnate. Indeed, we are in such a sorry state that it is rare to find anyone willing

[47] Klaus Wagner, *El doctor Constantino Ponce de la Fuente: el hombre y su biblioteca* (Seville, 1980), p. 24. Jacobo Cromberger's stepson, Tomás Ungut, imported 'international books' into Seville from abroad (see Hazañas, i. 105–6).

[48] For the huge imports of books into Spain from Lyons see Christian Péligry, 'Les Éditeurs lyonnais et le marché espagnol aux XVIe et XVIIe siècles', in Díaz y Díaz *et al.*, *Livre et lecture*, pp. 85–93.

[49] On the economic weakness of 15th-c. printing at Barcelona, where bookselling nevertheless flourished, see Madurell and Rubió, *Documentos*, pp. 55*–8*. D. W. Cruickshank, '"Literature" and the Book Trade in Golden-Age Spain', *Modern Language Review*, 73 (1978), 799–824 (p. 816), suggests that after 1570 undercapitalization of the Spanish presses led to the existence of a large number of small offices,

to print a book in Latin, even one which would sell well, let alone set a Greek manuscript up in type. This barbarism is shameful and unworthy of any civilized man.[50]

Admittedly Vergara was here complaining particularly about scholarly editions, but a quarter of a century later the prosperous Lyonese publisher and merchant-bookseller, Guillaume Rouillé, prefaced his Spanish-language edition of a more popular work, Alciati's *Emblems*, which was destined for the Spanish market, by addressing his readers thus:

When I considered (dear reader) all the books—both those already in print and those still in manuscript—of which you are deprived in Spain because of the shortcomings of the presses, I realized how great a service I could render you by sending you beautifully printed and carefully corrected editions.[51]

Although this could be a conventional use of *captatio benevolentiae* by a foreign publisher anxious to curry favour with his potential customers, it does suggest that demand in Spain was, indeed, outstripping the capacity of the domestic presses. This indication is confirmed by the ready market found in Spain by Rouillé and other foreign publishers like him, who began to specialize in editions printed for export to the Peninsula. Rouillé even considered it worth his while employing a learned Spaniard as a proof-reader to ensure that the Spanish-language books he published were linguistically accurate.[52]

During the first two decades of the sixteenth century it was unusual for a Spanish book in the vernacular and intended for sale in Spain to be printed abroad.[53] As the century wore on, however, the percentage of imported books increased and in about 1540 it appears that production of books printed in Spanish outside the country overtook domestic output.[54] This trend was accompanied by a decline in the quality of Spanish printing, which was probably both a cause and an effect of the large numbers of imported editions. It is possible that imports increased to such an extent that, even if demand was growing, Spanish printers were competing with each other for a reduced slice of the market; quality would then fall as they endeavoured to cut costs.[55]

Sir Henry Thomas's explanation for the deterioration in the quality of the

none of which could afford to undertake major projects. They were obliged to rely on editions which brought in rapid returns on limited investment, leading to the predominance of ephemeral and popular printing in Spain. For Seville, at least, Cruickshank's date of about 1570 is late.

50 Cited by Gil Fernández, *Panorama social*, pp. 572–3.

51 Andrea Alciati, *Los emblemas* (Lyons, 1549), p. 3.

52 Martin and Chartier, *Le Livre conquérant*, p. 259.

53 Norton, *Printing in Spain*, p. 135.

54 Theodore S. Beardsley, Jr., 'Spanish Printers and the Classics: 1482–1599', *Hispanic Review*, 47 (1979), 25–35 (p. 30).

55 This is one of the reasons adduced by D. W. Cruickshank for the decline of Spanish printing in the 17th c. (see his 'Some Aspects of Spanish Book-Production in the Golden Age', *The Library*, 5th ser., 31 [1976], 1–19).

Spanish printed book, which he believes was well under way some twenty years earlier, is somewhat different. He maintains that, while the unfavourable economic climate in Spain and the demise of the original German master-printers were of importance, it was the increased output by the Spanish presses, with the consequent exhaustion of supplies of good-quality paper and the employment of large numbers of unskilled workers in the printing-shops, which caused the rot to set in.[56] The economic crisis of the mid-century and the restrictive legislation on printing would, then, merely have hastened a process which was already under way. Whatever the reasons for the general deterioration—and the repercussions of this at Seville in particular will be discussed more fully in Chapter Four—the decline of the Cromberger press in that city can be dated to the mid 1540s.

In the second half of the century other factors combined to complete this decline, amongst which the most important were the legislation applied to the presses as Spanish intellectual life closed in, and the granting of printing monopolies and privileges on sales. In 1572 Philip II gave the Antwerp printer, Christopher Plantin, the monopoly for the printing of breviaries, missals, and other liturgical books destined for Spain and the American colonies, and in the following year granted the Hieronymites of the new monastery of El Escorial a privilege on their sale. By these actions he seriously undermined the livelihood of Spanish booksellers and printers. At the turn of the new century the granting of a sale-privilege to the Hospital General in Madrid on Nebrija's *Artes* had a similar deleterious effect, for this grammar had by then been decreed the only one permissible for the teaching of Latin in Spanish universities.[57]

Printing in Seville during the Fifteenth Century

By the time printing arrived in Spain, Seville was the most populous and prosperous city in Castile. Situated in a rich agricultural region and, owing to its position on the banks of the Guadalquivir, at the pivot between the Mediterranean and Atlantic maritime routes, it had long been the major Christian city in Southern Spain. Its imports included slaves, gold, wheat, and cloth, while it exported agricultural products, wool, dyestuffs, and fish to Northern and Southern Europe. It boasted soap and textile industries of its own, but its importance was securely founded on trade and, ever since its reconquest from the Moors in the mid thirteenth century, a large number of foreign merchants, many of them Genoese, had settled in the city.[58] In that same century Alfonso the Wise, building on the early missionary

[56] Thomas, 'The Output', pp. 89–90.

[57] Christian Péligry, 'Les Difficultés de l'édition castillane au XVIIe siècle à travers un document de l'époque', *Mélanges de la Casa de Velázquez*, 13 (1977), 257–84 (pp. 264–5).

[58] Miguel Ángel Ladero Quesada, *La ciudad medieval (1248–1492)*, Historia de Sevilla, 2, 2nd edn. (Seville, 1980), pp. 75–103.

work of the Dominicans, had founded a *studium* at Seville for the study of Latin and Arabic, and this was recognized by Alexander IV as a *studium generale* in 1260. It is probable that the short-lived *studium generale* had its own *scriptorium*. The intellectual life of the city during the Middle Ages has been dismissed by some historians as insignificant, but recent research suggests that, despite the lack of a permanent medieval university there (the *studium generale* had a precarious existence, although a long-lived *studium* for the teaching of grammar was founded well before 1365, and possibly even as early as 1261[59]), Seville was blessed by a cultured clergy and fostered a lively trade in books, possessing scribes, illuminators, and binders who were capable of producing manuscripts commercially. By the 1470s, Seville was the major Castilian supplier of manuscript books to a growing lay readership. Fine choir-books were produced there, but the city was especially strong on the more 'popular' end of the market: secular books written in the vernacular, the authors and patrons of which were associated with Seville. This points to a flourishing intellectual life there. Painter's claim that the state of the book-trade in Spain is reflected by Spanish incunables and the formats in which they were published is particularly true of Seville. The active printing industry which sprang up there so soon after the arrival of the presses in the country provides further evidence for a pre-existent trade in manuscripts in the city.[60] The emphasis of this local manuscript trade on vernacular literature helps to explain the printers' predilection for that sort of book, for printers everywhere tended to produce the kind of works which were already circulating in manuscript in the centre in which they had set up their presses.

It is unclear exactly when printing came to Seville, but the city's importance as a commercial centre would have made it attractive to the earliest printers. Indeed, the first book certainly printed there bears the date 1477, thus making Seville the first printing-site in Castile with the exception of Segovia, which had an ephemeral press whose dates are doubtful. Although firm proof is wanting, there are indications that there could have been a press at Seville even earlier than 1477. In that year when the court was resident in the city, a royal licence was granted to one 'Theodorico Alemán'—possibly the Thierry Martens of Aalst who had been printing at Louvain four years earlier—to import printed books without paying taxes on them.[61] Although the licence was addressed to the authorities of

[59] José Sánchez Herrero, 'Centros de enseñanza y estudiantes de Sevilla durante los siglos XIII al XV', in Miguel Ángel Ladero Quesada (ed.), *Estudios dedicados al profesor D. Ángel Ferrari Núñez*, En la España medieval, 4, 2 vols. (Madrid, 1984), ii. 875–98 (pp. 875–9).

[60] Lawrance, 'Nuño de Guzmán', pp. 90–4.

[61] All trace of Martens is lost in the Low Countries from 1474 to 1486 (see Wytze and Lotte Hellinga, *The Fifteenth-century Printing Types of the Low Countries*, 2 vols. [Amsterdam, 1966], ii. 78–9). Some bibliographers claim that two indulgences were printed at Seville as early as 1473, but I do not find their arguments wholly convincing (see Carlos Romero de Lecea *et al.*, *Historia de la imprenta hispana* [Madrid, 1982], pp. 133–4).

Murcia, the statement included in it that Martens had been harassed by local officials at Cádiz and San Lúcar de Barrameda suggests that his activities had been centred on Seville.[62] This licence was the first royal favour granted to printers in Spain. Three years later the monarchs not only made this tax-exemption applicable to all importers of books, but also wrote to the Seville authorities expressing their gratitude for the good treatment which had been given there to a German printer and publisher called Miguel Dachauer.[63] Dachauer helped to finance at least one edition printed at Seville, although there is no evidence that either he or Martens worked there as printers.

At least one historian of Spanish printing suspected that it was German capital and skill which lay behind the activities of the first press known to have operated in the city.[64] This was run by three men: Antonio Martínez, Bartolomé Segura, and Alfonso del Puerto. Unusually for a Spanish city, the first known printers of Seville were therefore native Spaniards, although Hazañas suggests that they had received their training from German master-printers and that Martínez was the only experienced craftsman among them, Segura and del Puerto merely acting as his assistants and financial partners.[65] Like many early printers, Segura may originally have been a scribe, for he was almost certainly a member of the family of public notaries which operated one of Seville's notarial offices, of which the Crombergers were to be frequent clients in the following century. In 1477 the three men signed two books, the much-reprinted *Sacramental* of Sánchez de Vercial (Haebler 597), and Díaz de Montalvo's *Repertorium quaestionum super Nicolaum de Tudeschis* (Haebler 210); these are the first two editions known for certain to have been printed in the city. The surviving output of this press is small and indicates only intermittent activity: Martínez disappears from sight for a time and, in 1480, Segura and del Puerto print an edition of Rolevinck's *Fasciculus temporum* (Haebler 583) in close imitation of the one printed by Georg Walch at Venice in the previous year. Lyell maintains that this was the first illustrated book produced in Spain.[66] Two years later del Puerto prints one book on his own, the *Crónica de España abreviada* of Diego de Valera (Haebler 654), and then, with the exception of one book which Vindel attributes to him and dates to 1483, nothing more is heard of the press until 1486, when Antonio Martínez reappears and prints a Spanish translation of Domenico Cavalca's *Specchio della Croce* (Haebler

[62] W. I. Knapp, *The Earliest Decree on Printing or Thierry Martin of Spain (Seville, December 25 1477)* (New Haven, 1881).

[63] Document dated 10 Feb. 1480 partially transcribed in Hazañas, i. 13.

[64] Stephen Gaselee, *The Early Spanish Printing Press: A Lecture Delivered Before the Anglo-Spanish Society of Great Britain and the Spanish-speaking Countries on Wednesday, December 19th, 1923* (London, 1924), p. 4. Diego de Valera's *Crónica de España abreviada* (Seville, Alfonso del Puerto, 1482) was jointly financed by Dachauer and García del Castillo.

[65] Hazañas, i. 6.

[66] James P. R. Lyell, *Early Book Illustration in Spain* (London, 1926), pp. 3–5.

144) apparently without the help of either of his former colleagues.[67] Evidence for printing then disappears from Seville until the beginning of the following decade, when, in response to a royal invitation possibly extended on the advice of Fray Hernando de Talavera, two firms of foreign printers settled in the city.[68]

One of these consisted of four Germans, Paulus of Cologne, Johannes Pegnitzer from Nuremberg, Magnus Herbst of Fils, and Thomas Glockner. Together, they styled themselves the 'Compañeros alemanes', or 'German partners'. Their typographical material suggests that they had worked in Venice before being called to Seville by Queen Isabella, for whom in 1490 they printed their first book, an edition of Alfonso de Palencia's *Vocabulario universal* (Haebler 510). They printed some forty incunable editions, which bear witness to a high standard of craftsmanship, and in 1492 produced the first printed notes of music seen in any Spanish book.[69] As Hazañas points out, a large proportion of their editions were published by booksellers, not all of them resident in Seville.[70] Although Paulus of Cologne and, later, Thomas Glockner dropped out of the company, Pegnitzer and Herbst were still printing at Seville in the early years of the sixteenth century.

The other firm of immigrant printers had been summoned from Naples, where they had worked in Mathias Moravus' office. They were Meinardo Ungut, probably a German or Fleming, and Stanislao Polono, a Pole. They initially established their press at the centre of the book-trade in the city, the Calle de Génova.[71] Their first book, the *Defensiones sancti Thomae Aquinatis* written by the future Archbishop of Seville, Diego de Deza, was signed at the beginning of 1491 (Haebler 203). In the nine years of their joint activity Ungut and Polono printed the remarkably large number of over seventy known editions, making their press, whose mantle was later to fall on Jacobo Cromberger, the most prolific in fifteenth-century Spain.[72] So active was it that Ungut and Polono appear to have had their own workshop for wood-engraving, producing the many blocks cut in the local style which illustrate their books.[73] Nevertheless, despite this large output, Ungut and Polono were as dependent as the 'Compañeros alemanes' on commissions from booksellers. For a time these two firms of immigrants appear to have been rivals, for otherwise their printing in 1491 of the same book, the

[67] Vindel, *El arte tipográfico*, viii. 245–52.

[68] Carlos Romero de Lecea, 'Hernando de Talavera y el tránsito en España "Del manuscrito al impreso" ', in Colombás *et al.*, *Studia Hieronymiana*, i. 315–77 (p. 356).

[69] In their edition of Durán, *Lux bella* (Haebler 237).

[70] Hazañas, i. 70.

[71] The booksellers Juan and Alfonso Lorenzo, who financed the printing of incunable editions at Seville, lived in this street, as did the Genoese family of booksellers, the Labezaris, with whom the Crombergers later had important dealings. For the Calle de Génova in the 16th c. see below, pp. 27–9.

[72] Sheppard and Painter, 'Introduction to the Presses', p. lix.

[73] Hazañas, i. 35.

Siete partidas of Alfonso the Wise, is difficult to explain. Yet, only a few years later, when Talavera required a press at Granada after this last Moorish kingdom in Spain had fallen to the Christians, it was Pegnitzer and Ungut who travelled there and co-operated in the production of several books on the first Granadine press.

The fourth press at Seville was of only minor importance. In 1492 Pedro Brun, a Genevan who had previously worked in Catalonia, and Juan Gentil, possibly an Italian, printed their first book at Seville, an edition of Fernando Mexía's *Nobiliario* (Haebler 411). By the time that Brun issued his next known edition in *c.*1499 (Haebler 674), he was working alone, and he continued to print intermittently at Seville until 1507 or 1508, although it seems that his output was very small.

A brief analysis of the titles printed by the four incunable offices at Seville illustrates what was said in general terms above about the manuscript book-trade in the city. The proportion of vernacular editions is remarkably high even for Spain: only about one-third are in Latin, and a good number of these are service books. Apart from the jobbing printing of indulgences and vernicles in which at least Ungut and Polono are known to have engaged, there was, as we should expect from the major centre of Castile, a substantial production of 'utilitarian' editions: school-books, medical works in the vernacular, liturgical books and manuals for priests, as well as the many compilations of laws in which Ungut and Polono specialized.[74] More interesting, however, are the many volumes of translations, doubtless intended for the increasingly literate lay readership. These are exemplified by such editions as Alfonso de Palencia's translation and adaptation of Plutarch's *Lives* ('Cuatro compañeros', 1491) (Haebler 550), the vernacular version of Curtius Rufus' history of Alexander the Great (Ungut and Polono, 1496) (Haebler 186), Josephus' *Jewish War* translated by Alfonso de Palencia (Ungut and Polono, 1492) (Haebler 344), Juan García de Castrojeriz's version of Giles of Rome entitled *Regimiento de príncipes* (Ungut and Polono, 1494) (Haebler 156), the *Caída de príncipes* translated from Boccaccio (Ungut and Polono, 1495) (Haebler 53), Boethius' *De consolatione* translated by Antonio Ginebrada (Ungut and Polono, 1497 and 1499) (Haebler 59 and 60), *Cinco libros de Séneca* in Alonso de Cartagena's version (Ungut and Polono, 1491) (Haebler 621), and the *Proverbios*, a collection of *sententiae* erroneously ascribed to Seneca, which had been translated and glossed by Pedro Díaz de Toledo (Ungut and Polono, 1495; 'Dos compañeros', 1500?) (Haebler 618 and 620). Intended for a similar readership were works of Castilian history like Diego de Valera's *Crónica de España abreviada* (Alfonso del Puerto, 1482), López de Ayala's *Crónica del rey don Pedro*

[74] Ungut had also printed a large number of indulgences and vernicles for the Bishop of Jaén in 1493 or earlier (see Gestoso, pp. 5–6).

(Ungut and Polono, 1495) (Haebler 38), the popular *Crónica del Cid* ('Tres compañeros', 1498) (Haebler 173), Pedro del Corral's *Crónica del rey don Rodrigo* (Ungut and Polono, 1499) (Haebler 174), and Hernando del Pulgar's *Claros varones de España* (Polono, 1500) (Haebler 566).

Devotional works in the vernacular form a large proportion of the presses' output: Cavalca's *Specchio della Croce* translated into Castilian by Alfonso de Palencia (Antonio Martínez, 1486; Ungut and Polono, 1492) (the second of these being Haebler 145), Juan Alfonso de Logroño's translation entitled *Contemplaciones sobre el rosario* of a work by Gaspar de Gorricio (Ungut and Polono, 1495 and 1497) (Haebler 301 and 302), a vernacular version of the *Scala celi*, a work traditionally attributed to St Jerome (Ungut and Polono, 1496) (Haebler 342), Jiménez de Prejano's *Lucero de la vida cristiana* (Ungut and Polono, 1496) (Haebler 717[8]), Fray Íñigo de Mendoza's *Vita Christi* (Ungut and Polono, 1499), and a Castilian version of the *Imitatio Christi* attributed at the time to Jean Gerson (Ungut and Polono, 1493) (Haebler 294 *bis*). Devotional works in the vernacular would form one of the mainstays of the Crombergers' output in the following century, and many of their books were reprints of these incunable editions.

A final category of literature of entertainment (also including some Castilian works of popular morality) comprises titles frequently reissued at Seville in the sixteenth century: Diego de San Pedro's *Cárcel de amor* ('Cuatro compañeros', 1492) (Haebler 603), Pérez de Guzmán's *Coplas* (Ungut and Polono, 1492) (Haebler 534), Jorge Manrique's *Coplas* (Ungut and Polono, 1494) (Haebler 392), López de Mendoza's *Proverbios* (Ungut and Polono, 1494; 'Tres compañeros', 1499) (Haebler 425 and 426), Mena's *Trescientas* (Ungut and Polono, 1496; 'Tres compañeros', 1499; 'Dos compañeros', 1499) (Haebler 412, 414, and 413), and his *Coronación* (Polono, 1499) (Haebler 416), and a Castilian translation of Boccaccio's *Decameron* entitled *Cien novelas* (Ungut and Polono, 1496) (Haebler 54).

The books mentioned above, several of which were illustrated, indicate both the taste of the readership for which they were intended and the sort of titles which the Seville presses of the fifteenth and sixteenth centuries repeatedly issued. As we shall see in the next chapter, Jacobo Cromberger was to carry on the traditions of the early Spanish and, in particular, German printers who had made Seville the most active centre of book-production in Spain.

CHAPTER 2

THE HOUSE THAT JACOBO BUILT (1504–1528)

Jacobo Cromberger

LIKE so many of the early printers in Spain and elsewhere in Europe, Jacobo Cromberger, or Jácome Alemán as he was better known in Seville, was a German.[1] He was born in 1472 or 1473, but his place of birth is unknown.[2] His surname suggests that his family may originally have come from Kronberg or Cronenberg, but even if this is correct, neither was necessarily still the family home by the time he was born.[3] More suggestive is the other name by which he was sometimes called in notarial documents drawn up in Seville: 'Jácome Nurenberge' or 'Moranberga'.[4] These forms may be attributable to confusion with the name of his son-in-law, Lázaro Nuremberger. The two often accompanied each other on their frequent visits to the notaries' offices and, indeed, Lázaro occasionally appears in documents as 'Lázaro Cromberger'.[5] Yet the various forms of Jacobo's surname indicate that he was probably from the Southern German city of Nuremberg.[6]

Although it was not among the first cities in Germany to boast a printing-office, when the craft did reach Nuremberg in 1470 it soon became a major centre for

[1] Jacobo was normally referred to as an 'alemán' or German, but this term is vague. In notarial documents Stanislao Polono was also called an 'alemán', as were several merchants who were definitely Flemings. On the other hand, Lázaro Nuremberger, who was a German, was on occasion referred to as a 'mercader flamenco' (Flemish merchant). Juan Cromberger was called an 'alemán' although born and bred in Seville.

[2] The date of his birth can be calculated from a document of 1527 or 1528 in which he claimed to be 55 (see Hazañas, i. 75, 149).

[3] Kronberg is situated about 10 miles north-west of Frankfurt am Main; Cronenberg is about 20 miles north-east of Cologne. Not surprisingly, there were families called Cromberger living in Cologne in the 16th c. For further conjectures on Jacobo's origins see Enrique Otte, 'Jakob und Hans Cromberger und Lazarus Nürnberger, die Begründer des deutschen Amerikahandels', *Mitteilungen des Vereins für Geschichte der Stadt Nürnberg*, 52 (1963–4), 129–62 (pp. 131–2).

[4] For example 'Moramberga' in 1514 (see document dated 9 May 1514 [APS, Oficio 4, Libro 2 of 1514, fols. 361^r–364^r]); 'Nurenberge' in 1525 (see Hazañas, i. 129–31); 'Nuremberger' in 1528 (see document dated 1 Aug. 1528 [APS, Oficio 4, Libro 3 of 1528, unfoliated]) and 'Moranberger' (in a second document drawn up on the same day and preserved in the same place).

[5] José Toribio Medina, *La imprenta en México 1539–1821*, 8 vols. (Santiago de Chile, 1908–12), i. p. lviii.

[6] Gerald Strauss, *Nuremberg in the Sixteenth Century* (New York, 1966), pp. 129–30, like many before him, states that Cromberger was from Nuremberg, but has kindly confirmed that he has no proof for this claim. I am indebted to Dr Hermann Kellenbenz for a copy of his 'Die Beziehungen Nürnbergs zur Iberischen Halbinsel: besonders im 15. und in der ersten Hälfte des 16. Jahrhunderts', in Gerhard Hirschmann *et al.*, *Beiträge zur Wirtschaftsgeschichte Nürnbergs*, 2 vols. (Nuremberg, 1967), i. 456–93, where he asserts (p. 466) that Cromberger came from the city but adduces no evidence to support this claim.

the production of printed books and remained so in the sixteenth century.[7] The most important local industries were mining and metal-working; it was the seat of several substantial international trading companies; and local citizens were heavily engaged in business with foreign countries, including Spain, during the late Middle Ages. The development of the allied craft of printing in Germany led to a number of Nurembergers' being found among the first printers in Spain: one of the pioneers of printing in Seville, Johannes Pegnitzer, was a native of Nuremberg, as were some of the earliest book-merchants in Catalonia.[8] Nevertheless, Nuremberg archives shed no light on Jacobo's origins, although it is true that artisans by the name of Cromberger, particularly cobblers and cabinet-makers, lived in the city in the late fifteenth and early sixteenth centuries.[9] Any suggestion that Jacobo, too, was a native of the city must, therefore, remain conjectural, but is supported by the close family and business relations he enjoyed with the famous merchant from Nuremberg, Lázaro Nuremberger. What can be denied with certainty is the assumption made by some historians of Spanish printing that Jacobo was a scion of the great printing family of Nuremberg, the Kobergers, who also had interests in Spain.[10]

If Jacobo's origins are uncertain, little can be added about his early life or training before he arrived in Seville in his mid-twenties. The first reference to him in the local archives is dated February 1503, by which time he was already a denizen (*vecino*) living in the parish of Santa María la Mayor.[11] Nowhere in later documents is there any mention of when Jacobo came to Seville, or even Spain, nor where he came from, but certain fragments of evidence allow us to piece together a plausible, if unsubstantiated, early biography.

Taking up residence in Seville was not enough to qualify the immigrant for denizenship; as Jacobo was referred to as a *vecino* in 1503, it is possible that he had arrived there at least as early as the middle of the previous decade.[12] Jacobo

[7] Klaus Wagner, 'La reforma protestante en los fondos bibliográficos de la Biblioteca Colombina', *Revista española de teología*, 41 (1981), 393–463 (pp. 395, 399).

[8] Rubió, 'Integración', p. 106.

[9] Staatsarchiv Nürnberg, *Nürnberg Bürgerbüchern*, Rep. 52b, No. 307, fols. 131, 158; No. 306, fols. 40, 79. I am grateful to the archivists of the Staatsarchiv, Stadtarchiv, and Landeskirchliches Archiv of Nuremberg, and to Dr Benjamin Arnold of the University of Reading for their searches on my behalf for a record of the Cromberger family in that city.

[10] Vicente Barrantes, 'Apuntes para un catálogo de impresores desde la introducción del arte en España hasta el año 1600', *Revista contemporánea*, 26 (1880), 385–407, and 27 (1880), 43–72, 421–36 (26, p. 403); Haebler, *The Early Printers*, p. 60.

[11] Gestoso, p. 12. In 1525 Jacobo was said to have been a *vecino* of Seville for 'about 25 years' (royal decree dated 26 July 1525 [AGIS, *Indiferente General*, Legajo 420, Libro 10, fols. 36ᵛ–37ʳ]).

[12] Pike, *Aristocrats and Traders*, p. 14, implies that before 1597 the residence requirement for denizenship was seven years. She does, however, point out that it was often obtained fraudulently before such a qualification was fulfilled. Lázaro Nuremberger settled in Seville at some time after Mar. 1520; by Feb. 1523 he was already being referred to as a *vecino*.

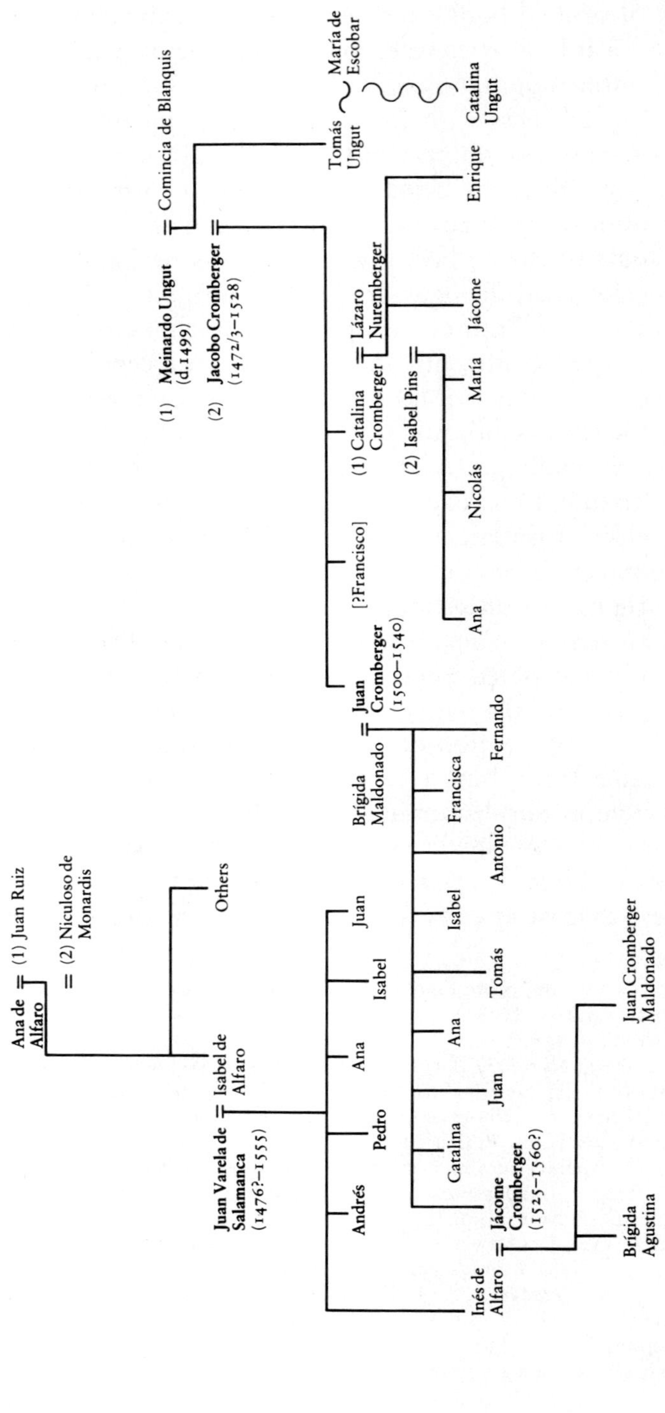

Fig. 1 Cromberger family tree (simplified; printers in bold type)

may have been employed as an assistant of Ungut and Polono; he was certainly called upon by Ungut's widow to witness certain facts concerning the press, and the other witnesses were Ungut's former partner, Stanislao Polono, and a printer who possibly worked with them, Pedro de Mendieta.[13] If this were the case, he would have been associated with the press at least by the late 1490s. He may even have worked in Moravus' office at Naples and have come with Ungut and Polono to Seville from Italy in about 1491.[14] He would not have been impossibly young at that stage and, indeed, could already have served an apprenticeship in Germany or elsewhere before going to Naples. The normal age for a youth to be apprenticed to a master, at least in Seville, was thirteen, and the case of the Spanish type-founder and punch-cutter, Antonio de Espinosa, reveals that by his late teens or his early twenties, a man could have served an apprenticeship and become a skilled craftsman even in a trade demanding mastery of complex technology.[15] It is equally possible that Cromberger had gone independently to Seville in the 1490s and joined the firm either as a skilled worker or a raw apprentice.[16] Spanish presses did not always recruit either foreign employees who were already skilled or native workers who needed training; in 1511 Jacobo Cromberger took on 'Juan de Basilea, the German' as a pressman for two years and agreed to teach him from scratch to 'ink the type, pull the bar, and everything else an apprentice should learn'.[17] What can be stated with confidence is that by 1503 Jacobo was a craftsman whose skill could match that of any contemporary in Spain, for in that year he printed books of a very high quality.

If he was an employee in Ungut and Polono's shop at the end of the fifteenth century, his status was too humble for him to be mentioned in the colophons of their editions. However, death was to change all that and Jacobo replaced one of his masters in more ways than one. Between 24 October and 2 November 1499, Meinardo Ungut died leaving a young widow, Comincia de Blanquis, and a son, Tomás, who was about nine years old.[18] The funeral baked meats must

[13] Gestoso, p. 15.

[14] It is a moot point whether Moravus had travelled to Seville with Ungut and Polono. In her will of 1511 Comincia de Blanquis mentioned a 'Master Matía, one of my husband's partners'. It is not certain that this is the same Mathias Moravus, nor does the document make it clear whether the husband referred to was Ungut or Cromberger (see Gestoso, p. 15).

[15] Gestoso, p. 117.

[16] Otte, 'Jakob und Hans Cromberger', p. 131, surmises that Cromberger was called to Seville in about 1500 by his compatriot Johannes Pegnitzer, but this is unlikely because Cromberger worked with Polono, not with the 'Compañeros alemanes', of whom Pegnitzer was by then one of the two surviving members. Agustín Millares Carlo, *Introducción a la historia del libro y de las bibliotecas* (Mexico City, 1971), p. 112, also makes Cromberger an associate of the 'Compañeros', but with no authority.

[17] Hazañas, i. 84–5. Juan de Basilea is probably the same man as Hans Henschel of Basle, who in 1535 went to the Indies as the factor of Juan Cromberger, Lázaro Nuremberger, and Christoph Raiser (see below, p. 79).

[18] Francisco Collantes de Terán, 'Un taller alemán de imprenta en Sevilla en el siglo XV', *Gutenberg-Jahrbuch* (1931), 145–65 (p. 156).

have coldly furnished forth the marriage tables, for Jacobo's first child by Comincia, Juan, was conceived, presumably in wedlock, early in 1500.[19] As Comincia was later to point out, Jacobo brought no wealth at all to their marriage; she, on the other hand, contributed the not inconsiderable estate left by her late husband.[20] There was, for instance, the debt of 100,000 *mrs* owed to her and Polono by the Bishop of Jaén for some missals printed in 1499; there was also her furniture and her clothes. More importantly, there were the 'great sums of *maravedís*, and goods, and jewellery both silver and gold' which formed Tomás Ungut's inheritance but which had passed, in part at least, to his mother until he was old enough to receive it.[21] Most important of all was Meinardo's share of the printing firm and the separate business of bookselling in which he and his partner had engaged: three presses with their equipment, types, instruments for the process of casting letters, and all the bed-linen and crockery which had been used by the printing-shop's employees.

Unless he were a person of considerable private means or able to attract generous patronage, an aspiring printer setting up an office would from the outset incur crippling debts. The presses themselves were not expensive, but matrices, types, moulds, and type-metal were costly.[22] To this would be added the rent of premises spacious enough to house the vats used in dampening the paper, the presses themselves, the printing material, the compositors' benches and their large cases, storage space for paper and books, and accommodation for the employees who served the presses and who would lodge with the master-printer. In addition, capital had to be available to pay for wages, food, paper, and ink and, if the printer was financing his own editions, this would be recovered only slowly when those booksellers who had neither died nor gone bankrupt between receiving the printed books and paying for them honoured their debts. It is therefore apparent why so many printers found themselves at the mercy of creditors. A constant theme of Thomas Platter's account of the foundation of his printing-office in sixteenth-century Basle was the whirlpool of debt into which he was sucked.[23] In

[19] In Sept. 1525 Juan legally accepted the deed of gift of the Cromberger press and must therefore have been at least 25 years old.

[20] See her will of 1511 (Hazañas, i. 143–8). Francesco Guicciardini, the Florentine ambassador to Ferdinand's court in 1512 and 1513, thought Spanish customs regarding widows' rights to their late husband's estates strange enough to warrant comment: 'In Spain women are treated well . . . for they not only recoup their dowry after their husband's death, but his estate at the time of marriage is calculated and, if it has grown during their married life, the increase is divided, one-half being entirely at the widow's disposal—whether she remarries or not—even if she has children by the deceased husband' (see *Viaje a España de Francesco Guicciardini Embajador de Florencia ante el Rey Católico*, tr. José María Alonso Gamo [Valencia, 1952], pp. 57–8).

[21] Comincia de Blanquis said that she had none of Tomás Ungut's inheritance when she married Jacobo, but this is seen to be untrue from subsequent documents (see n. 29 below and Tomás's request for the appointment of a *tutor*, or guardian, dated 29 April 1511 [Hazañas, i. 88–9]).

[22] See below, pp. 172–3, n. 37.

[23] Platter, *Autobiographie*, p. 108.

between his bankruptcies, the hapless François Estienne, who worked at Geneva from the middle of the sixteenth century, had to barter his printed books against a fresh supply of paper and even on occasion deliver to his cautious supplier sheets printed one day as security for the paper he would use on the next.[24] Examples of printers who went to the wall, often after producing only a handful of editions, are legion in Spain as elsewhere.

Given these harsh conditions, it is not surprising to find young assistants making their way in the world of printing by turning the demise of their masters to their own advantage. They often married the widows, gaining control of the expensive equipment which would otherwise have been beyond their grasp. On their side, the widows would, through these unions, suffer the minimum of disruption to their lives in a world in which women were especially vulnerable if they had no husband to support them. Such matches were therefore frequent: Guyone Viart, the widow of the Parisian printer Johann Higman, married Henri Estienne I, who had possibly worked in Higman's press, and then, when Estienne died, became the wife of one of his partners, Simon de Colines; Johann Froben's widow married the Strasburg printer Johann Herwagen, who then went into partnership with her sons by her first marriage; Jaqueline, the widow of the London printer Thomas Vautrollier, married her first husband's apprentice, Richard Field, famous in Spanish Protestant printing as 'Ricardo del Campo'.[25] Such marriages were not unknown in Spain. For example, the first printer at Burgos was Fadrique de Basilea. His daughter Isabel married Alfonso de Melgar, who took over the press in 1517 or 1518 but must have died in 1525.[26] Some five years earlier Giovanni Giunti, a member of the famous printing family of Florence, Venice, and Lyons, had left Italy for Salamanca, where he acted as the agent for the family's interests in Spain.[27] In 1526, now known as Juan de Junta, he married Isabel and in the following year issued his first signed book from the press at Burgos. Similarly, Miguel de Eguía married Brocar's daughter, María, and took over his father-in-law's famous press at Alcalá in the early 1520s.[28] It would, however, be incorrect

[24] Robert M. Kingdon, 'The Business Activities of Printers Henri and François Estienne', in G. Berthoud *et al.*, *Aspects de la propagande religieuse* (Geneva, 1957), pp. 258–75 (p. 264).

[25] Elizabeth Armstrong, *Robert Estienne, Royal Printer: An Historical Study of the Elder Stephanus* (Cambridge, 1954), p. 5; Susan V. Lenkey, 'Printers' Wives in the Age of Humanism', *Gutenberg-Jahrbuch* (1975), 331–7; Denis B. Woodfield, *Surreptitious Printing in England 1550–1640* (New York, 1973), p. 35.

[26] F. J. Norton and Edward M. Wilson, *Two Spanish Verse Chapbooks: Romançe de Amadis (c.1515–19), Juyzio hallado y trobado (c.1510): A Facsimile Edition with Bibliographical and Textual Studies* (Cambridge, 1969), p. 2.

[27] In a royal privilege for the printing of the *Siete partidas* dated Jan. 1520, he and two compatriots were styled book-merchants of Salamanca (see Beltrán de Heredia, *Cartulario*, ii. 414). See also William A. Pettas, *The Giunti of Florence: Merchant Publishers of the Sixteenth Century* (San Francisco, 1980), p. 109.

[28] Norton, *Printing in Spain*, p. 43.

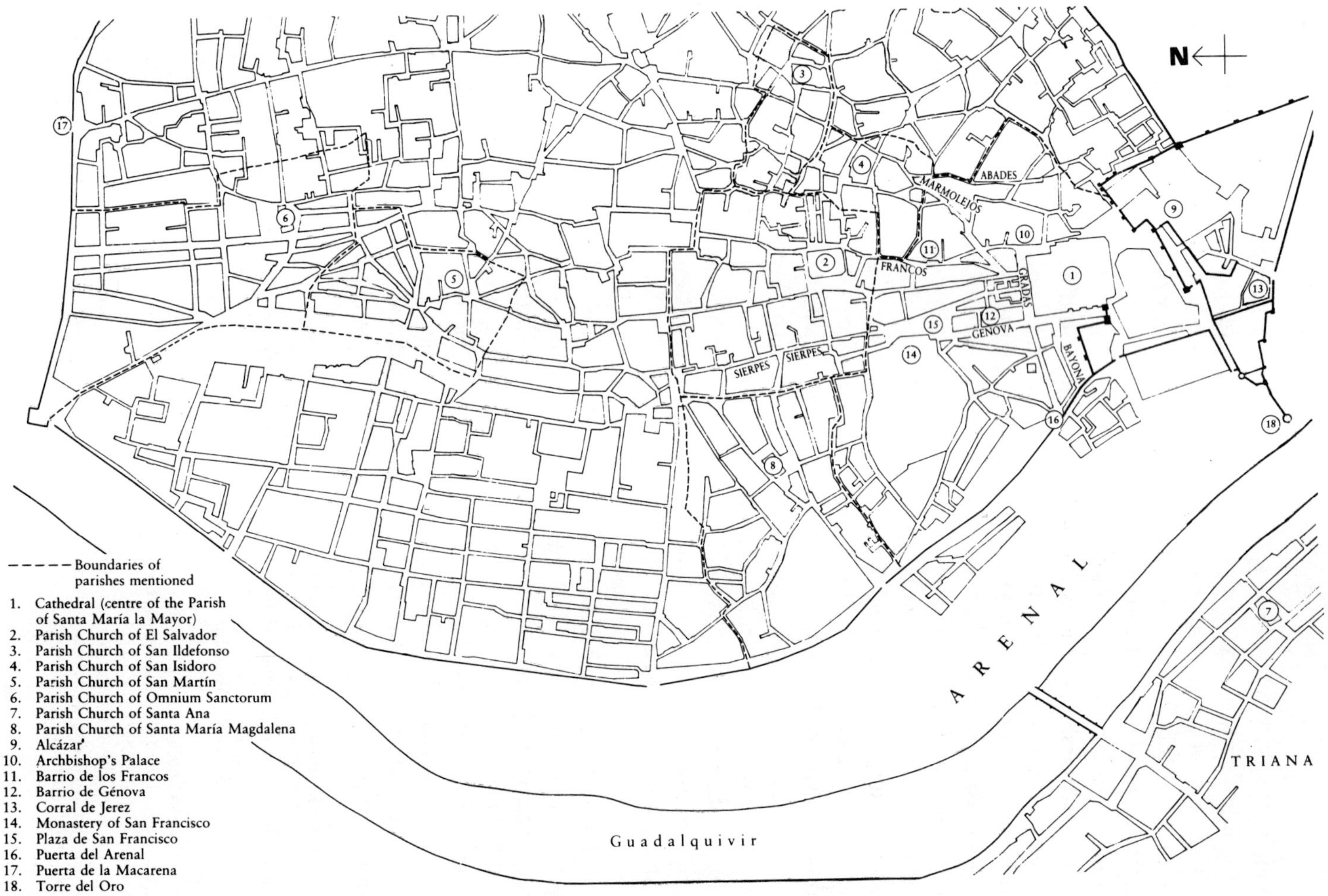

Fig. 2 Street plan of Seville *c.*1500

to conclude that such marriages of convenience were exclusive to the printing trade; rather, they were customary in the sixteenth century from the royal houses down to the commercial classes. In Seville they were frequently used to weld together fortunes or mercantile interests. The Crombergers were particularly careful and, as shall be seen, successful in using these alliances to further their twin activities of commerce and printing, but it was Jacobo's marriage to Ungut's widow which provided the first step in the family's ascent to prosperity.

In 1500, at the age of 27 or 28, he had risen from the status of a penniless immigrant to that of a partner in Seville's most thriving press and was also in possession of the capital to finance his operations.[29] He was then living with his wife and stepson in Ungut's former house in the Calle de Bayona, which fell within the boundaries of the parish of Santa María la Mayor. The most populous parish of the city throughout the century, Santa María occupied a large area surrounding the cathedral; it was the hub of Seville's commerce and was always the richest district in the city.[30] As was normal in European cities at this period, the residents of each street specialized in selling particular wares or were engaged in particular occupations. Santa María boasted the Calle de Génova, which was famous for its silversmiths and its numerous bookshops, many of the latter being somewhat primitive, as they were set up in the doorways of houses.[31] Bookselling in Seville, as elsewhere in Spain, was largely in the hands of immigrants, particularly the Germans, Flemish, and Genoese. Indeed, so many foreign merchants had traditionally lived and worked in Santa María that two of its quarters were known as the 'Barrio de Génova' and the 'Barrio de los Francos', and in the last third of the fifteenth century the Genoese colony at Seville had grown substantially.[32] From surviving notarial documents it is possible to piece together a history of much of

[29] On his marriage, Jacobo took possession of 120 gold florins and a silver cup which were part of Tomás Ungut's inheritance (see Gestoso, p. 15). Tomás's inheritance was invested, at least in part, in the Cromberger press until he came of age (see document dated 29 Apr. 1511 [APS, Oficio 4, Libro 2 of 1511, fols. 1322^v–1326^r]: 'certain sums of money and other possessions belonging to the said Tomás Ungut . . . have been invested in the said printing-office'; also the document dated 4 Apr. 1514 [APS, Oficio 4, Libro 2 of 1514, fols. 124^v–128^r] which mentions the profit gained on the money invested in the press on Tomás's behalf).

[30] Alonso Morgado, *Historia de Sevilla* (Seville, 1587), fol. 110^r: 'this parish includes the quarters and streets where the most prosperous residents of the whole of Seville live'.

[31] Morgado, *Historia*, fol. 57^r, calls them 'the very rich bookshops of the Cal de Genova'. The Calle de Génova remained an important centre of printing in 17th-c. Seville (see Edward M. Wilson and Don W. Cruickshank, *Samuel Pepys's Spanish Plays* [London, 1980], pp. 16–34).

[32] Ladero Quesada, *La ciudad medieval*, p. 132. At the beginning of the 16th c. the largest foreign community in Spain was that of the Genoese of Seville (see Ruth Pike, *Enterprise and Adventure: The Genoese in Seville and the Opening of the New World* [Ithaca, 1966], p. 1). Guicciardini noted the plethora of foreign craftsmen working in Spain: 'The Spaniard is often considered a wily and intelligent man, but he does not bother with any mechanical or liberal art; almost all the craftsmen at court are from France or elsewhere. Neither does the Spaniard devote himself to business' (see *Viaje a España de Francesco Guicciardini*, p. 56).

the property in the Calle de Génova during at least the first half of the sixteenth century, and the picture which emerges is one of a closely-knit community of artisans, merchants, apothecaries, and retailers linked by marriage or common commercial interests. Jacobo Cromberger's relations or associates signed many contracts for the purchase, sale, lease, or renting of property in the street. For example, in 1514 Cromberger acted as the agent of the Portuguese bookseller Afonso Lourenço for the rent of a house which the latter owned in the Calle de Génova; the same bookseller also provided one of Jacobo's retail outlets in Lisbon.[33] In the same year a local bookseller who was a frequent client, Luis Méndez, collected rent due on a house which had been let to a colleague there.[34] Many documents have come to light concerning the house and shop in the same street in which the Genoese bookseller, Niculoso de Monardis, had lived, and how they were gradually acquired over the years by his son-in-law Juan Varela de Salamanca, Cromberger's frequent partner in local affairs, who also lived and worked in the street.[35] When Jacobo's son, Juan, sent the young Genoese bookseller Guido de Labezaris to Mexico as his factor in 1536, he took over Labezaris's house and bookshop in the street and let the latter out to a silversmith.[36] This shop had previously been owned by Guido's father Sebastiano, who was also a bookseller.[37] It comes as no surprise, then, that when a Toledan client of the Crombergers, Juan López de Pastrana — who was the stepfather of the famous Seville doctor and author of medical treatises, Nicolás de Monardis — decided to return to the city, he should choose to live in the Calle de Génova.[38] By 1556, when Juan Varela de Salamanca's estate was divided among his heirs, one of whom was Jacobo's grandson, Jácome Cromberger, Varela had acquired at least four shops in the street, which were all rented to booksellers, and one house in which a silversmith lived.[39]

However, the most suggestive glimpse we are given of the Calle de Génova's residents is in 1520 on the day after the outbreak of the short-lived *comunero* riots in Seville. A large number of them met in the house of a local *converso* silversmith, Juan de Córdoba, and there formed a vigilante group to protect their lives and property from the mob, which they feared would make the street one of their first ports of call. Five of the signatories were booksellers, one a printer, and at least two were silversmiths. Three were immigrants, and the names of most

[33] Gestoso, p. 17.

[34] Document dated 7 May 1514 (APS, Oficio 4, Libro 2 of 1514, fols. 364^{v}–365^{v}).

[35] Hazañas, i. 231, 245–7.

[36] Two documents dated 10 July 1536 (APS, Oficio 1, Libro 2 of 1536, fols. 68^{r}–69^{r}) and one dated 1 Aug. 1536 (APS, Oficio 10, Libro 1 of 1536, Cuaderno for July, fols. 24^{r}–25^{r}).

[37] Gestoso, p. 33.

[38] Hazañas, ii. 221.

[39] Document dated 7 Aug. 1556 (APS, Oficio 5, Libro 2 of 1556, unfoliated).

of the others suggest a *converso* origin which in some cases can be proved.[40] All were prosperous and half of them were connected in some way or other with Jacobo Cromberger, one being his stepson, Tomás Ungut, and another Varela. They were right to suspect that, had the *comunero* movement gathered the momentum in Seville which it had done in other Spanish cities, the Calle de Génova, peopled as it was by foreigners and *conversos*, would have been an obvious target for the rebels; pickings for looters would have been rich into the bargain.

But there were other important streets in the parish which contributed to its wealth. There was the commercial Calle de Francos, where Justo Alemán, a printer and Comincia de Blanquis's brother-in-law, was living in 1512.[41] Most important of all, however, were the Gradas, the steps which still run alongside the Patio de los Naranjos of the cathedral and where the Genoese bankers had their stalls, slaves were sold, and several of the city's notaries public did a thriving business drawing up the contracts between the bankers and merchants in this most active site of Seville's booming commercial quarter. During his stay in the city in the spring of 1526, the Venetian ambassador, Andrea Navagero, visited the Gradas and was later to describe them in the following terms:

> Around the cathedral . . . there is a sizeable area paved with marble and enclosed by chains . . . All day long a host of gentlemen and merchants stroll up and down in this, the most beautiful corner of Seville. It is called the Gradas, and the street and square which face it are constantly thronged with people. It is a sort of market where numerous auctions take place.[42]

The parish in which Jacobo Cromberger had settled was, then, the commercial centre of a city where many merchants were, like him, of German or Genoese

[40] Hazañas, i. 159, 211–13. The Calle de Génova had long been associated with *conversos* (see Claudio Guillén, 'Un padrón de conversos sevillanos (1510)', *Bulletin hispanique*, 45 [1963], 49–98 [pp. 52–3, 71]). *Conversos* were 'new Christians' who had converted, normally under duress, from the Jewish faith.

[41] Document dated 28 Feb. 1512 (APS, Oficio 1, Libro 1 of 1512, Cuaderno 13, fols. 312^{v}–313^{v} [*olim* 363^{v}–364^{v}]). Justo Alemán (or 'Cancios', 'Canaos', or 'Canan') had probably worked in Ungut's or Cromberger's printing-shop. In 1511 he became Tomás Ungut's guardian and Tomás referred to him as his uncle, while Justo called Tomás 'my nephew' (see Hazañas, i. 88-9). He died in 1524 and Jacobo became guardian to his children, Melchior and Catalina, no doubt investing their inheritance of about 600,000 *mrs* in his own business until they came of age. When Jacobo died, Juan Cromberger became their guardian and held Melchior's money when the latter went to live with his uncle and aunt in Flanders. Justo's wife, Catalina, was a native of Bruges (see document dated 30 May and 1 June 1530 [APS, Oficio 3, Libro 1 of 1530, fols. 558^{r}–561^{v}]). It is therefore possible that Comincia de Blanquis or Meinardo Ungut was also from that city which was the centre of the 15th-c. book-trade of Flanders and a flourishing commercial centre, as Caxton found in the same century. Their son, Tomás Ungut, was frequently referred to as a 'Flemish' merchant, but, as has been noted, this adjective is vague. Catalina Alemán, Justo's daughter, married the rich Florentine banker and merchant, Federigo Alborgo; Lázaro Nuremberger was godfather to their first offspring in 1539 as was the Lucchese banker and slave-trader, Cristoforo Francesquín (see Hazañas, i. 174).

[42] Andrea Navagero, *Il viaggio fatto in Spagna et in Francia* (Venice, 1563), fol. 13^{r-v}.

origin. Its tightly-knit group of foreign and native residents, many of the latter being *conversos* and all living near the cathedral, fairly reflects the world in which Jacobo moved. It was a society of businessmen, artisans, and retailers in which the trades of silversmith and bookseller were frequently represented.[43]

López de Pastrana has already been mentioned as a Toledan bookseller, but when he settled in Seville he did not abandon his other livelihood, becoming the apothecary for the Hospital de las Cinco Llagas. His case was not untypical, for several men combined the two activities, and it is not surprising that apothecaries were the third group (the other two being booksellers and silversmiths) constantly found in documents concerning the Crombergers. Indeed, this fact may also help to account for the large number of medical works which the family press printed in Spanish during the whole of the first half of the century; these editions were designed for apothecaries and surgeons rather than for the university-trained doctors, who would normally have used text-books in Latin.[44] Although not belonging to the professional classes, members of the guilds of apothecaries, and particularly silversmiths, enjoyed considerable social prestige in sixteenth-century Seville; together with printers they formed the highest class of artisan and it may also be something more than a coincidence that they were frequently of *converso* origin.[45]

Whether by luck or by judgement, Jacobo Cromberger found himself at the beginning of the century ideally placed for a successful career as a printer. Not only was he by then the owner of presses, their equipment, and enough capital to avoid having to take out loans to finance editions, but Seville was an excellent location for his activities.[46] In an age when transport by land was slow and expensive, the major centres of printing required easy access to rivers or the sea — a factor which weighed in favour of important sites like Lyons, Paris, Venice, and Antwerp. Seville was situated on a navigable river; it had long-established trade links with Mediterranean countries through the Straits of Gibraltar and with Portugal and Northern Europe via the Atlantic sea-route. This was particularly important for Spanish printing because the industry required large shipments of paper and this had to be imported in the main from Italy and France.[47] On the other hand, Seville printers soon looked to export their products

[43] In Sept. 1516 Cromberger was asked on two occasions to be a godparent. On the first, the child was the daughter of the bookseller, Juan Francisco; the other godparents were Nicolás Alemán, an immigrant silversmith, Juan de Oñate, another silversmith who was active in trade with the New World, and Martín de Oñate, presumably one of Juan's relations. On the second occasion the baby was the daughter of another bookseller, Fernando Díaz (see Hazañas, i. 104).

[44] See below, p. 161, for a discussion of editions of medical works.

[45] Pike, *Aristocrats and Traders*, pp. 133, 141, 143.

[46] In contrast to most 16th-c. printers, there is no record of Jacobo's having borrowed money to finance the production of editions. On the contrary, the Crombergers frequently lent small sums to others.

[47] Spain boasted the first paper-mill in Europe but, by the 16th c., the major centres of the industry

as well as to supply the demand in their own city, and it was important that books be distributed rapidly to the major fairs and other centres if returns on investment were not to be dangerously slow.[48] Jacobo was fortunate in his choice of adopted home, for the major cities in Southern Spain and Portugal which were his nearest potential market either had no presses at all or very inactive ones. Most of these cities were bishoprics and commissioned large editions of liturgical works, while they also provided outlets for more popular books. Whereas over 300 recorded editions were produced at Seville from 1501 to 1520, only 39 were printed at Lisbon, 16 at Granada (of which 12 were produced by the Seville printer Varela), two at Murcia, and none at all at Évora, Córdoba, Badajoz, Cádiz, Málaga, and Jaén.[49] Moreover, in Seville itself there was little competition from other printers for these markets during Jacobo Cromberger's lifetime. From 1504 until his death in 1528, he dominated the industry: over two-thirds of the books printed in the city came from his presses, a further fifth came from those of his friend Varela, and the insignificant remainder were produced in unidentified or ephemeral offices.[50]

Another feature of Seville made prospects bright for Cromberger. When Meinardo Ungut had died in 1499, the inventory of the equipment in his printing-office included the following entries: 'item, the partnership has approximately 600 *libras* of type-metal which has been cast into letters; item, a black slave; item, three presses with their equipment and iron chases'.[51] The appearance of a black among the tools of the trade is suggestive, for in 1540, when an inventory

were in France and Italy. In the mid 14th c. Italian paper had first appeared in Spain, and it gradually replaced Spanish paper, which was of lower quality (see Jean Irigoin, 'L'Introduction du papier italien en Espagne', *Papiergeschichte: Zeitschrift der Forschungsstelle Papiergeschichte in Mainz*, 10 (1960), 29–32. The Crombergers bought paper from the Doria, Spinola, Adorno, Soprani, Negrón, Grimaldi, Rondinelli, Gentile, Camilioni, Centurione, Calvo, Fantoni, Bonguillermini, León, Cano, and Burón, few of whom were of any other nationality than Genoese (see Gestoso, p. 12; Hazañas, i. 77–83, 90, 94–5, 101, 103–4, 109, 112; documents dated 11 June 1555 [APS, Oficio 23, Libro 2 of 1555, fol. 346^{v}] and 18 Mar. 1556 [APS, Oficio 23, Libro 1 of 1556, fols. 921^{r-v}]). For the dispatch of French paper to Spain from 1552 onwards see Henri Lapeyre, *Une Famille de marchands les Ruiz: contribution à l'étude du commerce entre la France et l'Espagne au temps de Philippe II* (Paris, 1955), pp. 563–5, 571. Paper was France's second-largest export to Spain after textiles. In the late 1550s it even became the largest. French paper was particularly suited to the printing of missals, while Spanish paper was used for cheaper books (see Manuel Abizanda y Broto, *Documentos para la historia artística y literaria de Aragón procedentes del Archivo de Protocolos de Zaragoza*, 3 vols. [Zaragoza, 1915–32], i. 309, 341).

[48] Antonio Domínguez Ortiz, *Orto y ocaso de Sevilla*, 3rd edn. (Seville, 1981), p. 51. The two settings of the *Passiones* or *Lectiones* printed by Polono and Cromberger in 1503 suggest an interest in this export market (see Norton, *Printing in Spain*, p. 133).

[49] Norton, *A Descriptive Catalogue, passim.*

[50] Most of the Crombergers' surviving books were signed despite Norton's observation that printers tended not to sign books in cities where they faced no serious competition (see Norton, *Printing in Spain*, pp. 20–1).

[51] Nicolás Tenorio, 'Algunas noticias de Meinardo Ungut y Lanzalao Polono', *Revista de archivos, bibliotecas y museos*, 5 (1901), 633–8 (p. 636).

of Juan Cromberger's estate was made, the slaves he employed in his workshop were also listed: three blacks were *batidores*, or beaters, and one black and a 'white slave'—a North African or *morisco*—were *tiradores*, or pullers.[52] Unfortunately, although there is evidence that Jacobo also owned both black and 'white' slaves, it is not known in what capacity he employed them. Nevertheless, it is most probable that, like so many artisans engaged in a wide variety of trades in the city, he took advantage of the large slave population for which Seville was renowned in the sixteenth century and used them to reduce his labour costs. The wages of pressmen were among the highest paid to any artisan, at least in Lyons.[53] So printers anxious to cut costs would naturally look for cheaper labour if guilds did not protect the interests of the journeymen.[54] In Seville there was no such guild for the printing industry during at least the first half of the sixteenth century and so employment restrictions did not exist.

There has recently been considerable debate among historians about the relative profitability of free and slave labour in the antebellum South of the United States, but it is unlikely that there was much doubt in sixteenth-century Seville that the purchase of slaves to work in the printing-shops made sound economic sense. Admittedly, if only free labour were employed, when business was slack expenses could be reduced by not hiring journeymen and running the office at half-press, but it is equally the case that free workers would have to be paid for the remainder of the contract which the master-printer had made with them, and that this could vary from the time it took to print a single edition to a period of three years. On the other hand, slaves could be employed in other sorts of work if the printer had, like the Crombergers, diversified his activities, or they could be hired out. Of course, slaves would normally be fed by their masters, and the cost of food rose sharply during the frequent famines to which Seville was subject, but the contracts drawn up with free pressmen also included the clause that the master-printer would provide board and lodging for his workforce. Even taking into consideration the cost of buying a slave, there is no doubt that slave labour was cheaper than free employment in Seville at this time: the purchase price of a slave was approximately equivalent to only one year's wages of a free pressman.[55]

[52] Gestoso, p. 74. The *moriscos* were descendants of the North African invaders of Spain, and had been forcibly converted to Christianity.

[53] Natalie Zemon Davis, 'Strikes and Salvation at Lyons', *Archiv für Reformationsgeschichte*, 56 (1965), 48–64 (pp. 52–3).

[54] See A. C. de C. M. Saunders, *A Social History of Black Slaves and Freedmen in Portugal 1441–1555* (Cambridge, 1982), p. 223, n. 12. A 'beardless Ethiop youth' was employed in a press at Fossombrone in Italy in 1513 (see F. J. Norton, *Italian Printers 1501–1520: An Annotated List with an Introduction*, Cambridge Bibliographical Society Monographs, 3 [London, 1958], p. xxii).

[55] Wages in the printing industry throughout the period are difficult to assess. In 1519 pressworkers hired in Seville were earning between 3 and 3.5 *ducados* per month (about 1,100 to 1,300 *mrs*) while in 1528 a puller received only 2.5 *ducados* (about 950 *mrs*) (see Hazañas, ii. 243–4, 277). Apprentices

It is true that there were certain risks involved in the purchase of a slave. In 1526 Jacobo Cromberger invested 17,000 *mrs* in a North African called Barca. Employing the usual formulae found in sale contracts for slaves, the vendor guaranteed that Barca was of good character, but this proved to be somewhat short of the truth.[56] When Jacobo realized the true nature of his slave, he decided to cut his losses and sold him to one Francisca Fernández for a price considerably below what he had paid. His apprehensions were well founded. In 1529 the unwitting purchaser sued Jacobo's heirs on the grounds that their father had guaranteed the slave's good character, but that she had been cheated of her investment: Barca had been arrested and hanged for an undisclosed crime and Jacobo's heirs were obliged to return the purchaser's money.[57] Nevertheless, such cases appear to have been rare, especially among blacks—the *morisco* and North African slaves were always considered less reliable—and there is no evidence that free labour was any less troublesome.[58] In 1550 the Seville printer Sebastián Trugillo had a white apprentice imprisoned for running away before he had fulfilled his contract.[59] At least slaves were unable to organize and bring the presses to a halt as the printing journeymen did at Lyons and Paris.[60] Some of the workers in Cromberger's shop would always have been free, especially the skilled compositors, but the heavier manual tasks were frequently performed by slaves. As free labour was expensive, this use of slaves may account in part for the profitability of his operation.[61]

Seville thus offered good prospects for an enterprising printer: it was a large

were clearly cheaper. In 1513 the average price in Seville for a mature male black was approximately 11,000 *mrs* (see Alfonso Franco Silva, *Regesto documental sobre la esclavitud sevillana (1453–1513)* [Seville, 1979]). In 1526 Jacobo Cromberger paid 17,000 *mrs* for a mature 'white slave' and in 1530 his son bought four mature 'whites' for 200 *ducados* (or 18,750 *mrs* each) (see document dated 6 Mar. 1530 [APS, Oficio 4, Libro 1 of 1530, fols. 859ʳ–860ᵛ]). Similar prices are quoted for 1528 by Pike, *Enterprise and Adventure*, pp. 63–4. When Juan sent his black slave, Pedro, to Mexico in 1539 he valued him at 100 *ducados* or 37,500 *mrs*, but Pedro was probably a skilled craftsman and would therefore have been untypically expensive (see Gestoso, p. 66). Pike maintains that by the middle of the century a mature black fetched a price of 80–90 *ducados* or 30,000–33,750 *mrs* (*Aristocrats and Traders*, p. 176).

56 Hazañas, i. 136.

57 Document dated 2 June 1529 (APS, Oficio 4, Libro [2] of 1529, unfoliated). Gestoso, p. 35, mentions this document but does not realize its implications.

58 Pike, *Aristocrats and Traders*, pp. 177, 180–1. The different attitude towards black and Moorish slaves was similar in Portugal at this time (see Saunders, *A Social History*, pp. 108–9).

59 Gestoso, p. 114.

60 Davis, 'Strikes and Salvation at Lyons'; ead. 'A Trade Union in Sixteenth-century France', *Economic History Review*, 2nd ser., 19 (1966), 48–69. There is a hint that pressmen at Zaragoza in the 1570s were not averse to striking for higher wages (see Abizanda y Broto, *Documentos*, i. 359).

61 In a different country and at a different time, but ones in which complete accounts were kept, a third of the cost of printing an edition was wages and another third paper (see L. Voet, 'Production and Sales Figures of the Plantin Press in 1566', in S. van der Woude (ed.), *Studia Bibliographica in Honorem Herman De La Fontaine Verwey* [Amsterdam, 1967], pp. 418–36 [p. 421]).

commercial centre situated on major trade routes; there was a potential market for books in the city and outside it; there was no fierce competition to supply this market; slave labour was plentiful; and, finally, there was in Seville an active group of scholars—the 'Hispalensis academia' about which more will be said later—whose help as editors and proof-readers could be solicited. But one further advantage of the city was to prove decisive for the Crombergers. In 1503 Seville was granted a monopoly of trade with the newly-discovered Americas, the Casa de Contratación was founded, and Jacobo, like many other artisans, gradually became aware of the possibilities for profit which the New World provided. Seville in the first half of the sixteenth century was therefore a city which afforded unique opportunities for the twin activities to which members of the family were to devote their lives: printing and trade.

Printing, Publishing, and Selling

Initially Jacobo limited himself to printing. The first edition which he signed was Johannes Versor's fifteenth-century exposition of a thirteenth-century work, *In magistri Petri Hispani Logicam indagatio*, the colophon of which bears the date 15 April 1503. As with all the books which he subsequently printed in that year, his name is preceded by that of Polono, who was the senior partner in the company. It is likely that Cromberger had been working with him since at least the time of Ungut's death, and he had taken over effective control of the Seville office from the end of 1502 when Polono went to Alcalá. Polono there produced two fine editions of Montesino's famous translation of the *Vita Christi* by Ludolph of Saxony, which must have taken many months to print. He stayed in Alcalá until March 1504. By that time the Seville office had passed into Jacobo's hands and Polono's name appears in no further colophons. He had not, however, died, but must merely have retired, for in April 1514 he was still involved in the affairs of Tomás Ungut, his former partner's son.[62]

For the remainder of the decade Cromberger printed a wide variety of works, of which an average of almost seven titles survives for each year. There is, though, a great disparity between the number of books printed in individual years, and it is likely that the disappearance of editions has been considerable. Outside factors probably also affected production. For example, the small number of titles known to have been printed in 1507 may well reflect the devastating plague and famine which struck Seville that year, while the years 1506 to 1508 were a period of depression for printers all over Spain.[63] An examination of contracts for Jacobo's

[62] Gestoso, p. 17.

[63] Norton, *Printing in Spain*, p. 55. Cromberger survived the depression. Other Seville printers like Pegnitzer and Brun did not.

purchases of paper during this decade indicates that the largest amounts were bought in 1506, which was also the year from which the highest number of editions printed exclusively by him in those years survive. A study of the titles appearing in that year gives some idea of the sort of books he was producing. There are ten surviving titles, including two folio editions of considerable size, which would have each taken many weeks to print as both contain over 100 sheets. One was Nebrija's Spanish–Latin dictionary, the other an edition of the Epistles and Gospels. At the other end of the spectrum there were two medical works in Spanish, Alfonso Chirino's *Menor daño de medicina* (18 sheets) and Diego Álvarez Chanca's *Tratado en que se declara de qué manera se ha de curar el mal de costado pestilencial* (2.5 sheets); two short books and a longer one of Spanish verse, *Coplas de Mingo Revulgo* (5 sheets), Fray Íñigo de Mendoza's *Coplas de vita Christi* (9 sheets), and Fernán Pérez de Guzmán's *Las setecientas* (12.5 sheets); two short works of devotional verse in Latin, Aegidius Delphus' *Septem psalmi penitentiales exametro metro exarati* (1.5 sheets) and Adam de Montaldo's *Passio domini nostri Jesu Christi carmine heroico composita* (5 sheets), and a final devotional prose work in Spanish and Latin, Luis de las Casas's *Tratado de la santa concepción de nuestra abogada la virgen María* (9 sheets).

The contrast between the slight and substantial editions printed in 1506 is one which will be seen throughout Jacobo Cromberger's career and those of his successors. It suggests a policy of keeping the presses occupied with short works in between the printing of more ambitious books, and of using cheap editions which would provide rapid income to offset longer-term investment in more expensive ones. In Paris, at least, the capital tied up in a single edition of a substantial book could be greater than the cost of equipping the printing-shop itself; it can therefore be appreciated how important it was for the printer to ensure that some money was continually flowing into his business.[64] The practice of combining slight and substantial editions was common to the early printers: Caxton seems to have engaged in it, while in Spain it was alluded to by Luis Ortiz in his *Memorial* (1558) to Philip II when he talked of, 'the ABCs, catechisms, books of rhymes, and other slight works which are needed by the Republic and which printers have to produce in slack periods between printing more ambitious editions'.[65] Ephemeral chap-books, certificates, indulgences, and jobbing printing must have played an important part in such a strategy, but most of the evidence for these publications has disappeared. The odd volumes composed of ballad chap-books and the occasional liturgical or devotional *pliego suelto* now to be found in a handful of the world's major libraries are merely the tantalizing remains

[64] Febvre and Martin, *L'Apparition*, p. 166.

[65] Blake, *Caxton*, pp. 66, 72; Luis Ortiz, *Memorial del contador don Luis Ortiz a Felipe II*, transcribed by Jaime Fernández Laville (Madrid, 1970), p. 49.

of what must have been one of the mainstays of the press's production. They have been preserved in the main by foreign travellers to Spain, who noticed that *pliegos* were typical of Spanish printing and collected them as curiosities.[66] It is only in the inventories of contemporary bookshops that evidence of the size of editions of such *pliegos* emerges. In 1528 Jacobo's shop contained 50,500 sheets of rhymes, 21,000 sheets of prayers, over 10,000 copies of devotional woodcuts normally of one sheet each, 3,000 'Rosaries of Our Lady' (two sheets), and smaller quantities of ABCs and pamphlets setting out the rudiments of plainsong.[67] When Juan Cromberger died eleven years later, the shop contained 10,000 ABCs, over 5,000 sheets of rhymes, over 3,000 copies of a 'Life of Our Lady', and various other broadsheets.[68] Of most of these not a single copy has survived.

The mention of a strategy for keeping the presses busy is based on the assumption that Jacobo printed largely what he chose to produce rather than what a publisher had ordered. The distinction between printer and publisher is an important one, for many printers never undertook publication, merely working under contract or on commission, while many publishers never dirtied their hands at a press. The publisher, were he author, bookseller, merchant, capitalist, or representative of an institution, ran the risks of being left with an unsold edition; the printer would merely be paid for his skill as a craftsman and took no share in whatever profits were made. Printers who did not issue their own editions tended to remain poor artisans while their counterparts, the booksellers and publishers, became prosperous, for it was in the trade that profit was to be gained.[69] If a printer did not have access to the capital required to publish editions on his own behalf, his presses would stand idle during a period when orders were slack, and he would incur losses.[70] If he did have the capital and chose his titles wisely, he would be able to use his presses efficiently and would also make large enough profits to enable him to continue financing editions of his choice rather than be beholden to publishers who placed their orders with him.

When Cromberger and Polono had worked as partners they had printed some editions for booksellers: the *Passiones* of 1503 was partly financed by the Seville

[66] Carlos Romero de Lecea, *La imprenta y los pliegos poéticos*, Colección de estudios y ensayos Joyas bibliográficas (Madrid, 1974), p. 31.

[67] Two documents dated 7 June 1529 (APS, Oficio 4, Libro [2] of 1529, unfoliated). Gestoso, pp. 37–42, reproduces part of one of these lists (book list: 1528) but his transcription is plagued by errors.

[68] Document dated 20 Sept. 1540 (APS, Oficio 17, Libro 2 of 1540, unfoliated). Gestoso, pp. 86–98, reproduces this inventory, but again with errors of transcription.

[69] The profits in the book-trade at Barcelona, for instance, went to the booksellers, not the printers (see Norton, *Printing in Spain*, p. 132, and Madurell and Rubió, *Documentos*, pp. 55*, 81*–3*). The same situation probably explains the relative prosperity of booksellers at Coimbra and Lisbon in the 1560s and the poverty of printers in those cities (see Jorge Peixoto, 'Aspectos económicos do livro em Portugal no século XVI', *Gutenberg-Jahrbuch* [1965], 142–9). For a comparison between the wealth of booksellers and printers at Lyons see Martin and Chartier, *Le Livre conquérant*, pp. 260–1, 269–70.

[70] Febvre and Martin, *L'Apparition*, p. 166.

bookseller Juan Lorenzo (or Lorencio), and the edition of Versor's *In magistri Petri Hispani Logicam indagatio* of the same year was printed for Gracia de la Torre and Alfonso Lorenzo, the latter probably being the Seville bookseller of that name who lived in the Calle de Génova. In 1504, the first year of Cromberger's independent career, he printed at least two works for local booksellers: Nebrija's edition of Persius' *Satires*, again for Juan Lorenzo, and the missal of León use for Lázaro de Gazanis.[71] Soon, however, like many printers in Castile, he began to publish works in his own right and subsequently printed only three surviving books on commission for a bookseller: the 1511 edition of the *Vida y excelencias y milagros de Santa Ana*, printed for the Valladolid bookseller Pedro de Villalta, and the manual of baptism and the breviary for the diocese of Évora in Southern Portugal which he printed in 1528 for António Lermet. Lermet possessed a royal privilege for the publication of liturgical works—a privilege which had presumably been granted by the Infante Dom Afonso, who had close links with the bishopric. The colophon of the breviary implies that Lermet not only financed the edition, but also had some supervisory role in its printing. It is particularly noteworthy, however, that Cromberger did not rely on such commissions, for printers at Seville in the previous century had been dependent upon that sort of work.[72]

Jacobo was clearly an enterprising publisher with a close knowledge of what would sell. One of his first ventures was the publication in 1505 of Padilla's *Retablo de la vida de Cristo*; this was so successful that the press issued at least four subsequent editions (1510, 1512, 1516, and 1518). Thereafter he was to publish many reprints of popular works, the first editions of many of which had originally come from his own office. He did, however, complement his publishing activities in two ways. First, as has been seen, he did some printing under contract, notably of liturgical works which had always been a staple for printers. He became justly famous in Spain for these editions. Service books, especially missals, would have been expensive to set up and slow to print. Not only did they contain a variety of types which needed to be harmonized, and a large number of pages, but they required careful proof-reading. Moreover, both sides of each sheet were printed in black and red inks, thereby more than doubling the work of the pressmen because each colour had to be printed separately—a skilled and time-consuming process. If the substantial print-runs and the fact that some of the copies were of expensive vellum are also taken into account, it will be apparent that the printing of such books would have been a costly undertaking. The capital tied up in these editions was too great for the majority of printers to afford, and so it was customary for the ecclesiastical authorities to make a large advance in cash, vellum, or paper before printing began. As these sorts of editions

[71] Norton, *Printing in Spain*, p. 11.
[72] Hazañas, i. 70.

were not normally published speculatively but printed under contract, they were an attractive proposition to Jacobo Cromberger, just as they had been to his predecessors Ungut and Polono.[73] They kept some of his presses occupied for a long period, ensuring a continuity of work and, once the edition was completed, he did not have to wait several years for individual booksellers to pay their bills and thus enable him to recoup his investment. In theory, at least, as soon as the edition was delivered he would be paid for his work.[74] This sort of printing, with its particular financial arrangements, complemented the publication of more rapidly produced books or ephemera. Cromberger may in this way have ensured that his presses operated without long periods of idleness and that there was a steady inflow of money to the office so that there were no times when he would be forced to take out expensive loans.

Second, he appears to have founded his business on secure enough an economic base to be able to branch out from his activities as a printer and publisher. By 1508 he owed money to one Juan Jiménez, a Seville bookseller, for binding books for him.[75] As it was normally the bookseller rather than the printer who bound books or had them bound for him, the document suggests that Cromberger was by this date engaged in retailing his products.[76] From then until the end of his life there is abundant evidence that Jacobo was a bookseller on a large scale and that he built up an extensive network of outlets. These ranged from the humble street-vendor who sold *pliegos* in the flea-market of Seville among the city's brothels (the 'very swarthy fellow who used to sell books and haberdashery at the Puerta del Arenal') and the peddlar, like 'Cortés who goes from village to village selling his wares', to the large merchant-booksellers who received shipments of several thousand books at a time and ran up bills of 200 *ducados* or more.[77] At the time of his death this network was recorded in the list of debts he was

[73] An exception to the printing of liturgical works under contract was the Cordoban missal which Montesdoca was to print speculatively with a colleague. But he got cold feet in 1559 and backed out of his commitment, selling the contract to Simón Carpintero (see Klaus Wagner, *Martín de Montesdoca*, pp. 123–4).

[74] In 1556, however, Varela's heirs were still owed 226,000 *mrs* by the chapter of Jaén Cathedral for the missals which he had printed for it, presumably at some time before 1539, when he appears to have retired from the craft (see document dated 7 Aug. 1556 [APS, Oficio 5, Libro 2 of 1556, unfoliated]). For further evidence of delays in payment for liturgical editions, see Beltrán de Heredia, *Cartulario*, ii. 416–17, 423–5. Cromberger did not always automatically hand over all copies of such editions to the chapter concerned, as his own stock reveals.

[75] Hazañas, i. 80.

[76] Binding-tools were part of the stock-in-trade of the 16th-c. bookseller. When the bookseller Guido de Labezaris sold his books to Juan Cromberger, his equipment included a set of such tools (see document dated 10 July 1536 [APS, Oficio 1, Libro 2 of 1536, fols. 70^{v}–71^{r}]). When Francisco de Monardis died, Pedro de Vargas rented his bookshop in the Calle de Génova and took over his stock including the binding-tools (see document dated 28 January 1530 [APS, Oficio 4, Libro 1 of 1530, fol. 386^{r-v}]). There is similar evidence from other countries (see, for example, Martin and Chartier, *Le Livre conquérant*, pp. 250, 261).

[77] Gestoso, pp. 43–54.

owed by numerous booksellers, and this list enables us to discover his major markets.

The largest was, naturally enough, that of Southern Spain, both Seville itself and towns in its vicinity: Ayamonte, Córdoba, Écija, Jerez [de la Frontera?], Llerena, and Zafra. Further afield in Andalusia and Extremadura there were Granada, Jaén, Málaga, and Mérida. Presses in the neighbouring kingdom of Portugal were few in number and very unproductive, so this market was supplied by printers in Spain as well as France and Italy. Large numbers of books which must have produced a substantial income for Jacobo frequently followed the sea-route from Seville to Lisbon; indeed, at least one bookseller went to a watery grave while transporting Jacobo's products to the Portuguese capital.[78] But his books were also dispatched to Évora and as far north as Oporto, while others sailed west to the Azores. The third large market to which his products travelled was Northern Castile. Although they would have had to make at least some of the slow journey by land, numerous copies were sold in the bookshops of Burgos, Salamanca, Toledo, and Alcalá de Henares. As might be expected, large sales were made to booksellers attending the great fairs at Medina del Campo which took place in May and October each year. Notarial documents show that members of the Cromberger family were frequently absent from Seville during these two months and suggest that Jacobo or his son, Juan, regularly attended the fairs. Juan was certainly at Medina in 1529 when he signed an agreement with Antonio de Guevara, Bishop of Guadix, concerning the printing by the Cromberger press of the bishop's *Libro áureo de Marco Aurelio* and his *Relox de príncipes*. In that agreement Juan implied that he also visited the fairs at Villalón and Medina de Ríoseco.[79] Substantial dealings with the booksellers of the nearby city of Valladolid provide further evidence of the importance of Medina for the Crombergers' distribution network. It is possible that whole editions were dispatched to Northern Castile as soon as they were printed, for Hernando Colón (Columbus), the natural son of the explorer and an ardent bibliophile, bought more copies of Cromberger editions in Valladolid and Medina than he did in Seville, where they had been printed and where he lived and built up his magnificent library. Some of these books were bought by Colón soon after they had come off the presses: Cromberger's editions of Boccaccio's *Fiammetta* and of Cortés's *Carta tercera* printed in 1523 were bought by Colón the following year at Valladolid. But the most rapid export of a Cromberger edition to Northern Castile of which a record survives was that of Cortés's second letter, which Jacobo finished on 8 November

[78] Hazañas, i. 152–3, and two documents dated 15 June 1528 (APS, Oficio 4, Libro 3 of 1528, unfoliated).

[79] E. García Chico, 'Documentos referentes a la imprenta en Medina del Campo', *Castilla: boletín del Seminario de Estudios de Literatura y Filología de la Universidad de Valladolid, Facultad de Historia*, 2 (1941–3), 233–98 (pp. 233-40).

1522 and Colón bought at Valladolid for one *real* three weeks later on 2 December.[80]

Copies of Cromberger books occasionally travelled further afield in Europe. In 1514 he printed Álvarez Chanca's edition and commentary of Villa Nova entitled *Commentum novum in Parabolis Arnaldi de Villa Nova*. Colón paid thirty pfennigs for a copy of this edition when he was in Cologne eight years later.[81] It can therefore be seen that Jacobo distributed books over a wide area and his editions would have been available in most of the major cities of the Iberian Peninsula.

But he did not limit his activities to those of a bookseller handling exclusively his own products. In 1511 there is already a suggestion that he was selling volumes which were so expensive that they were unlikely to have been printed on his own presses; either they would have been weighty printed tomes imported into Seville from abroad, or costly manuscripts.[82] The inventory of goods which formed part of the dowry of Jacobo's daughter contained an entry—'a table and chest from the shop'—which reveals that the family owned a bookshop; further entries confirm that this was attached to their house and printing-office in the Calle de Marmolejos, to which they had moved in 1511.[83] In the shop they sold the products of other Spanish and foreign presses. As had been seen, Jacobo arranged with the Alcalá printer and bookseller, Eguía, to exchange at a discount the books they printed and to sell each other's editions in their respective towns. The inventory of Jacobo's estate provides further evidence of his dealings with many of Spain's leading printers and suggests that he not only retailed their books in Seville and elsewhere, but also sent his own products to them. This was, of course, a frequent practice among contemporary printers and booksellers elsewhere in Europe, and barter was a common form of payment among them.[84] There are entries for Juan de Junta of Burgos, Ramón de Petras of Toledo, and Jorge Costilla of Valencia. On his death in 1528 Jacobo's shop also contained books printed by his friend Varela.[85]

Therefore, within ten years of embarking on his career as an independent printer, Jacobo had published his own editions, built up a comprehensive distribution network, and begun to sell the products of presses other than his own. Several different occupations were therefore combined in the same person, and this gave him a certain measure of security against fluctuations in the market as well as

[80] *Catalogue of the Library of Ferdinand Columbus; Reproduced in Facsimile from the Unique Manuscript in the Columbine Library of Seville by Archer M. Huntington* (New York, 1905), Nos. 4011, 4094, and 272. In the last of these Colón mistakenly dated his purchase to 2 Dec. 1521 when he must have meant 1522.

[81] *Catalogue of the Library of Ferdinand Columbus*, No. 260.

[82] Hazañas, i. 89–90.

[83] Document dated 7 June 1529, (APS, Oficio 4, Libro [2] of 1529, unfoliated).

[84] Febvre and Martin, *L'Apparition*, pp. 340–1.

[85] Gestoso, pp. 45, 46, 50, 51.

various opportunities to build upon his initial capital.[86] He was, then, in a position to branch out into other commercial ventures and in this way to diversify even further his business activities, especially in the 'boom-town' atmosphere of Seville at this period.

Frequently in the history of early printing rich merchants became interested in the potential profits to be gained in the industry. Indeed, without investment from capitalists, most presses could not have operated at all.[87] As was noted previously, the first printers of Valencia had been called there by a German merchant. The first printer in England, William Caxton, was an important mercer who turned to printing only when he was already wealthy and when times were no longer propitious for his mercantile activities; he then invested the money he had made as a merchant in his press.[88] Cromberger's career, however, shows a very different development. It was his press which provided the basis of his prosperity, and profits made from printing were then channelled in other directions. But this did not lead to his neglecting the quality of his craftsmanship. His books could match almost any others being issued from Spanish presses at the time and, if modern bibliographers have been lavish with their praise of his work, sixteenth-century commentators were no less laudatory.[89] So high was Jacobo's reputation in Spain that a near-contemporary talked of him as first among 'the most famous printers to come to this kingdom'.[90] His editions became so well known that they were imitated not only in other cities of the Iberian Peninsula, but also in Italy.[91] Yet this reputation was achieved without either the patronage enjoyed by a printer such as Brocar, who renewed and added to his material at the expense of Cardinal Cisneros, or the wealth of a Caxton.

The Cromberger Press and Portugal

Jacobo's reputation as a fine printer was firmly established only a few years after he began to sign books independently. In a royal letter of February 1508 King Manuel of Portugal invited him to settle in his realm and print legal and liturgical

[86] Gestoso, p. 41.

[87] Febvre and Martin, *L'Apparition*, pp. 173–92.

[88] Blake, *Caxton*, pp. 44–5.

[89] For example Barrantes, 'Apuntes para un catálogo de impresores', p. 404: 'The bibliophile who manages to make a "Cromberger collection", bringing together the editions printed by Jácome [i.e. Jacobo] and Juan, will be the envy of princes and millionaires for he will possess the finest gothic books ever produced in Spain.'

[90] Jerónimo Román, *Republicas del mundo divididas en XXVII libros*, 2 vols. (Medina del Campo, 1575), ii. fol. 214, cited by Cristóbal Pérez Pastor, *La imprenta en Medina del Campo* (Madrid, 1895), p. 196.

[91] Ruth Mortimer, *Harvard College Library, Department of Printing and Graphic Arts, Catalogue of Books and Manuscripts: Part II: Italian 16th Century Books*, 2 vols. (Cambridge, Mass., 1974), i. 24–5, 245–6; Norton, *Printing in Spain*, pp. 136–7, 152–3.

works for his subjects.[92] There is no trace in Seville archives of Jacobo's having been in the city in February or March of that year and it is likely that, as this letter suggests, he had gone to Portugal to discuss with the king the printing of an edition of the *Ordenações manuelinas*. At least two other presses had been operating sporadically in Lisbon shortly before this date: those of Valentim Fernandes of Moravia and Giovanni Pietro (João Pedro) Bonhomini of Cremona, but Fernandes may have decided in 1505 to abandon printing temporarily, while Bonhomini printed nothing from 1507 to 1512.[93] Even if Dom Manuel had issued his invitation because there was a danger that printing in Portugal would cease altogether if he did nothing to encourage it, it is nevertheless some measure of Jacobo's standing at that time that the king mentioned only him by name in what was an open invitation to foreign printers to settle in the kingdom. It also suggests that, as Cromberger already enjoyed a reputation in Portugal for his work, his books had by this early date found their way from Seville to Lisbon. This suggestion is confirmed by a power of attorney signed by Jacobo in May of 1508 asking a German merchant in Portugal to collect on his behalf a debt owed by a Lisbon bookseller whom he had been supplying.[94] He may even have already issued at least one edition specifically destined for the Portuguese market: the *Tractatus de indulgentiis* by Afonso de Portugal, Bishop of Évora, had come from his presses in about 1505.[95]

Despite tempting promises of high social status made by the king—printers who settled in Portugal were to be made gentlemen of the royal bedchamber—Cromberger did not accept the invitation. Modern *converso*-hunters, mindful of the origins of many of Jacobo's contacts in Seville, would doubtless attribute his refusal to the condition laid down in the royal letter that foreign printers would be welcome only if they 'were Old Christians without a trace of Jewish or Moorish blood'. King Manuel made a further proviso which shows that he was aware of the high cost of establishing a printing-office: any newcomer would have to possess the large capital of 2,000 gold *dobras*, and it could be argued that Jacobo did not take up the king's offer because he could not meet that requirement. But reliable evidence exists that he did have a large amount of money tied up in his equipment and was buying property in Seville just after his return from Portugal,

[92] Venâncio Deslandes, *Documentos para a historia da typographia portugueza nos seculos XVI e XVII*, 2nd edn. (Lisbon, 1888), pp. 12–14.

[93] Norton, *A Descriptive Catalogue*, p. 498. In the 15th c., Portuguese presses had been unproductive and few in number. Painter and Sheppard record only 30 Portuguese incunables compared with 856 for Spain (see 'General Introduction', p. ix, and 'Introduction to the Presses', p. lxxiii). The weakness of Portuguese printing is surprising when the wealth and size of Lisbon at this time are considered.

[94] Hazañas, i. 80.

[95] Cromberger was not the first Seville printer to look to Portugal; Pegnitzer had printed a primer for the Portuguese market (see Norton, *Printing in Spain*, p. 8).

so poverty is unlikely to have been his motive.[96] More convincing than either of these explanations for his reluctance to transfer his loyalty to Portugal was his financial acumen: his awareness of the unique opportunities offered by Seville as a centre for printing and trade.[97]

This does not mean that he turned his back on the neighbouring kingdom. When the *Ordenações manuelinas* were eventually printed at Lisbon by Fernandes in 1512 or 1513 Cromberger gave him some practical help by lending or hiring (printers seldom did something for nothing) the main type in which they were set (see the Notes for Type 7 in Appendix Two).[98] Evidence of numerous debts owed by booksellers in Portugal to the Cromberger press show that Jacobo and his son, Juan, found a lucrative outlet for their books there. Although they specialized in printing works in Castilian, their success in Portugal is not surprising because there was a considerable degree of bilingualism among sixteenth-century readers in that country. Indeed, the Crombergers even had a factor working for them at Lisbon as a distributor, one 'Luis Rodrigo'.[99] The family's products found their way into noble libraries in that kingdom during the sixteenth century and later: for example, Teotínio de Bragança (1530–1602), Bishop of Évora, acquired no fewer than nine books printed by Jacobo, Juan, and the latter's heirs.[100] The Visconde da Trinidade's library (now in the Biblioteca Geral of the University of Coimbra) and the Manizola Collection (now in the Biblioteca Pública e Arquivo Distrital in Évora) both contained copies of Cromberger editions, while the library of the Portuguese kings possessed many of the press's products.[101] Indeed, the Biblioteca Nacional at Lisbon and the library at Évora boast the richest collections

[96] Hazañas, i. 80.

[97] Otte, 'Jakob und Hans Cromberger', p. 132, claims that Nuremberg merchants were known for their caution; Jacobo's reluctance to risk his business in Seville for only the promise of riches in Lisbon was, then, perhaps characteristic of a Nuremberger.

[98] No document records Cromberger's passing this type to Fernandes but a close examination of it confirms that it was the same as the one owned by Jacobo. There are, admittedly, other ways in which Fernandes could have acquired it, but there is no evidence for the tempting possibility that there was direct contact between him and Moravus. They were both natives of the same area in what is present-day Czechoslovakia: Moravus was sometimes known as 'Mattias de Olmutz' implying an origin in Olomouc; Artur Anselmo, *Origens da imprensa em Portugal* (Lisbon, 1981), p. 149, suggests that Fernandes could have learnt his trade in the nearby town of Brno, of which, as Dr Oldřich Kašpar of the University of Prague kindly informs me, Fernandes was a native.

[99] Gestoso, p. 41. This 'Luis Rodrigo' was probably Luís Rodrigues, the royal bookseller, who began his own career as a printer at Lisbon in 1539.

[100] López de Mendoza, *Proverbios* (1516); Mena, *Las trescientas* (1517); Fernández de Enciso, *Suma de geografía* (1519); Mena, *La coronación* (1520); Rodríguez de Almela, *Valerio* (1527); Jiménez de Prejano, *Lucero* (1528); Guevara, *Libro áureo de Marco Aurelio* (1533); Aesop (1533); and Cazalla, *Lumbre del alma* (1542). Dr Ivo Castro of the University of Lisbon generously provided me with this information.

[101] Some idea of the holdings of the royal library, now dispersed, can be gained from an examination of the Biblioteca da Ajuda, the Biblioteca Nacional at Rio de Janeiro, King Manuel II's library at Vila Viçosa, and the Biblioteca Nacional at Lisbon.

of Cromberger editions in Europe after the Biblioteca Nacional at Madrid and the British Library. Arguments for the distribution of books in the sixteenth century based upon their present location must be treated with caution, for unwary use of this evidence would suggest that nowhere did the Crombergers have more active factors than in Bloomsbury and Manhattan. Nevertheless, the large number of their books which belonged to private or institutional libraries in sixteenth-century Portugal and were subsequently assimilated into major collections in that country does indicate that there was always a significant market in the kingdom for the Crombergers.[102]

They not only exported their Latin and Castilian books to Portugal, but also produced editions specially intended for sale there. In 1528 Jacobo printed for Évora use the two liturgical editions mentioned previously.[103] When he died later in that year his bookshop contained 414 Portuguese ABCs, 353 large Portuguese books of hours, and 283 little books of hours in Portuguese, some or all of which had come from his own office.[104] Earlier, in 1522, and probably on his own initiative, he had issued the *Horas de Nossa Senhora* in Portuguese. He might have been encouraged in this by his printing in the previous year of his monumental edition of the *Ordenações manuelinas*, his first venture into printing in the Portuguese language.

This work is the first edition of the second compilation of these laws, the first compilation being the one printed at Lisbon by Fernandes in 1512 or 1513 and reprinted by Bonhomini in 1514. We have already seen that the contract to print this book was probably offered to Cromberger in 1508 but that he had been unwilling to transfer his allegiance to Portugal. Cromberger's edition of 1521 raises the important question of whether he owned an itinerant branch office in Portugal, as has frequently been assumed. The colophons of Books I and IV of the *Ordenações* clearly state that they were printed by Jacobo at Évora; those of Books II, III, and V claim that he printed them at Lisbon, and the last gives a date of 11 March 1521. An entry in the inventory of Jacobo's estate shows that in 1528 two Portuguese merchants still owed him for some 1,000 copies of this edition; this is therefore the minimum number of copies printed.[105] An optimistic rate of production from one press in Cromberger's office was '3,000 sheets a day', which I take to mean 1,500 sheets printed on both sides (for a fuller discussion

[102] This approach is taken from an unpublished paper given at Birkbeck College, London, in 1981 by Jean-François Gilmont, who made enlightening use of a study of the present location of books to prove Jean Crespin's interest in supplying the 16th-c. German market. See also below, p. 75, n. 17.

[103] My discovery and identification of copies of these works refutes Nuno J. Espinosa Gomes da Silva's theory of an agreement between Galharde and Cromberger for the latter to print these editions under the former's name (see his 'Algumas notas sobre a edição das *Ordenações Manuelinas* de 1521', *Scientia Juridica*, 30 [1980], 5–16 [pp. 11–12]).

[104] Documents dated 7 June 1529 (APS, Oficio 4, Libro [2] of 1529, unfoliated).

[105] Gestoso, p. 50.

of the daily output of a press see below, p. 86). The *Ordenações* contains over 250 sheets and, if the compositors kept pace with the pressmen, 1,000 copies could have been printed in an absolute minimum of 170 working days if only one press were used. If that press were to have been dismantled, transported to Portugal, reassembled in one city, dismantled again, reassembled elsewhere, and have been capable of printing books to the high typographical standard of the edition of the *Ordenações* as well as being supplied with ink, paper, and skilled workmen, it is evident that the Seville office would not only have been seriously depleted for the best part of a year, but would also probably have made a financial loss even if Jacobo were paid most of the 400 Portuguese *reais* (equivalent to one *ducado*) for which each copy was to be sold. More than one press could have been sent, but this would have reduced yet further the Seville office's capacity for production in a year when it nevertheless printed several important editions on what would have been a small remaining number of presses.

Documents signed by Jacobo prove that he was in Seville at least in September and October of 1520 as well as mid-January of the following year.[106] He would therefore either have been unable to supervise the entire printing of the *Ordenações* or would have travelled back and forth from Seville to different Portuguese cities at the very time when he was printing a beautiful and complex work in Spain: his edition of the Seville missal, which he finished in December 1520. All of this just to print one edition, however prestigious, is most improbable. It is made even more so by the absence of any other works printed by Jacobo in Portugal even although we know that in his Seville office he produced books specially for the Portuguese market.[107] It is also doubtful whether the contract for the sale of copies of this edition to the Lisbon merchants would have passed before a Seville notary if Jacobo had really finished printing it in Lisbon itself as the colophon claims.[108]

Redondo, who discussed the editions of the *Ordenações* in some detail, believes that the 1521 edition was really printed at Seville. He is probably correct, but his explanation for the misleading colophons is implausible. He surmises that King Manuel asked Cromberger to produce the edition because nobody then working in his kingdom was competent enough to do so, and adds that Cromberger may have been able to print it more cheaply than his competitors. This is quite possible, but his further suggestion that Jacobo and the king were in league to disguise the fact that an important contract had gone abroad, and that the unusual type and

[106] Hazañas, i. 112.

[107] Agustín Redondo, 'Les Éditions des *Ordenações manuelinas* faites par les imprimeurs sévillans de la première moitié du XVIe siècle Jacobo et Juan Cromberger', *Mélanges de la Casa de Velázquez*, 2 (1966), 103–25 and three plates, suggests that the Évora breviary and baptismal of 1528 were printed in Portugal. This is incorrect.

[108] I have not been able to locate this contract, but it was drawn up by the Seville notary public Pedro Tristán (Oficio 17) on 18 May 1522 according to information contained in the document of 19 Nov. 1529 in which the debt was eventually paid (APS, Oficio 4, Libro 3 of 1529, unfoliated).

misleading colophons were used to deceive local printers who would otherwise have complained to the king, is most improbable. A printer like the Frenchman, Germão Galharde (Germain Gaillard?), could scarcely be expected to remain ignorant of the presence or absence in Lisbon, where he lived and worked, of one of the most important Spanish printers of the day. Furthermore, when Juan Cromberger produced another edition of the *Ordenações* at Seville in 1539, Galharde did not raise any objection to setting up matter which was added to the printed sheets sent by Juan from Seville to Lisbon. Jacobo probably employed the distinctive Type 7 for his 1521 edition less because it would deceive his competitors than because the original edition of the first compilation of the laws printed by Fernandes had used it and so it had become familiar to the potential users of the work. He may also have felt that it was particularly attractive for the Portuguese market, for it appeared again in his Portuguese *Horas de Nossa Senhora* of the following year, on the title-page of which he specifically drew attention to the style in which they were printed saying that the books were 'Printed in a large type'—the only reference to a type to appear in any of his books printed in any language.

It is, nevertheless, difficult to account for the misleading colophons if his *Ordenações* were, as I believe, printed at Seville. However, Juan's 1539 edition suggests an explanation. That edition was printed in the same type as the one used in 1521 but, oddly enough, the final sheets containing the list of errata were set up in a completely different gothic fount. There are two distinct settings of these final pages, but a comparison of both with the type and ornamental initials of two books printed at approximately the same time in Lisbon, the *Capitulos de cortes. E leys que se sobre alguñs dellas fezeram* (3 March 1539) and the *Statutos y constituyções dos virtuosos y reverendos padres conegos azuys* (25 August 1540), reveals that the final pages of Cromberger's *Ordenações* were really printed by Germão Galharde.[109] These pages may have been appended because it was only when Juan Cromberger's finished sheets reached Lisbon that they were found to contain misprints in the Portuguese.[110] These final pages also include the

[109] The two settings of the final colophon are (i) as reproduced by ex-King Manuel, *Early Portuguese Books 1489–1600 in the Library of His Majesty the King of Portugal Described by H.M. King Manuel in Three Volumes: Livros antigos portuguezes 1489–1600 da Bibliotheca de Sua Majestade Fidelíssima descriptos por S. M. El-Rei D. Manuel em três volumes*, 3 vols. (Cambridge, 1929–35), i. pl. 226; and (ii) also in the types of Germão Galharde, but with different line divisions and, unlike (i), with no mention of this being the third edition or of its having been printed in 1539. The setting (ii) may have been replaced by (i) because it was necessary to stress that the third edition had superseded that of 1521. It was normal for earlier editions of the *Ordenações* to be called in and destroyed as soon as a revised edition was published. Redondo, 'Les Éditions', does not realize that the two pages reproduced on his Plate III were both printed by Galharde, but were different settings of the sheet he reprinted for Juan Cromberger's edition of 1539.

[110] This would not have been the first time that a printer was asked to reprint a section of another's work (see Norton, *Printing in Spain*, p. 53).

authorization which had to be signed in each copy by royal officials if it were to have any validity. Also, and possibly more importantly, they contain the colophon which, being copied by Galharde from Jacobo's earlier edition, stated that the books had been printed by Jacobo at Lisbon in 1521. Galharde then merely added, 'Terceyra impressam. M.D.xxxix. annos.' with no indication that the whole book had really been printed by Juan in Seville.[111] This no doubt gave the edition an authority which it would have lacked if Portuguese lawyers and officials were to see that a major instrument of justice in the realm had been imported from abroad. Indeed, Juan may have printed a colophon on the last folio of the final gathering of Book V containing a clear reference to Seville as the place of printing. Galharde not only appended the authorization and list of errata but went as far as to remove this final sheet from Juan's last gathering. In doing so he would have destroyed any offending reference to Seville on leaf M_{10}, but also, of course, its conjugate leaf M_1 which contained part of the text. Therefore, in addition to printing the authorization on M_{10}^a and the new colophon copied from Jacobo's edition on M_{10}^b, he had to set up M_1 and reprint it in his own types. If the printer went to all that trouble, it would appear that something appearing on the original leaf M_{10} needed to be eliminated. This may well have been Juan's colophon.

In the light of this evidence from the 1539 edition, it is reasonable to surmise that, in order to confer authority on this important legal work, King Manuel insisted that Jacobo omit all reference to the real place of printing of the 1521 edition, replacing it with the names of those towns where the itinerant Portuguese court was then known to be resident. This would offer an explanation for Jacobo's misleading colophons and refute the suggestion that he ever sent his presses to Portugal. Although his death in Lisbon bears witness to his continuing interest in Portugal—an interest which he bequeathed to his son—it was in Seville that Jacobo concentrated his efforts during the first two decades of his career.

Growing Prosperity

When he married Comincia de Blanquis, Jacobo took over the family home in the Calle de Bayona, but by 1505 he was resident in the neighbouring parish of El Salvador. Two years later he had moved back into the parish of Santa María la Mayor and was living in the Calle de Génova alongside his colleagues in the book trade. He frequently changed residence in these early years, for in April 1508 he and Juan Varela de Salamanca bought a house in the Corral de Jerez, again in the parish of Santa María, in which he must have installed his family by November of that year. Rather than restlessness, these moves are probably a

111 For more details see the description of this edition in App. I.

reflection of the improvement of his fortunes. By 1510 he had enough money to begin to invest in property, for in that year he had bought a house in Santa María Magdalena and rented it to a tenant.[112] This was a common form of investment at the time in Seville, especially for ecclesiastical institutions, which owned a large proportion of the property in the city; they rented houses and shops to respected and reliable burghers who would then sublet them at a higher rent but at greater risk. By the latter part of the century even the city's brothels were ultimately owned by the Church.[113] In addition to buying property outright, Jacobo Cromberger invested heavily in such rents as he became more prosperous.[114]

The improvement of his business also implied the expansion of his activities as a printer and bookseller, and he doubtless required more extensive premises, especially for the storage of books. In 1510 he found and rented some suitable buildings into which he moved in the following year, letting his half of the house in the Corral de Jerez to a canon of the cathedral, Pedro Núñez Delgado, who will be encountered again later.[115] His new home was located in the Calle de Marmolejos in the small parish of San Isidoro, still close to the commercial centre of the city. This house was to remain the press's address for more than thirty years. It would have been large enough to accommodate his presses, equipment, stock, his immediate family, his workforce of slaves, apprentices, and pressmen, as well as his domestic servants and household slaves. In addition there would have been room for Jacobo's extended family and for visiting businessmen. At some time before 1525 Justo Alemán, Tomás Ungut's uncle, died. Jacobo, as the guardian of Alemán's orphaned children, must have taken them in, for Juan Cromberger subsequently replaced his late father as their guardian and promised not to charge the children for their board and lodging, thereby revealing that they lived in the family house.[116] Several years earlier a Flemish merchant called Juan de Haarlem had fallen ill when staying in Jacobo's house in San Isidoro and dictated his will there, indicating that it was customary for traders, no doubt often German-speakers, to stay with the family while they conducted their business in Seville.[117] No full description of the house survives, but in 1529, by which time alterations had been carried out and the house belonged to the family rather than being rented, it was valued at 2,800 *ducados* and listed as

a dwelling with its principal rooms, chambers, yards, garden, and outbuildings, situated

[112] Hazañas, i. 78–83.

[113] Mary Elizabeth Perry, *Crime and Society in Early Modern Seville* (Hanover, NH, and London, 1980), p. 230.

[114] Pike, *Enterprise and Adventure*, p. 16: 'The increase of rents under the stimulus of the price revolution, and the rise of land values due to the building boom, encouraged all members of the Seville business community to invest in urban property'.

[115] Hazañas, i. 83–6.

[116] Document dated 25 May 1529 (APS, Oficio 4, Libro [2] of 1529, unfoliated).

[117] Hazañas, i. 107–9.

in the parish of San Ysidro in this city and between, on the one side, the house of Alonso Pérez de Madrid, the Treasurer, and, on the other, a house owned by the cathedral in which Gerónimo de Burgos lives.[118]

Twenty-six years later the house was valued at 8,800 *ducados* and described as comprising:

a hall, stables, rooms on the ground and first floors, and other chambers; . . . in addition two more dwellings which are built on to it . . . each one having its own entrance, central courtyard, principal rooms, other chambers, and outbuildings . . . They are situated between, on the one side, a house owned by the cathedral of which Gerónimo de Burgos is tenant for life, and, on the other side, the house of Antonio de Soria, Pedro de Soria's son.[119]

An examination of the incomplete inventory of Juan Cromberger's goods when he died in 1540 suggests that the house had large living-rooms and bedrooms on two or three floors, one of which was extensive enough to contain many books and household stores. There was also a shop, a large cellar, a kitchen, and stables.[120] No mention was made of the printing workshop itself, but there is clear evidence contained in other documents that this, too, was attached to the house.[121]

Thus by 1511 Jacobo was a successful printer. Indeed, in the previous year he even seems to have acted as the editor of his own publications, the edition of Andrés de Li's *Repertorio de los tiempos* of 1510 having been revised 'through the labours and studies of Jacobo Cronberger, the German'. He was now installed in his new house, was modestly prosperous, and had gained something of a reputation as a solid citizen. His trustworthiness is witnessed in 1507 when he was made an *albacea*, or executor, to the will drawn up by a printer of pictures called Maestre Nicolás who had left money for the foundation of a chaplaincy in the church of Santa Ana in Triana, a town which lay on the opposite bank of the Guadalquivir from Seville, but which was, in effect, a part of the city.[122] Cromberger was to oversee this chaplaincy and subsequently undertook a good deal of business associated with it in accordance with his late colleague's wishes. After this, testators frequently chose him as executor to their wills: in 1511 a bookseller from Zamora who had fallen ill while in Seville appointed him to act in this capacity; in 1515 his old associate, Pedro de Mendieta, and also a wood sculptor or woodcut engraver, Jácome Plotino, both turned to him as did Juan de Haarlem in 1518, Ana de Alfaro, widow of the Genoese bookseller Niculoso

[118] Gestoso, p. 48.

[119] Document dated 20 Apr. 1555 (APS, Oficio 15, Libro 1 of 1555, fols. 781^{r}–784^{v}).

[120] Gestoso, pp. 75–84.

[121] In a document dated 10 Dec. 1516 Jacobo mentioned some indulgences dispatched to the Bishop of Jaén which he had printed 'in my own home' (APS, Oficio 3, Libro 2 of 1516, fols. 706^{v}–707^{r}).

[122] Gestoso, p. 12.

de Monardis, in 1522, and Ana Díaz, widow of the Seville bookseller Cristóbal Alemán, in 1527.[123]

His reliability, and particularly his skill as a printer, drew him to the attention of the cathedral chapter of Seville. This is revealed by an agreement that he made in 1511 with Villalta of Valladolid, who had acquired the right to issue the *Ordenanzas sobre el obraje de los paños*, which was to be a frequently reprinted money-spinner. It was agreed that Cromberger would print an edition and that they would both sell copies and share the profit. This was an important contract and, although it was soon cancelled, Cromberger did print an edition. Its interest lies in a clause stating that, in exchange for the right to sell Villalta's *Ordenanzas*, he would share with him his own privilege on the printing of all liturgical books for the diocese of Seville.[124] It is not known when he was granted this privilege, but it is evident that by 1511 he had been recognized as a fine printer of complex and expensive service books for which he owned specialized printing material: suitable vignettes and large woodcuts, a variety of gothic types, and the means to print music. He had used this material in the missals he printed for León (1504) and Seville (1507) and also in a breviary and an edition of the Constitutions for the diocese of Jaén, both probably printed in 1511. He subsequently added to his stock of types a very large gothic fount (Type 1), which was employed on title-pages but was particularly useful for certain parts of the liturgy such as the canon. It is no coincidence that he acquired this type in 1511, doubtless as a consequence or in anticipation of the monopoly granted to him by Diego de Deza, Archbishop of Seville. An immediate consequence of the monopoly was the edition of the *Constituciones del arzobispado y provincia de Sevilla* of February 1512 (unless this edition should be dated to 1513; see the Introduction to Appendix One). Jacobo was to remain the sole printer of service books for the diocese until his death in 1528, after which Varela appears to have acquired the monopoly, for he then printed a series of editions of the Seville missal (1529, 1533, 1534, and 1537). Cromberger, and later his son, printed many beautiful liturgical editions both of general Roman use and for a number of specific dioceses, which were mainly located in Andalusia and Southern Portugal. They also printed books commissioned by religious orders such as the Dominicans and the Franciscans. But Jacobo's relationship with the Seville chapter was an especially close one, an emblem of which was a particularly appropriate woodcut which adorned several of his books: it depicts two local saints, Isidore and Leander, carefully holding a model of the cathedral complete with its famous Moorish tower, the Giralda (reproduced in Appendix Three as WC:876).

Kingdon's researches have shown that one way in which a sixteenth-century

[123] Hazañas, i. 92–3, 101–3, 117, 141.
[124] Hazañas, i. 90–2.

printer could become prosperous was to attract patronage. In Italy Aldus Manutius was able to obtain Papal backing for his projects. In France Robert Estienne turned to Francis I, while his son, Henri, who ran a press in Calvinist Geneva, obtained financial support from Ulrich Fugger. In Spain Brocar was subsidized by Cardinal Cisneros, and in the Low Countries Christopher Plantin was favoured by Cardinal Granvelle and Philip II.[125] The other main source of income even for printers of scholarly editions was to cater for the constant demand for service books. In Protestant Europe this meant the printing of works like the Huguenot psalter, while in the Catholic countries it was missals and, in particular, breviaries which brought in a steady income.[126] When the breviary was standardized after the Council of Trent, there was an undignified scramble to obtain licences to print it. Plantin was able to corner the market for the somewhat different breviary and missal which Philip II commissioned and he amassed a considerable fortune by supplying them to Spain and the Americas together with all other liturgical books for Philip's kingdoms.

Although Jacobo Cromberger was not dealing with such enormous editions—Plantin was to deliver at least 47,000 breviaries and other liturgical works to Spain—his monopoly of liturgical printing for Seville not only set the seal on his reputation, but must have brought him considerable profits, especially because, as has already been seen, payment was normally made on delivery, thus reducing the problems of financing new editions while the income on those already printed had not been received.

Another sort of work also carried out for particular dioceses and guaranteed markets was the humble but profitable printing of indulgences. Jacobo ran off many thousands of them, but, by their very nature, most have disappeared. Just two examples, both printed for Seville cathedral, one in *c.*1518 and the other in 1519, survived until the Spanish Civil War (reproduced in Appendix Three as WC:723–725). References to the many which have vanished can be found in notarial documents: for example, Jacobo printed over 20,000 for the diocese of Jaén in 1514 and another 16,000 for it two years later.[127] Anti-Catholic

[125] Robert M. Kingdon, 'Patronage, Piety and Printing in Sixteenth-century Europe', in D. H. Pinkney and T. Ropp (eds.), *A Festschrift for Frederick B. Artz* (Durham, NC, 1964), pp. 19–36. In the 15th c. Caxton enjoyed noble patronage for his press at Westminster, and struggled to retain it when his original patrons fell from power (see Blake, *Caxton*, pp. 86–100).

[126] Breviaries were such a saleable commodity that Comincia de Blanquis declared in her will of 1511 that she had been paid in 'fifty breviaries printed for the Dominicans' rather than in money (see Gestoso, p. 15). In 1510 the London merchant, William Bretton, even considered it worth his while having a breviary of Seville use printed in Paris to sell in Seville through Jacobo Cromberger, in whose house copies were still to be found in 1518 (see Norton, *Printing in Spain*, p. 136). The Alcalá printer, Francisco de Navascués, was so eager to print the Oviedo breviary in 1536 that he put up half the money himself (see José Luis G. Novalín, *El inquisidor general Fernando de Valdés (1483–1568)*, 2 vols. [Oviedo, 1968–71], i. 92–3). The market for service books in Spain was huge (see Lapeyre, *Une Famille*, p. 566).

[127] Hazañas, i. 99–100, 105.

propagandists were quick to point out that indulgences were a delectable source of income for all concerned except the purchasers. The Papacy, the Crown, individual dioceses, and the mere suppliers of these unremarkable little forms all made a considerable profit because the retail price was greatly in excess of the cost of materials and the labour involved, especially after the advent of printing had made their production so simple. Printers fought hard for privileges to produce indulgences such as those of the Santa Cruzada, or Holy Crusade, as is witnessed by Brocar's complaint to his patron, Cardinal Cisneros, in 1517 that a Valladolid bookseller had travelled to Flanders to petition Charles V for a privilege on them which Brocar felt was rightly his.[128] The bookseller knew that the expense of his journey would be more than offset by the income from such a privilege. Similarly Miguel de Eguía's son claimed that the six-year privilege on the printing of the Crusade indulgences had been one of his father's major sources of wealth.[129] Varela considered the production of Crusade indulgences to be lucrative enough to warrant his setting up a branch-office in Toledo where he printed them from 1509 to 1515. Handsome profits could be made by printers, especially if they were unscrupulous enough to run off more copies of the indulgences than had been ordered and then sell them illegally on their own behalf. There must have been cases of such sharp practice in Seville in 1514 and again in 1525 when suspect printers, Jacobo Cromberger among them, were interrogated about fraudulent indulgences.[130] If he was involved in these shady deals, it is a nice irony that his great-grandson, Licenciado Juan Cromberger Maldonado, who had received a university education and had risen above the artisan class of his origins, was sent to the Indies in 1577 to be an administrator and watchdog of the indulgences of the Santa Cruzada.[131]

Jacobo printed large quantities of such ephemeral material as well as his magnificent liturgical editions. In 1512 he sold some sheets of devotional woodcuts and two thousand ABCs to a Franciscan expedition which was setting off under the leadership of Fray Alonso de Espinar to evangelize islands in the Caribbean. This was his first recorded association with the Indies—and, indeed, the first known connection between any printer and the New World—but it was not followed up until well into the 1520s, by which time more promising markets for books and other merchandise had been opened up on the mainland by Cortés's

[128] Norton, *Printing in Spain*, pp. 46–7. Later, in 1574, Seville printers competed for the contract to print indulgences for the Santa Cruzada para las Indias given by the Escorial, which had a privilege on them. The desirability of the contract can be seen by the staggering print-runs (see Cristóbal Bermúdez Plata, 'Las impresiones de las Bulas de la Santa Cruzada para las Indias', *Archivo hispalense*, 2nd ser., 6, No. 15 [1946], 9–40).

[129] Goñi Gaztambide, 'El impresor Miguel de Eguía', p. 87.

[130] Hazañas, i. 100, 126–7.

[131] Licence to travel to the Indies dated 25 May 1577 (AGIS, *Contratación*, Legajo 5538, Libro 1, fol. 429^{r}).

conquest of Mexico. Only once during this period did he have any direct contact with the Indies, when he tried to collect a debt owed to him by a resident of Hispaniola.[132] Nevertheless, throughout the second decade of the century he prospered; while the average number of books known to be printed by him in each year of the first decade of the century was only just under six, in the second decade this figure doubled.

There is again at this time evidence that he was an astute judge of the sort of books which would sell. In 1512 the royal chronicler, Fernán Pérez de Pulgar, lent him a manuscript of a Spanish translation of Lucan's *Pharsalia*.[133] It is probable that Jacobo was considering printing an edition of the work and it would indeed have been a candidate for publication as it had long been known in Spain. The Alphonsine compilers had used it as an authority for the *Estoria de España*, and a Castilian translation had been included in the fifth part of the *General estoria*. More recently it had been drawn upon by two fifteenth-century poets of whose works the Seville printers produced numerous editions during the early sixteenth century: Íñigo López de Mendoza, Marquis of Santillana, and Juan de Mena. Many libraries of fifteenth-century Castilian nobles contained manuscript copies of the *Pharsalia* either in the original Latin or in vernacular translations. Santillana, who was no Latinist, had Spanish and Italian translations as well as the Latin original.[134] Nevertheless, no copies of any such printed edition of Lucan in the vernacular are recorded, and it is probable that he had decided against publication. If this is the case, he made the correct decision, for Lucan was used as a school-text in Latin, not in the vernacular, and the lack of a more general interest in a translation at this time is witnessed by the fact that it was not printed anywhere in Spain until Felipe de Junta issued a Castilian version at Burgos in 1578.[135]

At the same time Jacobo was increasing his investments in the city and developing his commercial interests. In 1513 he engaged in some minor deposit banking, while in 1516 he supplied the chapter of Jaén not only with books, but also with timber which he had imported from Flanders.[136] He invested his profits in property and agricultural land in and around Seville. In April 1518 he had extensive alterations made to his house in San Isidoro and four months later he bought

[132] Hazañas, i. 99. This document is dated 12 Jan. 1514 and not 12 Feb. 1514 as Hazañas states (APS, Oficio 3, Libro 1 of 1514, fols. 152^{v}–153^{r}).

[133] Hazañas, i. 96.

[134] Mario Schiff, *La Bibliothèque du Marquis de Santillane*, Bibliothèque de l'École des Hautes Études: Sciences historiques et philologiques, 153 (Paris, 1905), pp. 136–40; Miguel Ángel Ladero Quesada and María Concepción Quintanilla Raso, 'Bibliotecas de la alta nobleza castellana en el siglo xv', in Díaz y Díaz *et al.*, *Livre et lecture*, pp. 47–59 (p. 50); M. C. Quintanilla Raso, 'La biblioteca del marqués de Priego (1518)', in M. A. Ladero Quesada (ed.), *En la España medieval: estudios dedicados al profesor D. Julio González González* (Madrid, 1980), pp. 347–83 (pp. 357, 359, 364).

[135] The first Spanish-language edition was printed at Antwerp by Joannes Crintius in about 1540.

[136] Hazañas, i. 97, 105.

another house in Triana together with a small vineyard in the Vega de Triana.[137] In the following year he accepted some more houses in Seville as security against a debt and these were later to pass into his hands when the debtor defaulted on his obligations.[138] A Genoese merchant appears to have mortgaged two water-mills near the town of Osuna to him in 1520 and promised to make him an annual payment in flour.[139] Again the debtor seems to have defaulted on the payments because, by 1529, the mills had become Jacobo's own property and were valued by his heirs at 240 *ducados*.[140] These valuable mills passed to Catalina Cromberger on Jacobo's death and their history can be traced under her ownership.

Although these acquisitions indicate that Jacobo was engaged in building up the family's finances, his investments at this stage were still modest. Other successful Seville artisans also owned slaves, took advantage of the building boom, buying several houses each, and invested in land outside the city at a time when prices were rising owing to the increasing demand for shipments of agricultural products to the New World and Northern Europe.[141] In the 1520s, however, there are several signs in different areas of his activities which indicate that Jacobo was beginning to operate on a new scale. The possible reasons for this are numerous, but probably centre on his increased ambitions as a printer and his growing involvement in the commercial opportunities offered by the Indies.

1520 was a productive year for Cromberger's presses, and paper purchases are the highest recorded in the history of the press. With the exception of 1512, more editions have survived from 1520 than for any other year of Jacobo's production or, indeed, of that of his son or grandson. These editions included a wide range of printed matter from numerous chap-books to substantial books like the Seville missal, from three further liturgical editions and a bulky volume of the press's first edition of Ludolph of Saxony's best-selling *Vita Christi*, to a slight legal work, and an interesting little book by Erasmus. If the number of sheets contained in one copy of each of these books, rather than just the number of editions, is calculated, so that the slight chap-books are put in a proper perspective when compared with more substantial works, the surviving editions printed in 1520 still indicate that the presses were almost twice as busy as in 1506, which had been the most productive year of Jacobo's independent activity in the first decade of the century. These statistics are not entirely trustworthy as so many editions must have disappeared, but it may be due to more than mere chance that the

137 Hazañas, i. 107, 109.

138 Hazañas, i. 110. For the subsequent protracted dispute over the rightful ownership of this property see documents dated 12 and 23 Jan. 1531 (APS, Oficio 4, Libro 1 of 1531, unfoliated) and 20 Nov. 1531 (APS, Oficio 4, Libro 4 of 1531, fols. 423^r–426^r).

139 Hazañas, i. 111–12.

140 Gestoso, p. 48.

141 Pike, *Aristocrats and Traders*, pp. 137–8.

purchase of a large quantity of paper coincides with the high output of surviving editions in that year.

After 1520 contracts for paper suddenly disappear until 1555, significantly the very year in which Jacobo's grandson, Jácome, complained that paper was not readily available in Seville.[142] It is, of course, possible that contracts for the purchase of paper have been lost or that I have just not come across them in the labyrinthine Archivo de Protocolos, but it is suspicious that no record of a single purchase during this period should have come to light.

It was essential for a printer to have a regular supply of paper, especially if his press was not situated near a mill or in an area where there was surplus production. Most of the paper used by the Crombergers seems to have been imported and, until 1520 and after 1555, it was bought from merchants rather than direct from the producers. Geneva provides a comparable example of a printing centre which did not possess local mills, and it is significant that printers there went to great lengths to ensure adequate supplies. Jean Crespin arranged to buy up the entire annual production of one mill when he set up his press in that city in 1550.[143] Robert Estienne had a standing order with a mill, and when he died in 1559 his son, Henri, strengthened the agreement by purchasing the mill's complete production for eight years at a rate of one thousand reams each year.[144] Both Crespin and Henri Estienne were even prepared to invest capital in those mills to acquire a monopoly of their output. It is a strong possibility that in 1520 or soon after Jacobo made a similar arrangement for the supply of sufficient quantities of paper for his needs either from a merchant or directly from a mill. His mills at Osuna were not, incidentally, ever used for the production of paper and there is no evidence that the Crombergers ever invested in the manufacture of this basic requirement for their activities. Only three years later Charles V was to give Brocar and Eguía a licence to import paper from France on the grounds that 'in our kingdoms it is scarce and expensive'.[145] Unfortunately, no document survives which enables us to state with certainty that Jacobo negotiated a steady supply of paper in the way I have suggested, but the absence of any *ad hoc* purchases points to such an agreement having been drawn up. This would account for the large stocks of paper which he left on his death: in 1528 almost two thousand reams of various sorts of paper were included in the goods which formed part of his daughter's dowry payment.[146] Indeed, he may even have used his regular

[142] Document dated 17 Oct. 1555 (APS, Oficio 10, Libro 3 of 1555, fols. 74r–75r).

[143] Jean-François Gilmont, *Jean Crespin: un éditeur réformé du XVIe siècle*, Travaux d'humanisme et Renaissance, 186 (Geneva, 1981), pp. 75–6.

[144] Kingdon, 'The Business Activities', p. 262. For further evidence of investment in paper mills by printers, see Febvre and Martin, *L'Apparition*, p. 46.

[145] Beltrán de Heredia, *Cartulario*, ii. 422–3.

[146] Document dated 7 June 1529 [*sic*] (APS, Oficio 4, Libro [2] of 1529, unfoliated).

supplies to engage in the retailing of paper. Several of his debtors in 1529 were notaries public, and it is noteworthy that some of the paper which he stocked was of a sort not found in any of the many printed books which I have examined. As paper for printing and paper for writing were somewhat different, this may have been of a quality prepared for the quill. Although the watermarks of several hundred Cromberger editions are described in Appendix One, it has not been possible to identify with any reasonable degree of certainty the origin of most of the paper found in them. As is the case with so many aspects of Spanish printing history, much research remains to be done on the origin of the paper used in sixteenth-century Spanish books. Until a full working manual of watermarks to be seen in sixteenth-century Spain is published, any speculations about the Crombergers' paper supplies must be treated with caution.[147]

Jacobo Cromberger, Lázaro Nuremberger, and the Indies

Jacobo appears to have consolidated his position as a major printer at the beginning of the decade just at a time when Spain was suffering the turmoil of the *comunero* revolt, and bookselling at the important centre of Medina del Campo had received a serious blow with the razing of an area of the town where booksellers had their warehouses, an event which may have benefited printers and merchant-booksellers in other Spanish cities. Then in 1520 or 1521 there followed the important order for the edition of the *Ordenações manuelinas*. However, two events, one international and one personal, appear to have affected the balance of Jacobo's activities, which until then had been centred upon printing with only very limited involvement in other sorts of commerce.

On 21 August 1521, after a long and bitter battle, Cuauhtémoc fled the Aztec capital of Tenochtitlán, which finally fell to Hernán Cortés, thus heralding the end of resistance to the Spanish conquest of New Spain. The devastated city was

[147] It was common for booksellers to retail paper (see Madurell and Rubió, *Documentos*, p. 42*). In the 1529 inventory of Jacobo Cromberger's estate a large quantity of 'papel de buytre' ('vulture paper') was listed (see Gestoso, p. 39). This description probably referred to the design of the watermark, for the paper referred to as 'papel de mano' contained a watermark of a hand. I have not found a watermark of a bird in any book printed by the Crombergers. In 1555 Jácome bought 22 reams of 'papel del pelegrino' (see document dated 11 June 1555 [APS, Oficio 23, Libro 2 of 1555, fol. 346v]). This paper was again not used in any Cromberger production, but I have frequently encountered the watermark of a pilgrim in paper used for notarial documents held in Seville archives. This watermark appears to have originated in Italy and is seen in 16th-c. Spanish books, but not, as far as I am aware, in editions printed at Seville in the first half of the 16th c. (see Oriol Valls i Subirà, *The History of Paper in Spain*, tr. Sarah Nicholson, 3 vols. [Madrid, 1978–82], ii. 163–5). Valls i Subirà's book is the best history of the paper produced and used in Spain. However, its catalogue of watermarks is not particularly useful as it contains only a small number of illustrations, especially of the design most frequently seen in Cromberger editions—the hand with a star above it. This is the most difficult watermark to assign to a place of manufacture because, although it originated in Italy, it was copied by paper-makers in many centres outside that country (see Valls i Subirà, *The History of Paper in Spain*, ii. 151–9).

rebuilt by the colonizers and, as Spaniards emigrated to the American mainland, new markets were created which had to be supplied by merchandise dispatched from Spain. For humble Seville artisans and rich merchants alike, the Indies were to prove a magnet for both emigration and trade. In 1526 Andrea Navagero, the Venetian ambassador, remarked that so many men had left Seville with the fleets that it had been abandoned to the care of women. He also marvelled at the variety of goods which were being sent from the port to the new colonies.[148] The conquest of Mexico was an event of incalculable importance to Seville and so it is not surprising to find that the first editions of Cortés's second and third letters to the king, the *Cartas de relación*, dated respectively 30 October 1520 and 15 May 1522 (the first letter—if, indeed, it ever existed—has been lost), were printed there by Jacobo Cromberger in 1522 and 1523. These editions symbolize his two main concerns in the remaining years of his life: printing and commerce with the New World.

The personal event was the judicious marriage of Jacobo's daughter, Catalina, to the young and enterprising German merchant, Lázaro Nuremberger, who first appears in Seville documents in 1522.[149] By 1523 he was already described as Jacobo's son-in-law.[150] He had, however, arrived in the city in 1520 and by 1523 at least was already a *vecino* living in the parish of San Isidoro, probably in his father-in-law's house.[151] Although he was then only twenty-four years of age, Lázaro was an experienced merchant. He was a native of Neustadt an der Aisch, a small town some twenty-five miles north-west of Nuremberg. His parents, Heinz and Anna, lived in Neustadt in the quarter between the Nuremberg and Windsheim Gates, and it was from their home that their son set out in January 1517 for Lisbon, the first stage of a long and dangerous journey as a factor of the Hirschvogel trading company of Nuremberg.

The Portuguese voyages of discovery during the fifteenth century, in which Nurembergers like the navigator and cosmographer Martin Behaim had played such an important part, had broken the Arab and Venetian monopoly of trade with the Orient. In 1511 the vital commercial port of Malacca on the Malay Peninsula had fallen to the Portuguese, who by 1515 dominated the carrying trade between Asia and Europe, Lisbon supplanting Venice as the main market in Europe for Asian goods. German merchants seized the opportunity to exploit this traffic in spices, gems, and silks which were now coming to Europe by the sea-

148 Navagero, *Il viaggio fatto in Spagna*, fol. 15^{v}.

149 Lázaro had contributed in that year to the arming of ships sent to the Azores to protect the bullion fleet coming from the Indies from attacks by French corsairs (see Otte, 'Jakob und Hans Cromberger', pp. 135–6).

150 Hazañas, i. 118. Hazañas fails to mention that Nuremberger was already styled in this document *vecino* of Seville (APS, Oficio 4, Libro 1 of 1523, fol. 315^{r-v}).

151 Otte, 'Jakob und Hans Cromberger', p. 158, suggests that from 1525 Lázaro lived next door to Jacobo, but later evidence suggests that they already shared the same house.

route round Africa, and representatives of the great trading companies were based at Lisbon from the beginning of the sixteenth century for this purpose. By 1516 Lázaro had already completed one tour of duty for the Hirschvogel company at Lisbon where he had worked from the age of fifteen, and he had then returned to Germany. However, it was customary for the Lisbon factor of one of the German companies to join the fleets bound for India and there trade on behalf of the others. Although he was still very young, Lázaro was chosen for this task in 1517.[152] Returning to Lisbon from Germany he joined António de Saldanha's fleet of five ships bound for the East. They made for Mozambique via the Cape of Good Hope and then proceeded to the west coast of India engaging in trade in Goa, Calicut, and Cochin. The ships returned to Lisbon in the summer of the following year and Lázaro subsequently wrote to Willibald Pirckheimer—the famous Nuremberg humanist, counsellor of Charles V, and friend of Dürer and Erasmus—giving him a full account of the voyage. The autograph manuscript still exists.[153] It was no doubt on this journey that Lázaro acquired the expertise as a trader in pearls that is evident from his subsequent activities in Seville, for in his account he makes a special mention of the gems and pearls which were acquired at Bhatkal.[154] He returned to Germany, visiting Nuremberg and Neustadt, but was soon back in the Iberian Peninsula where he can be traced in Lisbon in March of 1520. He was to have travelled to India again in that year to continue trading on behalf of the German merchants of Lisbon but, rather than make Portugal the centre of his activities as many of his compatriots had done, he moved to Seville, which was to offer even more promising opportunities for overseas trade. He may have taken the decision himself to go to Seville or he may have been sent there by the Hirschvogel company. His activities as a pearl trader when in Seville also suggest that he was a factor of Jörg Herwart, the leading gem merchant of Lisbon. In his place another factor of the Hirschvogel company, Jörg Pock, went on the trading venture from Lisbon to India in that year, and Lázaro was henceforth based in Seville, although he frequently made business trips to Lisbon.[155]

The German merchants resident in Lisbon and Seville maintained close

[152] Otte, 'Jakob und Hans Cromberger', p. 134.

[153] *Lazarus Nurenberger: Being a Photograph of the Manuscript Account of his Voyage to India in 1517 and 1518 Preserved in the Central Library of the Slovakian Academy of Sciences in Bratislava, Czechoslovakia*, ed. with an introduction by Ronald Bishop Smith (Lisbon, 1974). A transcript and translation into French have been published by Anne Kroell, 'Le Voyage de Lazarus Nürnberger en Inde (1517–1518)', *Bulletin des études portugaises et brésiliennes*, 41 (1980), 59–87.

[154] In his will of 1564, Lázaro mentioned various of his transactions in gems, including the sale of some to Süleyman the Magnificent at Constantinople (see Theodor Gustav Werner, 'Zur Geschichte Tetzelscher Hammerwerke bei Nürnberg und des Kupperhüttenwerks Hans Tetzels auf Kuba', *Mitteilungen des Vereins für Geschichte der Stadt Nürnberg*, 55 [1967–8], 214–25 [p. 220]).

[155] Otte, 'Jakob und Hans Cromberger', pp. 133–4.

commercial contacts with each other, and it has already been shown that German and Flemish traders visiting Seville and dealing in Lisbon stayed in Jacobo Cromberger's house in the Calle de Marmolejos: Juan de Haarlem, for example, who was Jacobo's guest in 1518, said that he owned merchandise in both Seville and Lisbon at that time.[156] Jacobo and Lázaro probably knew each other, or at least had acquaintances in common, before Lázaro arrived in Seville. In any case, if he did not move into the Cromberger household immediately, he would soon have met Jacobo, who was a distinguished figure in the closely-knit community of German artisans in the city. Not long after his arrival Lázaro was familiar enough with Jacobo to marry his only daughter, Catalina, and he was subsequently a frequent business partner of three generations of the family despite Catalina's premature death in 1530 and his subsequent remarriage.[157] He died in about 1564 after a life devoted to trade, during which he amassed a considerable fortune. Merchants like Lázaro maintained the important trade from the rest of Europe, through Seville, to the New World: manufactured goods imported from Northern Europe and Italy were dispatched to the Indies and in return bullion, gems, sugar, drugs, hides, and cochineal were sent from the Americas to Spain and thence distributed to other countries.

There is plentiful evidence in the Seville archives of Lázaro's commercial activities, but a more tangible and appropriate monument still remains in the form of the magnificent chest with its complex mechanism of multiple locks—probably the 'green chest' so frequently mentioned in the accounts of the Casa de Contratación—which stands on the landing of the former Casa Lonja de Mercaderes, now the Archivo General de Indias of Seville. It was manufactured in Nuremberg and purchased from Lázaro in August 1537 by the Treasurer of the Casa de Contratación, who paid 52 *ducados* for it and said that it was to be used

> for the safe keeping of the remaining *coronas, doblones, escudos*, and silver *reales* from the sums collected for His Majesty which Licenciado Carvajal of the Council of the Indies left in my care, and also for any gold, silver, and coin which needs to be locked away in this Casa de Contratación.[158]

Considering Lázaro's career before his marriage to Catalina Cromberger, it is no coincidence that it was at the time of the union that Jacobo's interest in trade with the New World blossomed. His friend and fellow printer, Juan Varela de

[156] Hazañas, i. 107–9.

[157] Lázaro and Catalina had two children, Jácome and Enrique. Catalina had died by March 1530 (see document dated 28 Mar. 1530 in which Lázaro talks of 'Catalina Coramberger, deceased, may she be with God' [APS, Oficio 4, Libro 1 of 1530, unfoliated]). He had married his second wife, Isabel Pins, by 1534, for in that year he refers to Catalina as 'my first wife, deceased' (see document dated 4 or 5 Jan. 1534 [APS, Oficio 4, Libro único of 1534, unfoliated]). Lázaro and Isabel had three more children: Ana, Nicolás and María.

[158] Otte, 'Jakob und Hans Cromberger', p. 129 and photograph facing p. 133.

Salamanca, with whom he had been associated at least since 1506, would also have been an example to him of the fortune to be gained in commerce with the Indies.[159] Varela had entered this trade as early as 1507, when he had acquired a share in a ship. Subsequently, during a lifetime of intense activity as a merchant, he became a rich and respected burgher who was in receipt of large amounts of bullion from the Americas. He sent his son to Mexico as his factor, gave his daughters handsome dowries, was elected a *jurado*—a sort of councillor—of the city in 1530, and on his death left a fortune of over 40,000 *ducados*.[160]

Jacobo had long been a friend of Pedro de Mendieta, a former printer who later became a silk merchant.[161] Mendieta had died in 1515 but, as an executor to his will, Jacobo no doubt kept an eye on the welfare of his son, Diego. After he had been left a young widower, Diego was sued by his father-in-law in 1522 over the return of his late wife's dowry. He was on the point of being sent to prison when Jacobo intervened in the dispute and promised to pay whatever sum Diego was deemed to owe in respect of this dowry.[162] Mendieta was thus in Jacobo's debt, but early in the following year an opportunity arose for Diego to recompense him and for them both to earn a considerable amount of money. Jacobo decided to send a shipment of merchandise to the newly conquered colony of Mexico. Unlike Varela, however, he did not at that time have free access to this market because non-Castilians were excluded, in theory at least, from trade with the New World. It seems, nevertheless, that in the 1520s Charles V realized that foreign capital was required to develop the colonies and so, with a little transparent subterfuge, foreigners like Cromberger were able to send cargoes across the Atlantic. An agreement was drawn up that Diego, who was presumably a Castilian, would accompany the goods. Instead of being Cromberger's factor, he officially became their owner because he 'bought' the cargo from Jacobo just before he set sail.[163] In this way Jacobo could not be charged with engaging in illegal trade. It is, however, quite evident from the documents that this purchase was a fiction.[164] It was Jacobo who had bought the valuable cargo of clothes and other goods, which were worth about 1,000 *ducados*; he also paid the freight charges and insurance. Diego, for his part, agreed to sell the goods at the best price he was offered and send the money back to Seville. Profits were expected to be high, for in the New World goods fetched prices far in excess of those paid

[159] Hazañas, i. 78.

[160] Document dated 7 Aug. 1556 (APS, Oficio 5, Libro 2 of 1556, unfoliated). Some comparison is offered by the estate of one of Medina del Campo's most important publishers and booksellers, Benoît Boyer, who was considered by later generations a symbol of prosperity long since passed. In 1592 his estate was valued at only half that of Varela (see Péligry, 'Les Difficultés', p. 270).

[161] Document dated 14 Nov. 1513 (APS, Oficio 1, Libro 2 of 1513, fol. 694^{v}).

[162] Hazañas, i. 117–18.

[163] Hazañas, i. 119–21.

[164] The cargo was even 'bought' again, this time by one Alonso de Céspedes when, in 1525, Jacobo wished to conceal his illegal trade with the Indies (see Hazañas, i. 125–6).

in Seville; sales would also be rapid, for Diego promised to return to Spain within five months. As Jacobo had put up the capital for the venture, he would receive the lion's share of the profits, but Mendieta would retain for himself the not inconsiderable share of one-sixth. Indeed, profits may have been increased if Jacobo owned the ship into which the cargo was loaded. In the agreement between the two parties Diego mentioned that he would travel in a vessel named the *Santiago*. No mention is made of the identity of the owners, but in September 1525 Jacobo and Lázaro Nuremberger possessed a ship of the same name.[165] Ownership of a vessel was an ideal means of ensuring transportation of cargoes, and in England, at least, it was normal for more than one merchant to have a share in a single ship, thus reducing the risk of losing a large investment in the case of wreck.[166] If by 1525 Cromberger and his son-in-law had owned the *Santiago* for at least two and a half years and she was setting sail again, she would have already made a minimum of three or four voyages on the Atlantic run. This implies that they were beginning to trade on a large scale, especially as profits on merchandise would soon offset the cost of buying the ship in the first place. It is no surprise that Jacobo was frequently referred to in documents of this period not as a printer but as a merchant.[167]

In May 1523 Diego de Mendieta set sail from Seville carrying with him Jacobo's hopes of a successful career as a merchant in the Americas. Mendieta arrived safely in Mexico and sold the goods, but in February of the next year, long after he was due to return to Spain, he died in Vera Cruz. The expectation of handsome profits had not been misplaced, for Diego had increased the value of his cargo fourfold, and that amount of gold was found in his possession on his death.[168] However, the procedure for sending the treasure left by those who died in the Indies back to its rightful owners in Spain was cumbersome and corrupt.[169] Cortés had appointed one Hernán López de Ávila to the post of *tenedor de bienes de difuntos* ('holder of deceased persons' estates') in Mexico, but López de Ávila turned out to be a scandalous literalist in his interpretation of the word 'tenedor': he took possession of the gold, but then refused to give it up.[170] A long legal battle

[165] Gestoso, p. 22.

[166] G. V. Scammell, 'Shipowning in the Economy and Politics of Early Modern England', *The Historical Journal*, 15 (1972), 385–407 (pp. 400, 406).

[167] Gestoso, p. 28.

[168] For the inventory of Mendieta's estate and subsequent litigation see Cristóbal Bermúdez Plata, 'Relaciones entre Jacobo Cromberger y Hernán Cortés, con noticias de imprentas sevillanas', *Anuario de estudios americanos*, 4 (1947), 665–88.

[169] Ernesto Schäfer, 'La Casa de la Contratación de las Indias de Sevilla, durante los siglos XVI y XVII', *Archivo hispalense*, 2nd ser., 5, No. 13 (1945), 149–62 (p. 153). Varela's son, Pedro, died at Vera Cruz in 1543; his heirs were still claiming their inheritance over twenty years later (see Hazañas, ii. 100–8, 137–8).

[170] Jacobo was not the only victim of this corrupt official. In 1527 Alonso de Nebreda sued him (see Hazañas, i. 148–9), and even Cortés claimed in the following year to have been cheated by his appointee,

followed during which Jacobo attempted to sue him for the money and the *tenedor* used a series of ploys to frustrate his claims. Francesco Guicciardini's judgement based on his observation of the workings of contemporary Spanish law, 'Little justice is done, and civil cases are dealt with very slowly', is nowhere more apt.[171] López de Ávila died in Spain in 1527; Jacobo therefore changed tactics and sued Cortés himself when the Governor of New Spain returned from Mexico in the following year. Jacobo's claim was that Cortés had appointed the *tenedor* and was therefore responsible for his official's misappropriation of Mendieta's estate; he also alleged that Cortés had subsequently taken possession of López de Ávila's riches and brought them with him to Spain. Cortés denied everything and the case had still to come before the Council of the Indies when Jacobo died unpaid.[172] Juan Cromberger picked up the pieces and presented the *tenedor*'s father with a royal command signed by the Emperor and his secretary, Francisco de los Cobos, ordering him to repay the money. However, not only did all the papers connected with the case go missing in 1531 but, when they were recovered, López de Ávila's heirs dragged their feet, so that ten years after Mendieta had died in Vera Cruz, Lázaro Nuremberger was still suing the *tenedor*'s niece for the gold.[173] The whole venture had been a disaster. Although the amount lost was not great by the standards of the richest traders in gems and spices who had settled in Seville and Lisbon—1,000 *ducados* invested and over four times that sum expected in returns from Mexico—it must nevertheless have represented a blow to Jacobo's finances, for the whole stock of his bookshop a few years later was worth only slightly more than the bullion which Mendieta was due to send him from the New World.

Jacobo's first experience of transatlantic commerce was scarcely encouraging but, despite the reputation enjoyed by Nuremberg merchants for extreme caution, the lure of high profits from dealing with the Indies conquered any misgivings the family may have had, and they continued to dispatch goods to Hispaniola and Mexico. The small obstacle of the ban on non-Castilian merchants was soon overcome by Jacobo's employing his son as a 'front-man', for Juan had been born and bred in Seville. It is not known, for instance, what goods were sent to the New World by Jacobo in 1524 or whether news of Mendieta's demise had reached him by that date, but there is evidence that gold officially addressed to Juan arrived in Spain in the following year, doubtless the returns on a shipment of

who had stolen many of his possessions and fled to Spain with them (see Gestoso, p. 29). López de Ávila clearly made money out of his office: on his death he owned not only property in Spain, but houses, land, livestock, and slaves in Mexico (see Agustín Millares Carlo and José Ignacio Mantecón, *Índice y extractos de los protocolos del Archivo de Notarías de México D.F.*, 2 vols. [Mexico City, 1945–6], i. documents Nos. 61, 1080, 1474, 1707).

171 Guicciardini, *Viaje a España*, p. 44.

172 Gestoso, pp. 27–30.

173 Document dated 28 Jan. 1529 (APS, Oficio 4, Libro 1 of 1529, unfoliated); Medina, *La imprenta en México*, i. p. lviii; and document dated 4 or 5 Jan. 1534 (APS, Oficio 4, Libro único of 1534, unfoliated).

merchandise. On 21 May 1525 a ship which had sailed from Hispaniola, *La Magdalena*, docked at San Lúcar de Barrameda at the mouth of the Guadalquivir carrying '323 *pesos*, 3 *tomines* and 9 *onças*, all in a sack and addressed to Juan Combejer, the son of Jácome Alemán, the printer'; there is a marginal note in the account books of the Casa de Contratación showing that Juan collected the gold.[174] In the same month, an unnamed ship owned by Juan Cromberger and Lázaro was waiting to set sail for the Indies. Again the family was unfortunate: two royal agents appeared at the docks bearing a commission to requisition vessels capable of making the voyage across the Atlantic, for Charles V was in urgent need of his treasure in the New World to meet the expenses of his foreign policies in Europe.[175] The ship was seized but, after receiving a formal protest, the agents were persuaded to pay 460 *ducados* for it. Nevertheless, by September at least, another ship, the *Santiago* mentioned above, was being fitted out for its owners, Jacobo and Lázaro.

Their activities continued apace and Jacobo is found exporting hides to Santo Domingo even before the Mendieta affair had turned sour. Again he encountered problems. Only four or five years after the *comunero* revolt, there were still tensions between royal authority and local privileges. Although Jacobo had obtained permission from the Casa de Contratación to ship the hides, the Seville authorities confiscated them and prosecuted him for not having sought an export licence from them. A full record of the ensuing dispute survives.[176]

These documents concerning the family's commercial interests are less important in themselves than as evidence that the Crombergers had embarked in earnest on trade with the New World at a time when Jacobo was legally forbidden to engage in such activities. Nevertheless, the need for subterfuge soon came to an end. On 2 May 1524 he gave Juan permission to seek a favour from the Emperor. The result was a licence to travel to and trade with the Indies granted to Jacobo on 26 July 1525 despite his being a foreigner.[177] Lázaro had obtained a similar licence four days earlier, making them the first Germans in Spain to be accorded such a privilege.[178]

Printing and Trade

Jacobo's subsequent activities indicate that he became increasingly prosperous

174 AGIS, *Contratación*, Legajo 2439, unfoliated.

175 Otte, 'Jakob und Hans Cromberger', pp. 136–7.

176 AGIS, *Indiferente General*, Legajo 420, Libro 10, fols. 98^{v}–100^{r}; Otte, 'Jakob und Hans Cromberger', p. 136. It is strange that Jacobo should have wished to export hides to the Indies, for they were a commodity normally imported from the New World to Seville.

177 AGIS, *Indiferente General*, Legajo 420, Libro 10, fols. 36^{v}–37^{r}. This licence only concerned travel and trade; I can find no evidence to support Arthur Scott Aiton's assertion that Cromberger also secured at this time the right to contract printing for the New World (see his *Antonio de Mendoza*, p. 107).

178 AGIS, *Indiferente General*, Legajo 420, Libro 10, fols. 34^{v}–35^{r}.

and that commerce took up much of his time. In May 1525 he bought the house next door to the family home in the Calle de Marmolejos, probably with a view to expanding the printing-shop or to creating storage space for merchandise.[179] He also purchased a vineyard in Castilleja de la Cuesta, a village outside Seville.[180] He had already acquired vineyards, and supplies of cereals with the facilities to mill them. As shipping manifests of vessels bound for the Indies from Seville reveal that wine and wheat were regularly exported in huge quantities, these investments were probably linked directly with the American trade.[181] More certainly associated with the New World were his dealings with Fray Tomás Ortiz, Vicar-General of New Spain, concerning the export of cloth to Santo Domingo, and with Juan Francisco, an old colleague who had gone out to Hispaniola as the family's factor on the island.[182]

So active did Jacobo become in his role as merchant that he began to shed other duties. In August 1525 he handed over responsibility for the chaplaincy founded in Maestre Nicolás's will.[183] Much more important, however, was his decision to transfer the press to his son which is recorded in a deed of gift of the following month.[184] From then until his death in 1528, books were frequently signed by both father and son, but we may safely assume that some were entirely Juan's work. Juan was a skilled craftsman, as is witnessed by the books he printed alone after his father's death, and Jacobo clearly felt sufficiently confident to trust to his care the day-to-day management of the press even though Juan was only twenty-five years old at this time. Jacobo and Juan were close: Juan was the first-born and only son; his father was certain that he would not lose effective control of the press, for he gave as his reason for the gift to Juan, 'my great love for you and because you, my aforementioned son, are obedient to my wishes'.[185]

Jacobo subsequently seems to have divided his time between printing and

[179] Hazañas, i. 127–8. The houses were bought by Jacobo in Lázaro's name, but, as a later document dated 16 Dec. 1535 proves, Jacobo really acquired them for his own use (APS, Oficio 1, Libro 2 of 1535, unfoliated [5th document of this date]).

[180] The sale contract passed before the notary public of Castilleja, Álvaro de Ulloa, on 20 Dec. 1525; the notarial records for this village have disappeared, but the information is contained in a later document of 16 Dec. 1535 drawn up in Seville (APS, Oficio 1, Libro 2 of 1535, unfoliated [6th document of this date]). The 30 *arrobas* of wine from Castilleja listed in the inventory of Jacobo's estate no doubt came from this vineyard (see Gestoso, p. 48).

[181] Pierre Chaunu, *Séville et l'Amérique aux XVIe et XVIIe siècles* (Paris, 1977), p. 31.

[182] Hazañas, i. 129, 131; *Catálogo de los fondos americanos del Archivo de Protocolos de Sevilla*, v, No. 756.

[183] Gestoso, p. 21.

[184] Hazañas, i. 129–31.

[185] Jacobo possibly had another son, Francisco, who is a shadowy figure or even a scribal error. In her will of 1511, Comincia de Blanquis said that she had three children by Jacobo: Francisco, Juan, and Catalina (see Gestoso, p. 15). If he really existed, Francisco must have been dead by 1527, for he was absent from the list of her children in her will of that year (see Hazañas, i. 147). No other document records Francisco's existence.

commerce. Mindful of the enormous profits which could be gained in the spice trade, Jacobo and Lázaro invested in Sebastian Cabot's prospective voyage of discovery of the westward sea-route to the Spice Islands in 1526.[186] Lázaro put up 250,000 *mrs* while Jacobo, Juan Brunevedier, and Kasimir Nuremberger, who Otte maintains was a member of the family, formed a company, each investing 45,000 *mrs* and thus making the Crombergers' combined investment one of the largest in the expedition.[187] Again this venture was a disaster, for Cabot got no further than the River Plate and came home penniless in 1530. The Crombergers never again invested in exploration, but contented themselves with more straightforward commerce with the New World. Even before Charles V had granted him a trading licence, Lázaro had sent factors there and in 1526 he was the first merchant permitted to send agents of German nationality to the Indies. They were mainly engaged in setting up the Welsers' factory in Santo Domingo, for Lázaro represented the great Augsburg trading company at Seville, but he also sent one German factor, Bartel Blümel, to look after his own family's interests.[188] By the 1530s such permanent agents in the New World enabled him to become one of Seville's leading merchants with the Americas.[189]

Jacobo, for his part, cultivated high officials in the colonies: his dealings with the Vicar-General of New Spain have already been mentioned; he also built up contacts with Fray Juan de Zumárraga, Bishop-elect of Mexico, who acted on occasion as his agent in Mexico.[190] When Jacobo died, the bishop still owed him 15,000 *mrs*.[191] Lázaro's trading activities were always more important than those of his father-in-law, but Jacobo engaged in some independent ventures as well as frequently entering into partnership with Lázaro. In 1528 they had interests not

[186] Profits on one such venture from Spain to Malacca in 1512 or 1513 were so large that the return of a single ship covered the cost of fitting out the fleet of four vessels which made the journey and even gave an additional profit of 60–70 per cent. One ship which was lost at sea was bringing back a cargo valued at 300,000 *ducados* (see Guicciardini's letter to his brother from Valladolid, 17 June 1513, in *Viaje a España*, pp. 141-4).

[187] Gestoso, p. 23; Otte, 'Jakob und Hans Cromberger', p. 138. For an account of Cabot's voyage and of investment in it by merchants and the Crown see Pike, *Enterprise and Adventure*, pp. 103–17. Kasimir Nuremberger was only 21 but accompanied the voyage and later gave evidence in the litigation against Cabot (see José Toribio Medina, *El veneciano Sebastián Caboto al servicio de España y especialmente de su proyectado viaje a las Molucas por el Estrecho de Magallanes y al reconocimiento de la costa del continente hasta la gobernación de Pedrarias Dávila*, 2 vols. [Santiago de Chile, 1908], ii. 169–70, 507–9).

[188] It is worthy of note that several of the books printed by the Crombergers which are now in the Bayerische Staatsbibliothek, Munich, originally came from the Welsers' own library at Augsburg.

[189] Otte, 'Jakob und Hans Cromberger', p. 139. Lázaro Nuremberger was always associated with the trading companies of Nuremberg and Augsburg; he represented not only the Welsers, but also the Herwarts. One of the Fuggers' agents in Spain, Thomas Müller, married Lázaro's daughter María (see C. Collonge, 'Les Allemands dans l'Espagne du XVIe siècle', *Les Langues néo-latines,* 59 [1965], 1-22 [p. 6]). Lázaro's son, Nicolás, became an employee of the Fuggers (see Otte, 'Jakob und Hans Cromberger', p. 157).

[190] Hazañas, i. 154; and a document dated 1 Aug. 1528 (APS, Oficio 4, Libro 3 of 1528, unfoliated).

[191] Gestoso, p. 51.

only in Santo Domingo and Mexico City, but also in Yucatán; Jacobo was to say in the inventory of his goods: 'when we receive all our money from Yucatán, which may it please God to send us, my share can be calculated from the entries in my son-in-law's account books, for the greater part belongs to him'.[192] Jacobo's increasing investment in trade during the last years of his life does not mean that he neglected printing. In the deed of gift of the press to Juan, he was careful to retain the right to use it whenever he pleased and to take one-quarter of its publications. He seems to have exercised that right, for a number of books printed after the press had been handed over were signed by Jacobo alone. He was still thought of as a printer in 1526, when he and Varela were selected by members of the Seville book-trade as their most distinguished colleagues. Ferdinand and Isabella had exempted books from the *alhóndiga de pan*, a tax which required merchants to import a quantity of bread or grain into Seville equivalent to that of the goods they exported from the city. Nevertheless, by 1526 the local authorities were unwilling to respect this privilege and attempted to impose the levy upon the city's booksellers. In March of that year Jacobo and Varela were deputed by colleagues to present a formal protest to the Emperor.[193] In June the Cromberger press happened to be setting up an edition of Alfonso de la Torre's *Visión delectable de la filosofía*. It was oddly inappropriate in such a medieval compendium of knowledge to include a passage concerning a modern invention, but Jacobo took the opportunity to append a short history of printing, carefully emphasizing the positive contribution made by the craft to learning, science, and the Faith (for a full transcription of this history see the description of this edition in Appendix One). This is the only surviving piece of writing thought to have been by Jacobo and it is probably not out of character that, despite his elevated claims for the art of printing, the real purpose of his taking up his pen was to oppose a tax which threatened to reduce his profits. The campaign was successful and the exemption was confirmed by royal decree in the following month. When Varela printed a new edition of the *Ordenanzas de Sevilla* in 1527 he recorded in it his and Cromberger's petition and the happy result.[194]

Jacobo's printing was increasingly innovative in his final years. As usual, he printed fine editions of liturgical works under contract and also the sort of works of devotion and entertainment which past experience had taught him were best-sellers. But he was also willing to risk commissioning untried works for his press. In the prologue to the first edition of the Spanish-language *Moralia* of St Gregory which Jacobo printed in 1527, the translator, one Alonso Álvarez de Toledo, gave the credit for his labours to the 'continual encouragement and pressing demands

[192] Gestoso, p. 53.
[193] Hazañas, i. 135–6.
[194] Hazañas, i. 136, 141–2, 235–6.

of Jácome Cromberger, the German printer'.[195] Both the translation and the printing were major undertakings, for the edition contained over 1,000 folio pages. Nevertheless, it sold well for such a large and expensive book, only 124 copies remaining in the shop on Jacobo's death.[196] It had to be reprinted by Varela in 1534 and again by an unidentified Seville printer in 1549. So saleable were such books that when Varela gave his daughter, Inés de Alfaro, a dowry on her marriage to Jacobo's grandson, Jácome, 100,000 *mrs* consisted of the unromantic gift of 200 copies of the *Moralia*.[197] Jacobo was a shrewd judge of the market for his books, and anticipated by two years Juan de Valdés's exhortation to Christians to read the Spanish translation of the *Moralia*.[198] By 1527 Jacobo had learnt how well patristic works sold in translation. In January 1526 the Valencian printer, Jorge Costilla, had reprinted a Castilian translation of St Jerome's letters. These had previously been printed exclusively in Valencia, but Cromberger bought up some or all of Costilla's edition for his own clients. Indeed, it is open to conjecture whether the 1526 edition was printed entirely on commission for Cromberger, because Costilla's press was particularly dependent upon such work.[199] In the list of debts owed to Jacobo on his death appears the following entry: 'Jorge Costilla, printer, resident of Valencia, owes about 2,500 *mrs* which is the amount by which he was overpaid for the printing [or, 'for the edition'] of St Jerome's letters'.[200] The edition was a success; at the time of Jacobo's death only 64 copies remained unsold.[201] Juan Varela then reprinted it at Seville in 1532 and the Cromberger press reissued it in 1537, 1541, and 1548; during the second third of the sixteenth century it was one of the most widely read books in Spain.[202]

As well as the St Gregory and a large number of popular works, Jacobo issued

[195] *Los morales de sant Gregorio papa: dotor dela yglesia*, 2 vols. (Seville, Jacobo Cromberger, 1527), i. fol. +2^{a}; Aurora Domínguez Guzmán, *El libro sevillano durante la primera mitad del siglo XVI* (Seville, 1975), p. 80, incorrectly lists an edition of 1514.

[196] Gestoso, p. 38. Some 500 copies were probably printed. Part I was finished on 29 Apr. 1527; Part II on 15 June 1527. If 1,500 sheets could be printed each day on one of Cromberger's presses and a single press were used for this edition, about 460–500 copies of Part II which contained 140 sheets could have been printed in this time, allowing for rest days. Such calculations are, however, tentative for so many elements in the equation are unknowable.

[197] Document dated 29 Jan. 1551 (APS, Oficio 10, Libro 1 of 1551, unfoliated). For a similar case in France of books forming part of the dowry of a printer's daughter, see Martin and Chartier, *Le Livre conquérant*, p. 231.

[198] Bataillon, *Erasmo y España*, p. 359.

[199] Printers like Brocar subcontracted to presses in other cities. If this was also the case with Cromberger and the Valencian edition of St Jerome, Costilla nevertheless signed it himself. It was, however, quite normal in Spain for books printed on commission from a publisher to bear only the printer's name (see Norton, *Printing in Spain*, pp. 38, 131–2).

[200] Gestoso, p. 50.

[201] Document dated 7 June 1529 (APS, Oficio 4, Libro [2] of 1529, unfoliated); the entry is incorrectly transcribed by Gestoso, p. 37.

[202] Francisco López Estrada, 'Sobre la imprenta en Sevilla en el siglo XVI', *Archivo hispalense*, 2nd ser., 18, No. 57 (1953), 37–48.

in 1527 Fray Bernardino de Laredo's *Modus faciendi*, a book designed for apothecaries and, despite its title, largely in the vernacular. Laredo was an important figure in the medical and spiritual worlds of Seville and his popular *Subida del Monte Sión* was twice issued by the Crombergers. The printing of his *Modus faciendi* may, however, have been a result of the family's close relations with the author, for Juan later appointed him as an executor to his will.[203] Jacobo's surviving editions of 1528 were largely expensive folio volumes indicating that he was by this time prepared to invest substantial sums and wait for the relatively slow returns on such books compared with quarto or octavo editions. A calculation based on the number of sheets in a single copy of the surviving editions printed in 1528 (rather than merely on the number of editions) indicates that it was his most productive year. His ability to choose a potential best-seller is illustrated by his decision in that year to issue the first edition of Guevara's *Libro áureo de Marco Aurelio*, which the press subsequently reprinted at least nine times in this and its extended version containing the *Relox de príncipes*. Jacobo seems to have been no more distressed by the prospect of printing an unauthorized edition of this book at a time when it was certainly protected by the author's privilege than he had been some sixteen years earlier when he had pirated parts of Castillo's *Cancionero general* under the title *Cancionero llamado guirnalda esmaltada de galanes*.[204] He bequeathed to Juan the task of coming to an arrangement with Guevara enabling the press to print further editions of the *Libro áureo* legally. In the same year he produced numerous liturgical editions, one or two works by Erasmus, and the important Latin-language edition of Lucan's *Pharsalia*.[205]

It was at the apogee of his career as a printer that Jacobo decided to make another journey to Portugal. As has been seen already, that country provided a substantial market for his books; his son-in-law handled a thriving trade with Lisbon, especially in gems and pearls, where there was a large community of German merchants. Jacobo may have travelled there on general business for himself and Lázaro or to arrange more specifically for the sale of books. He may have wanted to size up the market for Latin works, especially if his *Pharsalia*, the surviving copies of which are both now in Lisbon, was the prototype for a new series of octavo editions of the classics rivalling those which were being poured out by French and Italian presses in imitation of the original Aldine series and imported into Spain. On the other hand, he had just printed two important

[203] For the Cromberger editions of Laredo's medical works see Teresa Santander Rodríguez, *Hipócrates en España: siglo XVI* (Madrid, 1971), pp. 9–17.

[204] Jack Gibbs, 'The Status of the Cromberger Editions of Antonio de Guevara's *Libro aureo de Marco Aurelio* and *Relox de principes*', *Bulletin of Hispanic Studies*, 54 (1977), 199–201.

[205] For a detailed discussion of this edition, see Clive Griffin, 'The Crombergers of Seville and the First Italic Book Printed in Spain', in Jean-François Gilmont (ed.), *Palaestra typographica: aspects de la production du livre humaniste et religieux au XVIe siècle* (Aubel, 1984), pp. 57–96.

liturgical works for Évora and might have been trying to win more orders for such books in Portugal. He also had to deal with a shipment of his books which had ended up in various towns of the Alentejo after his colleague, Diego Fernández, had been drowned off Tavira earlier in the year while taking them to Portugal. He had assured the widow and family that he would have the books brought back to Seville and would settle up with them there.[206] Whatever the reasons for his journey, on 18 August 1528 Jacobo gave his son a full power of attorney to act on his behalf during his absence.[207] By 8 October Comincia de Blanquis was a widow. In March of the following year his heirs said that their father had died in Lisbon seven months earlier—soon, therefore, after his arrival there.[208] The Lisbon earthquake of 1755 has destroyed any chance of his grave or any record of his death being found.

Jacobo was one of the finest printers who had worked in Spain. His output was not only prolific, but the quality of his editions, like those of other members of the first generation of German master-printers who settled in the Iberian Peninsula, was excellent. Starting out from a humble position as a printer's assistant, he had been able by a judicious marriage to begin a career as a major printer in a city which was ideally situated for working with little competition and a ready market for his products. He was a good judge of the sort of books that the market would absorb and he seldom printed editions which did not sell.[209] The marriage of his daughter to Lázaro Nuremberger and the opening of the New World meant that the experience he had gained as a bookseller with an extensive network of business contacts could be put to good use in more ambitious commercial ventures with the Americas and the East Indies.

The diversity of the works which came off his presses—from the products of jobbing printing to complex service books, from editions of the classics to humble *pliegos sueltos*; the balance between editions which brought in rapid or longer-term returns; and his spread of investments in printing, bookselling, land, urban property, and trade gave his business a great degree of stability and flexibility. Printers in Spain, as elsewhere, frequently fell victim to a failure to bridge the gap between investment in the production of an edition and the return on its sales. At the same time, merchants in Seville who relied on the arrival of a fleet in order to pay off expensive loans often went bankrupt when the fleet was delayed or their ships went down. Diversity of activities minimized the risk and effect of

206 Hazañas, i. 152–3; and two documents dated 15 June 1528 (APS, Oficio 4, Libro 3 of 1528, unfoliated).

207 Hazañas, i. 154.

208 Three documents dated 8 Oct. 1528 (APS, Oficio 4, Libro 3 of 1528, unfoliated), and Hazañas, i. 154.

209 For an analysis of the sort of books printed by Jacobo see below, pp. 145–64.

such misfortunes and while commerce could be financed from the profits of printing, so could the press be financed from the profits of trade.

On his death, Jacobo's estate showed all the signs of prosperity. He had a large number of investments, he was owed a good deal of money but had few creditors, he possessed both negro and North African slaves, and his house was furnished with all the necessities, and not a few of the luxuries, for a comfortable domestic existence. The stock of his bookshop numbered over 160,000 items and his estate was valued at a very respectable 12,000 *ducados*.[210] Most importantly, he had bequeathed to the son who had learnt his craft at his father's side, a flourishing and well-equipped press, an established network of outlets for his books, and a lively interest in the American trade. These were to be consolidated and expanded by the next generation of Crombergers.

[210] In 1548, after high inflation had been experienced in Spain, Pedro de Medina considered book-merchants of Medina del Campo who were worth 10,000 to 12,000 *ducados* to be very rich (see his *Obras*, ed. Ángel González Palencia [Madrid, 1944], p. 133).

CHAPTER 3

JUAN CROMBERGER: THE CONSOLIDATION AND EXPANSION OF THE PRESS (1528–1540)

The Seville Press

DURING a long life as Seville's leading printer, Jacobo Cromberger had founded one of the most successful presses in sixteenth-century Spain. On his death there was no question of who would take his place. Juan had worked with his father for at least three years and had grown up in the atmosphere of the printing-shop. He also knew many of his father's associates who were printers and booksellers, and it was only natural that on Jacobo's death his son should take over complete control of the press and foster relations with his father's old colleagues. But Juan had done more than acquire a knowledge of the workings of the press itself; as we have seen, by 1525 he was already engaged in commerce with the New World. Three years later, even his father owed him a substantial sum of money and he was by then being referred to in notarial documents as a merchant rather than as a printer.[1]

When Jacobo died, Juan and his wife were living in the family home in the Calle de Marmolejos, where they were to remain for the rest of Juan's life.[2] As their eldest son, Jácome, had been born by September 1525, they were presumably married by at least the end of the previous year. Brígida Maldonado was a judicious choice as a bride. Her origins are obscure, but a document which she signed in 1545 offers a clue to her background; in that year one Mateo Carón was travelling from Seville to the New World authorized by her to represent the Crombergers' interests in Mexico. He probably possessed specialized knowledge of printing and the book trade, for he was to deal with the 'press, books, and the privilege on the printing and sale of books'. Brígida added that this representative was her brother.[3] I know nothing more of Mateo, but he was almost certainly related to the important family of book-merchants, the Caróns of Salamanca, with whom

[1] Gestoso, pp. 30, 35. Throughout the remainder of his career he would be styled a merchant as well as a printer; on one occasion the word 'printer' was even deleted from a document (see Gestoso, p. 68). Already in 1524, Juan was being called a book-merchant (see Hazañas, i. 125).

[2] In her will of 1527 Comincia de Blanquis mentioned that she had installed the couple in a room in the family home and given them linen and clothes to the value of 130 *ducados* (see Hazañas, i. 146).

[3] Document dated 17 Mar. 1545 (APS, Oficio 17, Libro 1 of 1545, fols. 420^r–421^v [*olim* 704^r–705^v]). We should not necessarily expect children of a single marriage in 16th-c. Spain to have the same surnames; two of Juan's children, for example, were Jácome Cromberger and Ana de Maldonado; Martín de

Jacobo and Juan had frequent dealings. We find that another member of the family, Martín Carón, was in Seville in 1532 when he and Juan Cromberger settled an account for the purchase, or possibly exchange, of books.[4] Brígida Maldonado proved an ideal wife: she bore Juan many children and when he died in 1540 her family background stood her in good stead. She did not remarry like so many other printers' widows, but she managed the press and the family's overseas business competently until her son, Jácome, was old enough to take charge.[5] Unlike her mother-in-law, Comincia de Blanquis, who, although the wife of two printers, was illiterate, Brígida Maldonado was able to write. On her husband's death she could not only sign her name in a clear and confident hand, but followed the royal medieval practice of alluding to a widow's grief: she signed herself pathetically 'the wretched Brígida Maldonado'. Many months later she was still 'the grief-stricken Brígida Maldonado'.[6]

When his father died Juan inherited almost 6,000 *ducados* as well as control of the presses and equipment which he had been managing since 1525. A large part of this substantial sum was made up of debts owed to Jacobo by booksellers from all over Spain which Juan tried to collect during the 1530s. He also inherited the house in San Isidoro, its furniture and goods, several slaves, a large amount of paper valued at over 1,000 *ducados*, the vineyard in Triana, houses in Seville, and a variety of goods which debtors had left as sureties with his father.[7] Jacobo had clearly made arrangements for Juan to take over the press and its premises immediately so that continuity of printing could be maintained. An equal share of the estate fell to the other beneficiary, Juan's sister Catalina; this was made up of money, grain, and wine, all of which her husband, Lázaro Nuremberger, could use in trade with the New World. She also inherited the mills at Osuna, the vineyard at Castilleja de la Cuesta, which was larger than the one in Triana,

Montesdoca's brother and sisters were Alonso Núñez de Montesdoca, Inés Núñez, and Marina de Cortegana (sometimes called Marina de Consuegra) (see Klaus Wagner, *Martín de Montesdoca*, p. 28).

[4] One document dated 7 June 1529 (APS, Oficio 4, Libro [2] of 1529, unfoliated) and three documents dated 12 June 1532 (APS, Oficio 4, Libro 1 of 1532, unfoliated). Gestoso, p. 31, is mistaken when he makes Juan Cromberger's wife the daughter of Licenciado Luis Mexía Ponce de León and Doña Catalina Maldonado. It was Juan and Brígida's daughter, Catalina Maldonado, who married this Ponce de León. Their daughter was christened Brígida Maldonado after her grandmother.

[5] On 16 Sept. 1540 the offspring were listed as: Jácome (15 years old), Catalina (11), Ana (8), Tomás (7), Isabel, Antonio, 'Francisco', Fernando (all under 7), and Juan (over 7). Jácome, Juan, and Tomás were all able to sign their names (see Gestoso, pp. 69–71). As a daughter, Francisca, appears in later documents while 'Francisco' disappears immediately, the name was doubtless wrongly transcribed by Gestoso. Antonio does not appear in a list of the children in 1545 and had probably died by then (see document dated 17 Mar. 1545 [APS, Oficio 17, Libro 1 of 1545, fols. 420^r–421^v (*olim* 704^r–705^v)]). Juan does not appear in 1548 and was therefore probably dead by that date (see document of 3 Feb. 1548 [APS, Oficio 17, Libro 1 of 1548, fol. 811^{r-v}]). That document contains the last reference to Isabel and Fernando which I know.

[6] Gestoso, pp. 100, 101.

[7] Jacobo must, therefore, have given loans or credit.

property in Seville, and the enormous stock of the family bookshop or warehouse. Juan spent some two-thirds of his inheritance buying these books from his sister and thus taking control of the two sides of the family's interest in the press: printing itself and selling.[8]

Under Juan the press was productive. Although new printers set up in the city during the time that Juan was working on his own, none could approach the output of his office, which still issued almost 55 per cent of the surviving editions printed in Seville from the period 1529 to 1540. Varela had been the other major printer there during Jacobo's lifetime and he remained so until 1539 when he issued his last known books.

Varela had been born in about 1476 and was therefore no longer a young man when he gave up printing. In 1544 he described himself as 'old, infirm, and beset by the worries of serving His Majesty in this city'.[9] Nevertheless he lived on until 1555 and age is less likely to have been his reason for abandoning his craft than his involvement with more profitable activities and a consciousness that his social status was no longer compatible with the practice of a mechanical trade. As has been noted, he had long been an owner of ships and had regularly sent merchandise to the Indies. Records of the bullion sent from the New World to Seville are preserved in the Archivo General de Indias, and they contain entries for large sums remitted to him.[10] The wide range of Varela's activities which intensified in the 1530s and 1540s, meant that in the last decade of his work as a printer he did not increase his share of the city's total production of printed books; this share was a steady 20 per cent as it had been in the days of Jacobo.

The remaining 25 per cent of books printed in Seville in the 1530s came in more or less equal numbers from the presses of (i) Bartolomé Pérez, who began printing in the Calle Sierpes in 1529 and ceased in 1539; (ii) Domenico de Robertis, later a prolific printer, who began in the Calle Abades in 1533 and was to continue to work for at least fifteen years; and (iii) of unidentified printers.[11] The incalculable number of Seville editions which have disappeared makes these figures tentative, but it is likely that the preponderance of surviving books attributable to Juan Cromberger's press does indeed reflect a far greater output than that of his

[8] Juan must have given Catalina over half the value of the books soon after she inherited them. On 14 Aug. 1531 he paid the remainder of the debt to Lázaro, who was by then a widower (APS, Oficio 4, Libro 3 of 1531, fols. 195^r–196^v).

[9] Hazañas, ii. 104.

[10] For example, AGIS, *Contratación*, Legajo 2439 contains records of large quantities of precious metals remitted to Varela at Seville during the years 1533–45. An additional reference to gold being sent to him from Mexico by his son in 1533 is AGIS, *Contratación*, Legajo 689A, fols. $[4]^v$–$[5]^r$.

[11] Domenico de Robertis was in Seville by at least 1530 (see Hazañas, ii. 271). In 1532 he was calling himself both a bookseller and a book-merchant (see document dated 3 Oct. 1532 [APS, Oficio 1, Libro 2 of 1532, unfoliated]).

colleagues. There was, then, the same lack of serious competition to Juan's dominance of Seville printing in the 1530s that Jacobo had enjoyed for almost thirty years previously. The press flourished under his management: the very incomplete figures for book sales extracted from contracts with booksellers are considerably higher than those of Jacobo.[12]

Like his father, Juan was a publisher as well as a printer and the titles he chose were those he knew he could sell. An examination of surviving copies indicates that the range he published was similar to his father's, although there was a distinct decline in the number of scholarly Latin titles printed. Juan's production was dominated numerically by books of popular devotion in Spanish: from translations of the much-reprinted *Imitatio Christi* of Thomas à Kempis and the *Vita Christi* by Ludolph of Saxony, to the works of Spanish authors: Argomanas, Ciruelo, Osuna, Laredo, and Valtanás. He continued to print numerous liturgical works both of general use and specifically for named dioceses. Medical and legal works were still issued, but the number of these editions was small compared with those in the other main areas of the press's activity: works of popular morality or compendia of information, and literature of entertainment. In the former category Juan published many editions of Antonio de Guevara's *Libro áureo de Marco Aurelio*; in the latter he produced several editions of romances of chivalry, many of which would have been destined for the Indies despite the royal decrees of 1531 and 1536 banning their export there.[13]

Just as his father had printed school-books, so did Juan publish books for students. In 1535 he issued an edition of Michele Verino's *Liber distichorum*, a popular school-book in Spain, and he may even have printed many editions of this sort of text for those learning Latin. Evidence about this is, however, lacking because school-texts were eminently destructible, and this would have been one of their attractions for publishers: the more frequently they were pulled to pieces by young hands, the more copies the presses would have to print. It was doubtless for this reason that the Seville printer, Simón Carpintero, sought a royal licence in 1559 to reprint any school-text which had appeared in Spain during the previous fifty years.[14] Nevertheless, the fact that Juan had acquired a small roman type

[12] The source for these figures is numerous documents preserved in APS. Only prices and not numbers of books are recorded in the sale contracts. It is therefore impossible to calculate how much the higher figures are attributable to increased volume of sales and how much to price-rises.

[13] On 4 Apr. 1531 the Queen forbade the export of romances of chivalry (see Irving A. Leonard, *Books of the Brave: Being an Account of Books and of Men in the Spanish Conquest and Settlement of the Sixteenth-century New World* [Cambridge, Mass., 1949], p. 81). This ban and subsequent repetitions of it in 1536 and 1543 were ignored. I have examined numerous shipping manifests in the AGIS containing large quantities of romances bound for the Indies and which were approved for export by the inspectors with the phrase 'these are not prohibited works'. Presumably they were only interested in preventing the export of heretical or politically undesirable works.

[14] Klaus Wagner, *Martín de Montesdoca*, p. 124.

ideally suited to this sort of Latin text (Appendix Two, Type 19) and yet that it is found only in the unique copy of this one edition of the *Liber distichorum* is suggestive: he is unlikely to have invested in a new type only to use it in one book. It is more probable that the editions in which it appeared were soon thumbed to pieces and have disappeared.

To this list of the sort of books known from surviving copies to have been printed by Juan can be added editions which have disappeared but which were recorded in the inventory of his shop made in 1540. Those titles of which large numbers of copies were stocked are obviously likely to have been products of his own presses. In some cases the inventory positively states that the books were 'from our press'.[15] There were works of popular devotion ('611 large Flos sanctorum from our press', '202 Scala celi'); liturgical works ('a Roman psalter from our press', '2 Roman breviaries from our press', an edition of diurnals for the diocese of Badajoz);[16] works of Erasmus probably printed by Juan ('206 Psalms of Erasmus'); numerous romances of chivalry and chronicles; popular medieval tales like the *Doncella Teodor*; and the sort of publications which, by their very nature, cannot be expected to have a high survival rate, like cookery-books and chap-books. The diversity of material which Juan printed, from the prestigious edition of the five-volume *Ordenações manuelinas* already discussed in Chapter Two to the thousands of broadsheets and chap-books, coupled with the strategy of printing a mixture made up of works which gave long-term returns and those which produced an immediate income, mirrors Jacobo Cromberger's practice.

Juan also continued to supply the network of booksellers built up by his father. In the list of his debtors drawn up at the time of his death, booksellers appear who represent the three markets which Jacobo had supplied: Seville and the remainder of Andalusia—Écija, Jerez, Jaén, and Granada; Portugal, including Évora, Lisbon, and even a town as far north as Guimarães; and Northern Castile—Salamanca, Toledo, and, of course, Medina del Campo.[17] There is even a record of a debt owed by a Valencian bookseller, suggesting that Juan had found a retail outlet there. This is not as surprising as it may at first appear because Valencian had been in decline as a literary language ever since the beginning of the century and the reading public there had become accustomed to books printed in

[15] Document dated 20 Sept. 1540 (APS, Oficio 17, Libro 2 of 1540, unfoliated); this is reproduced with errors and omissions by Gestoso, pp. 73–99.

[16] The diurnals were printed in 1529–30, as is witnessed by a document of 1529 concerning payment by the Bishop of Badajoz for missals, breviaries, and diurnals which Juan printed (see Gestoso, p. 35). Copies of the missal and breviary survive; the diurnals have disappeared.

[17] Other documents indicate a flourishing trade with Portugal. In 1529 Juan loaded a shipment of books on a Portuguese caravel bound for Lisbon (see document dated 12 May 1529 [APS, Oficio 4, Libro (2) of 1529, unfoliated]). Several copies of Juan's books which I have examined contain contemporary annotations in Portuguese.

Castilian.[18] The printers at Valencia, who were themselves either Castilians or non-Spaniards, certainly produced books in Castilian in an effort to reach larger markets than their local one; Valencia, as the Empire's main Mediterranean port, was anyway a cosmopolitan city.[19] This network of suppliers and distributors was reinforced by personal contact; as we have seen, Martin Carón travelled from Salamanca to Seville and there drew up a contract in Juan Cromberger's house, and Juan himself travelled to Medina del Campo, where he met local booksellers. It was also strengthened by marriage: just as Juan had married Brígida Maldonado, so one of his major clients, Juan de Espinosa, a book-merchant of Medina del Campo, married his daughter, Ana de Maldonado.[20] Espinosa was an important figure who dealt with businessmen and printers of the stature of Andrés Ruiz, the Portonarii, and Guillaume Rouillé.[21]

Juan's publishing activities resembled those of his father in more than just the titles he selected and the network he used for distributing his books. It has already been seen that Jacobo published an unauthorized edition of Guevara's *Libro áureo de Marco Aurelio*. Guevara complained bitterly at the infringement of his royal privilege, claiming that Cromberger's edition was full of errors, but he sought no recompense and his anger did not prevent him from selling permission for subsequent editions to the very press that had ignored it in such a cavalier fashion—and was, incidentally, renowned for printing precisely those books which the good bishop considered a public scandal.[22] Juan continued this practice of publishing what he liked and then coming to an arrangement with the author or holder of the privilege if he complained. In 1519 Jacobo had printed the first edition of the *Suma de geografía* written by the explorer and cosmographer Martín Fernández de Enciso. The *Suma* was an important navigational treatise, which was one of the first works printed in Spanish to describe the New World. Charles V had granted Enciso a privilege for it and this was printed at the beginning of the first edition. When the first edition had sold out, Juan decided to reprint the book and did so in 1530 without consulting the author. It was a good deal easier, however, for Enciso to take action than it had been for Guevara, because he lived

[18] Philippe Berger, 'Contribution à l'étude du déclin du valencien comme langue littéraire au seizième siècle', *Mélanges de la Casa de Velázquez*, 12 (1976), 171–94.

[19] Joffre's edition of St Jerome's letters printed at Valencia in 1520 was destined for Castile (see Norton, *Printing in Spain*, p. 123).

[20] Juan Cromberger must have known Juan de Espinosa well, for Espinosa witnessed the agreement signed by Cromberger and Guevara in Medina del Campo in 1529 (see E. García Chico, 'Documentos', pp. 233–40). See also documents dated 11 July 1534 (APS, Oficio 4, Libro único of 1534, unfoliated) and 7 June 1548 (APS, Oficio 17, Libro 1 of 1548, fols. 102ᵛ–103ᵛ) in which Brígida Maldonado and Juan de Espinosa release each other from debts incurred in the book-trade by Juan Cromberger, Juan de Espinosa, and the latter's father.

[21] On Juan de Espinosa's business contacts see Lapeyre, *Une Famille*, p. 568.

[22] Guevara's condemnation of romances of chivalry, the *Cárcel de amor*, and the *Celestina* is cited by Maxime Chevalier, *Lectura y lectores en la España de los siglos XVI y XVII* (Madrid, 1976), p. 156.

only a few streets away from the press. He immediately had the distribution of the books prohibited and they were forcibly put in store. Many months later, in 1531, author and printer found a solution to their dispute and signed an agreement while they were, somewhat ominously, 'in the city council's prison which is in the parish of El Salvador'. Enciso granted Juan permission to sell the books, and Juan gave him in return two free copies and the small sum of five *ducados*.[23]

Business and Trade

Juan had also followed his father's example in other kinds of business both at home and in the Indies. There are many records of his renting out property in Seville and, in 1534, he had acquired a sizeable tract of agricultural land outside the Puerta de la Macarena, an area then famous for its market gardens.[24] His main commercial activity besides printing was, however, trade. For many years he maintained some sort of partnership with an old associate of his father, García Cerezo de Carranza, who bought large quantities of merchandise in Seville, doubtless intending to ship it to the Indies.[25] The New World was also the focus of Juan's own trading activities, and there is evidence that he was associated in this with his brother-in-law, Lázaro Nuremberger. This is not surprising, for Lázaro continued to live with Juan in the family home even after Catalina's death, although he may have moved out when he remarried; in 1539 he was resident in the neighbouring parish of El Salvador.[26] Each man frequently authorized the other to deal with his affairs, especially before leaving Seville on business trips. As has been seen, Juan attended the fairs of Northern Castile; Lázaro also visited his trading partners in Lisbon as well as certainly travelling further afield to Antwerp.[27] The commercial relationship between Juan and Lázaro is often obscure because most of their joint ventures must have been agreed verbally and no record of them remains. In many cases we are left with only hints of a common interest. Thus in 1529 Juan accepted jewels as a payment for books, the jewels doubtless being handled by Lázaro who specialized in this commodity as representative of

[23] Document dated 14 Nov. 1531 (APS, Oficio 4, Libro 4 of 1531, fols. 304^{r}–305^{r}). For a fuller discussion see Klaus Wagner, 'La *Suma de geografía* del bachiller Martín Fernández de Enciso: privilegio, honorarios del autor, tirada', *Estudios y notas*, 24 (1972), 67–9.

[24] In 1534 he had bought the 'Huerta de las Almenas' for 450 *ducados* from the noble Neyra family. Four years later, after considerable litigation, Juan returned the land and the Neyras returned his investment (see document dated 20 Dec. 1538 [APS, Oficio 1, Libro 1 of 1538, fols. (2)^{r}–(4)^{r}]).

[25] Documents dated 4 June 1529 (APS, Oficio 5, Libro 3 of 1529, fol. 125^{r-v}), 14 Aug. 1534 (APS, Oficio 5, Libro 5 of 1534, fol. 26^{r-v}), and 19 Apr. 1535 (APS, Oficio 5, Libro 2 of 1535, unfoliated [2 contracts]).

[26] A document dated 28 Mar. 1530 was signed 'in the home of the said Lázaro Morberger and Juan Coranberger' (APS, Oficio 4, Libro 1 of 1530, unfoliated). For Lázaro's new residence, see Gestoso, p. 32.

[27] Otte, 'Jakob und Hans Cromberger', p. 148.

the Herwart firm.[28] They both traded in slaves and on at least one occasion dealt jointly with the same slave-vendors; Juan and Brígida Maldonado purchased four 'white' slaves, while Lázaro bought a fifth.[29] It is possible that Juan acquired slaves to work in the press or on his land, but most would have been destined for the Indies. Lázaro dealt with 'white' and black slaves and became a trader of some importance, sending thirty 'white' female slaves to the Indies in 1536 and doing business with one of the major slaving families of Seville, the Jorges. Juan dealt on a much smaller scale, but bought and sold both blacks and 'white' slaves.[30] In 1537 he was granted licences to send seven slaves to the Indies, but had shipped only three by the time of his death. His step-brother, Tomás Ungut, dispatched the remaining four on Brígida Maldonado's behalf in 1541.[31]

Juan also appears to have participated in Lázaro's dealings with the great gem merchants of Lisbon and Augsburg, the Herwarts. As early as 1532 Christoph Pissinger, a factor of the firm, was a guest in Cromberger's house in Seville where he signed contracts with Juan, Lázaro, and Christoph Raiser, the latter a man who was, like Lázaro, employed as a representative of the Hirschvogel trading company of Nuremberg, but who also later worked as an agent of several other German companies including the Fuggers.[32] Lázaro and Raiser were quick to capitalize on the new conquests in the Americas and in the 1530s made ambitious plans to trade with the recently discovered kingdom of Peru. Indeed, at the remarkably early date of August 1532, some four months before Pizarro even took Atahualpa prisoner in Cajamarca, we find Lázaro demanding payment for goods which he had sent to Peru.[33] Juan's association with his brother-in-law in

[28] Many years after the purchase, on 10 Nov. 1529, the Lisbon merchant Gabriel Lopes paid Juan Cromberger for 1,000 copies of Jacobo's 1521 edition of the *Ordenações manuelinas* in 62 gold rings encrusted with diamonds and rubies (see Gestoso, p. 58 [Gestoso dates this document wrongly]).

[29] Juan's and Brígida's slaves were Juan Moreno, Berber (22 years old), Luis (19), Diego Herrera (23), and Diego de Cabrera (18); they cost 200 *ducados* (see document dated 6 Mar. 1530 [APS, Oficio 4, Libro 1 of 1530, fols. 859^r–860^v]). Lázaro's slave was Juan Platero, 'a white slave branded on his face' (22) who cost 80 *ducados* (see document dated 28 Mar. 1530 [APS, Oficio 4, Libro 1 of 1530, fols. 1050^r–1054^v]).

[30] Document dated 23 Aug. 1536 (APS, Oficio 1, Libro 1 of 1536, unfoliated). A black slave mentioned in this document was named, appropriately enough for the property of a printer of romances of chivalry, Lançarote (Lancelot).

[31] Otte, 'Jakob und Hans Cromberger', pp. 144, 154.

[32] Two documents dated 29 Apr. 1532 (APS, Oficio 4, Libro 1 of 1532, fols. 1237^r–1238^v and 1238^v–1240^v). Jörg Herwart owned a large diamond-cutting workshop in Lisbon and had a royal monopoly on gems. He dealt with rivals in an unorthodox fashion: in 1530 he denounced Christoph Raiser and Hans Tetzel of Nuremberg to the Portuguese authorities for eating meat on the eve of St Bartholomew. Raiser fled, but Tetzel was imprisoned (see Hermann Kellenbenz, 'Os mercadores alemães por volta de 1530', *Revista portuguesa de história*, 9 [1960], 125–40 [pp. 131–2]). Raiser and Nuremberger later formed a company together to trade with Mexico. Steinberg, *Five Hundred Years*, p. 46, maintains that Juan Cromberger was himself employed as an agent of the Great Trading Company of Ravensburg, but I have found no evidence to support this claim.

[33] Document dated 1 Aug. 1532 (APS, Oficio 4, Libro 1 of 1532, unfoliated).

this trade is only revealed by a document drawn up after his death in which his widow organized the collection of his property 'in the said city of Mexico in New Spain of the Ocean Sea, and in the provinces of Peru, Cartagena, and Santa Marta'.[34]

Some measure of the scale of Lázaro's involvement with the New World is found in a document of 1539, which records that he sought large loans to cover his expenses; in that year he promised to pay the Lucchese banker, Cristoforo Francesquín, and the Florentine merchant, Giovanni Battista Ridorsi no less than four million *mrs* (almost 11,000 *ducados*).[35] An indication of the close association of the press and the family's mercantile activities comes in 1535 when Juan, Lázaro, and Raiser were looking for a factor to represent their trading and mining interests in Mexico. They chose Hans Henschel of Basle to go to the New World for them; Hans was probably, as Otte asserts, the 'Juan de Basilea' whom Jacobo had contracted as an assistant in his press over twenty years earlier.[36]

Juan Cromberger does not seem to have been entirely dependent upon Lázaro to take the initiative in trading ventures. He sent at least one cargo to Mexico in his own name only. In March 1536 he was sufficiently prosperous to invest the large sum of 3,000 *ducados* in a shipment of general merchandise and he chartered a vessel called *Los Tres Reyes* to transport it to Vera Cruz, the port of entry to Mexico.[37] The contract drawn up between him and the master of the ship contained a provision that not only could he add some blacks to the cargo if he wished, but that a cabin would be reserved for whomever he might appoint as an agent to accompany the merchandise. Just as Jacobo had selected Pedro de Mendieta's son to accompany his first major cargo to the Americas, so did Juan choose to send the son of a family friend. Guido de Labezaris (not to be confused with his grandfather of the same name) was the son of an Italian bookseller of Seville, Sebastiano de Labezaris, with whom Jacobo, Juan, and Varela had enjoyed a long acquaintance.[38] Guido was a bookseller in his own right who had two shops in the Calle de Génova, and Juan would have known him well; he certainly considered him a reliable agent at that time. Juan required him to represent his interests in Mexico City: he was to distribute general merchandise sent to him by the Crombergers, and in particular would handle the books which were dispatched to him from Seville. As Juan put up the capital for the shipment and Guido's fare,

[34] Document dated 20 Sept. 1540 (APS, Oficio 17, Libro 2 of 1540, fol. 1365^r). It is not clear whether Juan and Lázaro were associated with the factory set up by Hans Ort of Antwerp in 1536 at Nombre de Dios to trade with Peru (see Otte, 'Jakob und Hans Cromberger', pp. 149–50).

[35] Gestoso, p. 32.

[36] Otte, 'Jakob und Hans Cromberger', p. 146.

[37] Document dated 18 Mar. 1536 (APS, Oficio 5, Libro único of 1536, fols. 1154^r–1156^v).

[38] Varela was godfather to one of Sebastiano's sons (see Hazañas, i. 204). The family was not of Basque origin as maintained in *Cartas de Indias*, published by the Ministerio de Fomento (Madrid, 1877), p. 782, but Genoese. Sebastiano's father, Guido de Labezaris, had been associated with the printers of incunables at Seville.

he was to take all the profit on the goods which Guido accompanied and was to sell in Mexico. However, Guido was to take one-third of the profits on books which Juan sent to Mexico. He was also to look after Lázaro's and Raiser's interests in the New World if anything happened to their factor, Hans Henschel. It seems that they had learnt from Jacobo's mistake of having Diego de Mendieta as his sole representative. Guido was to be absent from Seville for four and a half years and during that time he leased his shops to Juan Cromberger; he also sold him his own considerable stock of books.[39]

Guido de Labezaris was a colourful figure and it is clear from the fragmentary evidence concerning his career that he could not be relied upon to content himself simply with being a bookseller in Mexico City. Two years after Guido had sailed for the Indies, Juan Cromberger heard that history had repeated itself and that his factor had perished there. He therefore began making arrangements for other agents to take over Guido's responsibilities.[40] However, it subsequently became apparent that Guido was far from dead; nevertheless, he did not return from Mexico when his tour of duty ended, and for many years afterwards Juan asked other agents in the New World to settle up with him and send the profits back to Seville. It is not difficult to sympathize with Guido, for he seems to have found bookselling a mundane occupation, and the New World offered opportunities for adventure even more remarkable than those contained in the fictional pages of Cromberger's romances of chivalry which he was supposed to be handling there. As will be seen later, he soon became involved with silver-mining outside Mexico City, and his restless nature later led him still further afield. First, he left Mexico in 1542 on Ruy López de Villalobos's ill-fated westward expedition to the Islas de Poniente, which Villalobos was to christen the Philippines. When he eventually managed to return to Mexico via Lisbon and Seville some six years later after having virtually circumnavigated the world, he had lost everything and had to travel to the Americas once more as the Crombergers' employee. Fray Nicolás de Witte later wrote to the king in an attempt to attract some royal largess for Labezaris and recounted the hardships he suffered on this voyage:

> he contributed more than anyone to the service of Your Imperial Majesty on the voyage of '42 . . . to the Islas de Poniente. He brought back to New Spain diverse information about those parts and how to navigate in them; this was a great service to God, Your Majesty, your subjects, and particularly to those of us who reside in the Indies. What is more, during that expedition the wretched Labezaris not only sacrificed all his wealth, but the natives carried off his wife so that he was left utterly undone.[41]

[39] Seven documents dated 10 July 1536 (APS, Oficio 1, Libro 2 of 1536, fols. 66^{r}–74^{r} and 87^{v}–88^{r}).
[40] Document dated 12 or 13 Mar. 1538 (APS, Oficio 17, Libro 1 of 1538, fol. 463^{r-v}).
[41] *Cartas de Indias,* p. 119. This work maintains (p. 782) that he returned to Lisbon on a Portuguese vessel in 1549; this is incorrect, for in 1548 he was already back in Seville, where Lázaro Nuremberger

However, the desire to escape from his post as the Crombergers' representative became too strong for him again; when his experience at sea won him a commission from the Viceroy to undertake a voyage of discovery to Florida in 1558, he readily accepted. This expedition was repeated the next year, but this time he was fortunate to escape with his life from the hurricane which overwhelmed the fleet and caused the voyage to be abandoned.[42] Nothing daunted, on his return to Mexico he took on the important post of treasurer of the Philippines and of Miguel López de Legazpi's fleet, which eventually set sail for the islands in 1564. At last he seems to have found his vocation there. After Legazpi died in 1572 Labezaris took over command of the expedition, and so, when the newly-founded settlement of Manila was attacked by the forces of the Chinese corsair Li Mahong two years later, he was in charge of its defence. Although he was an old man by this time, he successfully defeated the pirates. He subsequently negotiated a peace with the Chinese, continued the construction and fortification of Manila, and carried on the task of pacification of the Philippines, where he eventually died.[43]

Interesting though Labezaris's adventures may be, his career serves more importantly in the context of this study as an illustration of the unreliability of factors at this time and of the very tenuous control their masters had over them. It will be important to bear this in mind in what follows. Guido's exploits in the Far East were, of course, of little help to the Crombergers, although it has to be admitted that he, or their other representatives in Mexico, do seem to have remitted considerable quantities of bullion to Juan before Labezaris set off on his first trip to the Orient. In 1539 alone Juan received 1,000 *pesos* of gold and over 1,000 *marcos* of silver.[44] Some of this money would later be invested in the project for

authorized him to be his representative in Mexico (see document dated 15 Sept. 1548 [APS, Oficio 15, Libro 2 of 1548, fol. 624r]).

[42] H. Ternaux-Compans, *Voyages, relations et mémoires originaux pour servir à l'histoire de la découverte de l'Amérique publiés pour la première fois en français*, 20 vols. (Paris, 1837–41), xx. 143–55, and J. Ignacio Rubio Mañé, *Introducción al estudio de los virreyes de Nueva España, 1535–1746*, 4 vols. (Mexico City, 1955–63), ii. 73–81.

[43] *Cartas de Indias*, pp. 782–3; J. Ignacio Rubio Mañé, 'La expedición de Miguel López de Legazpi a Filipinas', *Boletín del Archivo de la Nación*, 2nd ser., 5 (1964), 427–798 (pp. 652–3, 755–6, 776–9); id., 'Más documentos relativos a la expedición de Miguel López de Legazpi a Filipinas', ibid., 11 (1970), 82–156, 453–556 (pp. 103–5, 128–30, 146–9, 154–5, 491–5). At some time before 1560 Guido de Labezaris had been appointed *corregidor* and *alcalde mayor* (similar royal appointments empowering the incumbent to oversee political, economic, and judicial matters in Indian villages) of Tuspa and Zapitlán, Mexico; already by 1550 he was *corregidor* of half the village of Taymeo (see Luis Muro, *La expedición Legazpi-Urdaneta a las Filipinas (1557–1564)* [Mexico City, 1975], pp. 123–4, who also claims that Labezaris made another journey to Spain in 1552). It is worth noting that the Viceroy of New Spain wrote to Philip II in 1573 saying that when Labezaris was in the Philippines he was not respected by his men because 'they had known him here in Mexico when he was engaged in a lowly trade, for he was a bookseller' (Rubio Mañé, 'Más documentos', p. 155).

[44] Otte, 'Jakob und Hans Cromberger', pp. 161–2.

which the Crombergers are best known in the history of printing: the foundation of the first press in the New World.

The Foundation of the Mexican Press

Fray Juan de Zumárraga, the Erasmian bishop-elect of Mexico, had been well known to the Crombergers. As will be seen in Chapter Six, Jacobo was frequently associated with important ecclesiastical figures in Seville, and it is not surprising that this, coupled with his frequent printing for the Franciscan Order, should have led to his acquaintance with Zumárraga. There were certainly commercial dealings between the two men, for in 1529 Zumárraga owed Jacobo money and in the previous year Jacobo had authorized the bishop among others to collect the treasure left in Mexico by Diego de Mendieta.[45] Fray Juan had been sent to the Indies in 1528. He soon decided that the establishment of a press in the New World would help with the task of evangelizing the Indians and of promoting education in the colonies.[46] In 1533 and 1534 he was back in Spain where he was at last formally consecrated Bishop of Mexico, held discussions with the future viceroy, Antonio de Mendoza, about the organization of Spanish rule in Mexico, and may have mentioned to him the plan to invite a printer to the colony. While in Spain, Zumárraga wrote to the Council of the Indies floating the project of setting up a press and paper-mill in his diocese, and it is possible that while he was in Seville he discussed his plan with the obvious choice of printer to found such a press, Juan Cromberger.[47] Not only did the bishop know the family, but he would have realized that their press was ideally located for the project. Juan was the leading printer in the city, which was the hub of the Indies trade, and the family had a tradition of printing for the Archbishop of Seville, under whose authority the new Mexican see fell. The Crombergers also had experience of printing for the Indies, strong commercial interests in the Antilles and Mexico,

[45] Hazañas, i. 154. Among the others was one Licenciado Marroquín, presumably Francisco Marroquín, the future Bishop of Guatemala, a licence for whose *Doctrinas de la Provincia de Guatemala* was given to Juan Pablos in 1556 (see document dated 21 July 1556 [AGNMex., *Mercedes*, Libro 4, fol. 365r]) transcribed with minor errors by Agustín Millares Carlo, 'Algunos documentos referentes a tipógrafos de México en el siglo XVI', in id., *Investigaciones biobibliográficas iberoamericanas: época colonial* [Mexico City, 1950], pp. 111–42 [p. 136]).

[46] For evidence of his enthusiasm for educating the Indians, see Joaquín García Icazbalceta, *Don Fray Juan de Zumárraga primer obispo y arzobispo de México*, Colección de escritores mexicanos, 41–4, 2nd edn., rev. Rafael Aguayo Spencer and Antonio Castro Leal, 4 vols. (Mexico City, 1947), iii. 125–39.

[47] Joaquín García Icazbalceta, *Bibliografía mexicana del siglo XVI*, 2nd edn., rev. Agustín Millares Carlo (Mexico City, 1954), pp. 23–4, 42. The Council of the Indies suggested that the Emperor provide funds for the transport and setting up of the press as well as giving the printer who took it to Mexico a privilege, but this advice appears to have led to no immediate action (see García Icazbalceta, *Don Fray Juan de Zumárraga*, iv. 116).

and were prosperous enough to afford the investment which the founding of the new press would require.

The bishop reported to the Council of the Indies in 1533 that good libraries were essential in the New World.[48] As a consequence, in May of the following year the Emperor gave him permission to spend one-fifth of Mexico cathedral's income over a period of three years on the creation of the library.[49] His first action was to contract one Benito Martínez, a bookseller from Villalón who was then in Seville, to accompany him to Mexico and help set up the library.[50] They immediately went to see Juan Cromberger and bought from him what was to form the basis of the collection. Unfortunately, the sale contract gives no details of what these books were: the document merely records that the purchase was of 'a variety of books', but it must have contained a large number because the bill was for 100,000 *mrs*. Zumárraga and Martínez each paid approximately half of this sum.[51] The bishop clearly took a personal interest in the transaction, because he signed the agreement himself.

The relationship between Cromberger and Zumárraga thus strengthened, the bishop returned to Mexico with Cromberger's books, which were the core of the first library to be founded in the New World. It may have been the promise of a permanent interest in printing and bookselling in the Indies that subsequently led Juan to invest heavily in his trading venture with Guido de Labezaris in 1536. He was then able to count on at least two factors in Mexico, Labezaris and Hans Henschel, as well as calling on Zumárraga himself to represent his commercial interests in the Americas.[52] Certainly Juan Cromberger became ever more identified with Mexico at this period and was the person to whom the authorities automatically turned when they wanted to supply printed material to the New World. For example, in 1537 the Emperor authorized the printing and binding at his own expense of 500 copies of Fray Juan Ramírez's *Santa doctrina* in Castilian and a Mexican language. The officials of the Casa de Contratación asked Juan

[48] When he went to Mexico in 1528 he took with him a personal collection of over 200 volumes (see Carlos E. Castañeda, 'The Beginning of Printing in America', *Hispanic American Historical Review*, 20 [1940], 671–85 [p. 672]).

[49] Lawrence S. Thompson, 'The Libraries of Colonial Spanish America', in id., *Essays in Hispanic Bibliography* (s.l., 1970), pp. 3–18 (pp. 7–8). The royal decree is dated Toledo, 21 May 1534, and reproduced in Alberto María Carreño, 'La primera biblioteca del continente americano', *Divulgación histórica*, 4 (1943), 428–31, 488–92 (p. 428).

[50] On 8 July 1534 Martínez agreed to pay a young bookseller from Osuna, Diego de Salamanca, 12,000 *mrs* per year if he would go with him to Mexico to work as a bookseller there for four years (APS, Oficio 4, Libro único of 1534, unfoliated).

[51] Document dated 11 July 1534 (APS, Oficio 4, Libro único of 1534, unfoliated). Zumárraga may have bought at this time the *Vita Christi cartujano* of Ludolph of Saxony printed by Juan Cromberger in 1530–31, which he took to Mexico with him (see Carreño, 'La primera biblioteca', pp. 490–1).

[52] Documents dated 10 July 1536 (APS, Oficio 1, Libro 2 of 1536, fol. 73^{r-v}); 12 or 13 Mar. 1538 (APS, Oficio 17, Libro 1 of 1538, fol. 463^{r-v}); 14 Mar. 1538 (APS, Oficio 17, Libro 1 of 1538, fols. 469^{v}–470^{r}).

to print it and he agreed to do so, but demanded an advance payment before he began to set it up. There were considerable problems with the translation of the work into an Indian tongue and, although letters went back and forth between the Casa de Contratación and the royal authorities during 1537 and 1538, it is not known whether the book was ever printed.[53] There is some confusion between this work and the *Cartilla y doctrina en lengua de indios de Michoacán* which Cromberger agreed to print in 1538 and on which he was given a five-year privilege; this might have been the same work and was probably two items, an ABC and a catechism, both written in the Tarasco language. Although royal authorization was given to Juan to print some sample copies and submit them to the Council of the Indies for examination, no trace of the edition survives. Considering the sort of books they were and their destination, this is scarcely surprising.[54] What the fragmentary evidence does indicate, however, is that not only did Zumárraga turn to the Crombergers when he wished to provide books for readers in the New World, but that Juan was generally thought of as the Spanish printer who produced editions specially for this market even in Amerindian languages. Thus, he even had a woodcut made with an inscription in Tarasco; this appeared on the title-page of Pedro de Gante's *Doctrina cristiana* printed in Mexico in about 1547, but was probably cut much earlier for use in Seville.[55] It was, then, only natural that Cromberger should have been the person to be asked to send a press to Mexico, and he did so in 1539.

The printer he chose to set up his branch office in Mexico City was Giovanni Paoli, or Juan Pablos as he was known in Spanish. Pablos was an Italian from Brescia, and the works he printed in Mexico, especially in the 1550s, indicate that he was a craftsman of some skill. However, a natural pride in the first printer in the New World, especially among Latin American historians, has transformed this obscure figure into a scholar-printer. After all, the argument goes, how could he fail to have been a remarkable and learned man if he brought the means of spreading culture to the Americas? Thus José Toribio Medina asserts of him that 'he must have received a university education in his native Lombardy or elsewhere in Italy; . . . he knew Latin and could even write it with a degree of elegance'.[56] Documents which I have discovered in the Archivo de Protocolos in Seville are

[53] Francisco A. de Icaza, 'Miscelánea histórica', *Revista mexicana de estudios históricos*, 2 (1928), 5–112 (pp. 28–30); García Icazbalceta, *Bibliografía mexicana*, pp. 26–7.

[54] Nicolás León, 'Noticia de una obra en tarasco', *Anales del Museo Michoacano*, 1 (1888), 62–4. ABCs were often accompanied by a catechism (see Klaus Wagner, 'Los impresores sevillanos Estacio y Simón Carpintero', *Archivo hispalense*, 2nd ser., 58, No. 178 [1975], 135–42 [p. 138]).

[55] Henry R. Wagner, *Nueva bibliografía mexicana del siglo XVI: suplemento a las bibliografías de Don Joaquín García Icazbalceta, Don José Toribio Medina y Don Nicolás León* (Mexico City, 1940), pp. 8–9.

[56] Medina, *La imprenta en México*, i, p. lxiii. This claim is based on a letter addressed to the readers of a book printed by Pablos. The letter could, of course, have been written by somebody else.

strangely at odds with this contention. Medina guesses correctly that Pablos must have had some experience of working in Cromberger's printing-shop, but his role there was not that of an erudite editor or even that of a proof-reader. As early as June 1532 he witnessed a sale of books to a Portuguese bookseller in Juan Cromberger's absence; at that time he and a colleague were described as 'Pedro Alemán and Juan Pablos the Italian, "ympresores" of printed books'.[57] As there is no record of either having been a printer in his own right, they must have worked as pressmen for a master-printer, a frequent meaning of the word 'impresor' in Spain at this time. This is confirmed by a document of August of the same year in which Pablos gave as his place of residence Cromberger's house in San Isidoro and authorized a minor legal official to collect his wages from his employer, possibly indicating a dispute all trace of which has disappeared. What is remarkable is that he asked the notary who drew up the document to sign on his behalf because he said that he did not know how to write his own name.[58] This was not Lyons, where pressworkers called for apprentices to the trade to be literate, nor Paris, where apprentices' statutes even demanded a knowledge of Latin; in Seville it was normal for such workers and even for booksellers to be unable to sign their own names.[59] Nevertheless, Pablos must have been considered by Cromberger a competent craftsman and a reliable agent; he had been employed in the Seville office for many years and, by 1539, he had risen to the rank of compositor and had learnt to sign. His signature indicates that even now, however, he was still much more at home with the sturdy composing-stick than the delicate quill.

The foundation of the press in Mexico is well documented, for the contract drawn up between Cromberger and Pablos on 12 June 1539 still survives.[60] Pablos agreed to go to Mexico with his wife and stay there for ten years. He was to be the manager-cum-compositor, this being the usual role for the manager unless a printing-shop was large enough to hire all the necessary compositors. When there was work he was to contract labour and print 3,000 sheets a day. This figure has led bibliographers either to depict Cromberger as an imperious employer who exploited his defenceless workforce, or to assert that it is so high that there must already have been a printing-press in Mexico which, with the single one sent with Pablos, would enable him to achieve such a high output.[61] There are several objections to these conclusions. First, contracts were very seldom obeyed to the

[57] Document dated 16 June 1532 (APS, Oficio 4, Libro 1 of 1532, unfoliated).

[58] Document dated 31 Aug. 1532 (APS, Oficio 4, Libro 1 of 1532, unfoliated).

[59] Natalie Zemon Davis, 'The Protestant Printing Workers of Lyons in 1551', in G. Berthoud *et al.*, *Aspects de la propagande religieuse*, pp. 247–57 (p. 257); Febvre and Martin, *L'Apparition*, pp. 194–5.

[60] A photographic reproduction of this contract can be found in *Documentos para la historia de la tipografía americana*, ed. with an introduction by Emilio Valtón (Mexico City, 1936).

[61] Lawrence S. Thompson, 'Some Reconsiderations of the Origin of Printing in Sixteenth-century Mexico', in id., *Essays in Hispanic Bibliography*, pp. 34–9.

letter, whether between merchants and their factors, or printers and their employees or clients. For example, the Archivo de Protocolos in Seville is full of sale contracts containing heavy penalty clauses if time limits for repayment were not met, but it was a rare debtor who paid up on time and the penalties were seldom imposed. Moreover, Juan Cromberger should have been more aware than most that he could exercise very little control over factors or employees as far away as Mexico.[62] There would therefore have been little point in his imposing conditions which he knew could neither be fulfilled nor enforced. Second, the staff which Pablos took with him was just adequate to man a single press. Third, although the number of sheets which a single printing-press could produce varied greatly in the sixteenth century, pressmen in England, France, and the Low Countries usually contracted to print a certain number of impressions each day, the figures ranging between 2,500 and 3,350.[63] It has been thought that Pablos's agreement to print for Cromberger 'three thousand sheets each day, just as you do in your office', implies 6,000 impressions (3,000 sheets printed on both sides), but there is no reason to suppose that this is what the clause meant. It probably referred to 3,000 impressions (1,500 sheets)—what most printers thought of as a fair day's work for a single press. Pablos is unlikely to have accepted to print double this quantity if, as an experienced pressman, he knew that it was an absurd demand.[64]

The agreement stated that Cromberger was to supply the press with all its materials: ink, type, instruments, and paper—a clear indication that, in spite of Zumárraga's letter to the Emperor written some six years earlier, there was little hope of paper being manufactured in Mexico. We can also conclude from this contract that Pablos would not cut his own punches nor even cast his own types. He was also obliged to accept the supervision of a representative of Cromberger's interests in Mexico and to lodge him in his own house. There is no evidence that this clause was ever fulfilled nor, if it was, of who acted as his agent.

Pablos put no money into the enterprise and was to be paid no wages for his labour, but when the ten-year contract expired, he would receive one-fifth of all the profits to be calculated on his return to Seville. He was to maintain himself,

[62] Factors even swindled their blood relations in Seville once they were far away in the New World; this seems to have been accepted by Seville merchants of the period as a fact of life (see Pike, *Enterprise and Adventure*, pp. 72–4).

[63] Philip Gaskell, *A New Introduction to Bibliography* (Oxford, 1972), p. 140. Febvre and Martin, *L'Apparition*, p. 92, give a range of 2,500 to 3,500 impressions. Michael Pollak considers even 3,000 *impressions* per day to be an impossible output, and he suggests that Pablos built another printing-press when he arrived in the Americas (see his 'The Performance of the Wooden Printing Press', *Library Quarterly*, 42 [1972], 218–64 [especially pp. 239–43]). If this is true, there is no convincing evidence for it.

[64] Admittedly, at the time of his death, Pablos owned two or three presses, but he doubtless acquired the additional ones in the 1550s when he took out a large loan to re-equip the Mexican office (see documents dated 1 Feb. 1563 reproduced in Medina, *La imprenta en México*, viii. 387–8, and 18 May 1563, reproduced in Millares Carlo, 'Algunos documentos referentes a tipógrafos de México', pp. 137–8).

his wife, and his workforce on the proceeds from sales of the books he printed. Any surplus would regularly be sent back to Spain. He was to keep full account books of what he printed, including details of the size of editions, sale-prices, the number of unsold copies, the income from sales, and the outgoings on overheads. Copies of this information were to be sent on three separate ships whenever he remitted profits to the parent press at Seville. Such accounts, if they survived, would be an invaluable source for the study of the beginning of colonial printing, but we can hold out no more hope of their ever coming to light than we can of the Crombergers' own ledgers, which are frequently referred to in surviving notarial documents but must have been destroyed long since.

Cromberger was careful to look after his own interests: all books were to be signed in his name and licensed by the Bishop of Mexico. As in Seville at this time, the licensing of printed books lay within the jurisdiction of the ecclesiastical authorities. He ensured that Pablos followed the due formalities of licensing so that the press would not lose money by having an edition called in after it had been printed. Similarly, he forbade Pablos to enter into any business on his own behalf or on that of any other person, doubtless fearing that the press would be forfeit if Pablos got into debt. As will be seen later, he also protected the Mexican press from competition by other printers.[65]

This long and detailed contract has given rise to the belief that Cromberger exploited Pablos mercilessly, giving him a pittance in return for ten years' work and tying his hands so that he could not engage in any activities on his own behalf to supplement his meagre income.[66] Juan Cromberger has thus been branded, by Latin American writers in particular, a ruthless employer; this judgement owes not a little to a modern protest at the exploitation of the subcontinent by foreign capitalists. In Cromberger's case it is an anachronistic and probably unjust criticism. At a period when employers were entitled to imprison their apprentices or beat them to within an inch of their lives, and when pressmen, at least in Cromberger's office, were working alongside slaves who often did the same jobs as free workers, contracts between printers and their employees were unlikely to radiate liberality. Cromberger intended to make a profit from his investment and was to know from his experience with Guido de Labezaris that he would have little control over Pablos once he was on the other side of the world. He was a successful and rich businessman partly, no doubt, because he knew how to drive a hard bargain, but the terms of the contract, and particularly of the agreement over Pablos's share of the profits, were not unusually exacting. We may recall that Diego de Mendieta was to earn only one-sixth of the profits gained on his

[65] See below, p. 167.

[66] Medina, *La imprenta en México*, i, p. lx; José Torre Revello, *Los orígenes de la imprenta en la América española* (Madrid, 1927), p. 11; Emilio Valtón, *Impresos mexicanos del siglo XVI* (Mexico City, 1935), p. 14.

trading voyage to Mexico on behalf of Juan's father; Guido de Labezaris was to take one-third of the profits on books but nothing at all on the general goods which he took for Juan to the Indies. Neither does this mean that the Crombergers were unusually harsh on their employees; contracts which exist in the Seville archives show that these terms were typical of agreements drawn up by employers at the time. Whatever modern historians may think of the conditions of the contract, the fact is that Pablos accepted and then, predictably enough, failed to comply with them. He neither worked for the family for ten years, nor did he return to Seville as he was bound to do by the contract. Nevertheless, his relations with the Crombergers seem to have been amicable enough for them to negotiate the transfer of the press to him in the 1540s.

On the same day that the contract was signed, Juan Pablos recorded that Cromberger had supplied him with all the equipment to set up the Mexican office. It consisted of a single printing-press, ink, paper, and tools all to a value of 100,000 *mrs*. The cost of transporting the equipment to the New World was 50 *ducados*, while the fares for Pablos, his wife, a black slave called Pedro, who was valued at 100 *ducados*, and one Gil Barbero, who had agreed to work in the Mexican office for three years as a puller, was 50 *ducados*. The cost of food for them all during the journey was an additional 20,000 *mrs*.[67] Juan Cromberger's total investment in the project was therefore 195,000 *mrs* or 520 *ducados*.

A day later Cromberger acted as guarantor for Pablos's wife, Jerónima Gutiérrez, who was granted a licence to travel to the Indies.[68] She and her husband must have arrived in Vera Cruz later in 1539 and made the journey to Mexico City with the press and its equipment. It is not certain where the press was set up initially, but by April 1540 it was in the Casa de las Campanas, a property owned by Zumárraga just off the Zócalo, the central square of the city. It is unlikely that it would first have been installed elsewhere and subsequently moved to this location; the bishop's desire for a printing-press probably led to the offer to accommodate it as soon as it arrived in Mexico.[69] Although Zumárraga was mentioned only in passing in the contracts signed by Cromberger and Pablos, the installation of the press in one of his houses confirms his commitment to the project. He and Cromberger did not sever their association once the press had been sent to Mexico, and he continued to represent the printer's commercial interests in the New World.[70] The press operated from at least 1540, but it is not

[67] Gestoso, pp. 65–8.

[68] AGIS, *Contratación*, Legajo 5536, Libro 5, fol. 180^{v}.

[69] Valtón in his introduction to *Documentos para la historia de la tipografía americana*, p. 5.

[70] Gestoso, p. 68. A letter sent by Zumárraga in 1544 attests to his continuing connections with the Crombergers and his dealings with their Mexican press. It also suggests that Miguel López de Legazpi, the conqueror of the Philippines, had direct dealings with the Crombergers (see Richard E. Greenleaf, *Zumárraga and his Family Letters to Vizcaya 1536–1548: A Collection of Documents in Relation to the Founding of a Hospice in his Birthplace* [Washington, 1979], pp. 110–12).

known what was Pablos's first edition. A work entitled *Breve y más compendiosa doctrina cristiana en lengua mexicana y castellana* with a colophon which apparently read, 'This catechism was printed in Juan Cromberger's office on the orders and at the expense of Juan de Zumárraga, first bishop of this great city of Tenuchtitlán, Mexico, New Spain: 1539', was recorded in the *Cartas de Indias* published in the nineteenth century, but no source for the reference was provided and no copy is now known.[71] This may, nevertheless, be accurate information as it is just the sort of work the press would have been sent to the New World to print, obviating the need for the Seville office to produce such bilingual editions with all the attendant problems of having to send the manuscript to Mexico for revision by experts in the native languages before the book could be set up in Spain. In December 1540 Pablos printed a *Manual de adultos*, two leaves of which, including the colophon, survive. This is the first book known for certain to have been printed in Mexico.[72] Juan Pablos was granted denizenship of the city in February 1542 and spent the rest of his life working as a printer there.

Some historians of the origins of printing in the New World have maintained that, contrary to his own claims, Juan Pablos was not the first printer to have worked in Mexico.[73] Their evidence is that at some time between April 1536 and 1538 Cristóbal de Pedraza, later Bishop of Honduras, wrote to the Emperor that a 'master-printer wishes to serve Your Majesty with his craft, and to travel to New Spain to print service books there'. On 5 September 1538 one Esteban Martín, who was styled a printer, was granted denizenship of Mexico City; this was accorded to Pablos some two years after he had arrived in the Indies, although the normal residence requirement was five years, and so it is possible that Martín had been a resident of the city since at least 1536.[74] It has even been claimed that Juan Cromberger sent Martín to Mexico with a press in 1534 after discussions with Zumárraga, but there is no evidence at all for this assertion. Finally, at a date supposed by some to be 6 May 1538 Zumárraga complained that printing in Mexico had almost come to a standstill because of a lack of paper, thus implying that a press had been operating there before that date.[75]

The arguments against these historians' claims are that there is no evidence

[71] *Cartas de Indias*, p. 787.

[72] For reproductions see José Torre Revello, *Orígenes de la imprenta en España y su desarrollo en América española* (Buenos Aires, 1940), p. 163.

[73] In the colophon of his edition of the *Constituciones* of the Archbishopric of Mexico printed in 1556, he styled himself 'Juan Pablos, Lombard, the first printer in this great, splendid and loyal city of Mexico' (see García Icazbalceta, *Bibliografía mexicana*, p. 123).

[74] Valtón, *Impresos mexicanos,* pp. 9–10. He maintains (p. 38) that five years' residence was the normal requirement for denizenship of Mexico City.

[75] García Icazbalceta, *Bibliografía mexicana*, pp. 32, 42, 47. For a review of the arguments see Thompson, 'Some Reconsiderations'; and for the case for Martín's being the first printer see Alexander B. Carver, 'Esteban Martín, the First Printer in the Western Hemisphere: An Examination of Documents and Opinion', *Library Quarterly*, 39 (1969), 344–52.

that Pedraza's unnamed master-printer did anything more than state a desire to go to the New World; Esteban Martín may well have called himself a printer, but this proves neither that he was the one mentioned by Pedraza nor even that he ever operated a press in the Americas, for there were many craftsmen in Seville who continued to style themselves printers long after they had retired from the craft;[76] and, finally, the date of Zumárraga's complaint is uncertain: since its publication in the nineteenth century, the document has disappeared and it is therefore impossible to check.[77] It has been suggested that the date was incorrectly transcribed and should read '6 May 1548'.[78] This is not only convenient for those who would like to make Pablos the first Mexican printer, but is also rather likely to be true. There is, after all, no other archival evidence to suggest that a Mexican press was already operating by 1538 and no book printed before Pablos's arrival is known.[79] Indeed, there is strong evidence for the absence of such a press. One of the earliest books intended exclusively for the market in the New World was, as has been seen, sent to Cromberger to print in Seville. The manuscript of the translation into a Mexican language had subsequently to be sent to Mexico for checking and then come all the way back to Spain before it could be set up and the edition then exported back to the New World. This was a cumbersome and time-consuming process which would surely have been avoided had there been facilities for printing the book in Mexico itself. The suggestion that the date of Zumárraga's complaint should be 1548 is further strengthened by the fact that

[76] In 1548 'Maese Juan Zillo, printer, denizen of Seville' was constructing mills for grinding wheat, while in 1554 'Claudio Borgoñón, printer, denizen of Seville in the parish of San Ildefonso' was a prospector for mines (see Gestoso, pp. 109, 124). Indeed, Martín may have been in Mexico not as a printer, but merely in connection with the business of his relation, Andrés Martín, who was a bookseller there (see Luisa Cuesta Gutiérrez, 'La imprenta y el libro en la América hispana colonial', *Gutenberg-Jahrbuch* [1957], 160–7 [p. 161]).

[77] *Cartas de Indias*, p. 786.

[78] García Icazbalceta, *Bibliografía mexicana,* pp. 32–4. Zumárraga died on 3 June 1548.

[79] Henry R. Wagner, *Nueva bibliografía*, p. 9; Millares Carlo in García Icazbalceta, *Bibliografía mexicana*, p. 33. It has been claimed that Esteban Martín printed the first Mexican book, a Castilian edition of St John Climacus' *Escala espiritual para llegar al cielo* but, as no copy survives and the only reference to it comes from an unreliable account written as late as 1596, the claim is doubtful (see Carver, 'Esteban Martín', pp. 348–51). Other bibliographers have insisted on attributing to Martín books which are not even known to have been printed in Mexico, such as Toribio Motolinía's *Doctrina* (see Medina, *La imprenta en México*, i, p. xlvii). Francisco Vindel repeatedly asserted that a book in his possession had been printed on a small press sent to Mexico by Juan Varela de Salamanca in 1531 with his son Pedro, and that it could be dated as early as 1532–4. His arguments are, however, fanciful (see his *El primer libro impreso en América fue para el rezo del Santo Rosario (Méjico, 1532–34)* [Madrid, 1953]; the *Apéndice* to that study [Madrid, 1954]; *Réplica en 'Carta abierta' de Francisco Vindel al dictamen emitido por la Academia Mexicana de la Historia correspondiente de la Real de Madrid sobre su descubrimiento e investigación de 'El primer libro . . .'* [Madrid, 1954]; *En papel de fabricación azteca fue impreso el primer libro en América (apuntes que comprueban la falta de veracidad en un dictamen de la Academia Mexicana de la Historia)* [Madrid, 1956]). Alberto María Carreño, '¿El primer impreso en América?', *Boletín de la Biblioteca Nacional de México*, 5 (1954), 5–33, even accused Vindel of trying to perpetrate a fraud for base commercial reasons.

the late 1540s was the very period when we know that Juan Pablos's press was encountering difficulties and that he was faced with the choice of begging charity or seeing his family starve. It was also the time when the *Audiencia* of Mexico was complaining to the Emperor that only a trickle of books was arriving from Seville—another point made in Zumárraga's letter.[80] It is, therefore, prudent to assume, until any contrary evidence comes to light, that the bishop's letter dates from 1548 and that Juan Pablos printed the first books to come from an American press.

It has been shown that claims for Pablos's erudition and his exploitation at the hands of his employer have been greatly exaggerated. This conclusion invites enquiry into Cromberger's attitude to the foundation of his office in the New World, for this has also suffered from distortion at the hands of later writers.[81] Viewed by North and Latin American historians of printing in their continent and particularly through the lens of Mexican nationalism, the sending of Pablos across the Atlantic was a momentous event. There is no evidence that Cromberger or his contemporaries saw it in this way. On the contrary, although Juan left no account of his intentions or ambitions, there are several indications that the project did not enjoy the high priority in his affairs which later writers suppose.

If he did discuss the foundation of the press with Zumárraga when they met in Seville in 1534, and if he considered it an important undertaking, the delay of some five or six years before he managed to organize Pablos's voyage is surprising. It has even been suggested that the matter may have been broached by Zumárraga as early as 1527; if this were the case, the delay would be yet more striking.[82] When plans were eventually made, not only was Cromberger unwilling to supervise the installation of the Mexican press himself or to send a trusted relation in his stead, but he selected a non-Spaniard who was at best only semi-literate and who, although over thirty-two years of age when he was chosen for the job, had still not risen beyond the status of a mere employee in Cromberger's office.[83] The staff that accompanied him was scarcely impressive: Pablos's wife was to keep house, the black slave may have been going to work as a beater in the shop, and Barbero, an unknown pressman who could not sign his name, was to work as a puller. Neither did Cromberger envisage the office as a large operation, for he sent only one press. The material which accompanied it had not been specially commissioned: no new types appear to have been cut, and at least some of the woodcuts he provided were old stock from the Seville office. Indeed, the cost to Cromberger of the whole venture was only about one-sixth of the sum he had

[80] García Icazbalceta, *Bibliografía mexicana*, pp. 25, 47.

[81] Torre Revello, for example, claims that Pablos's press 'must have been as important as the one which its owner, Juan Cromberger, ran in Seville' (see *Orígenes de la imprenta en España*, p. 97).

[82] Carver, 'Esteban Martín', p. 346.

[83] In 1532 Pablos said that he was over 25 (see document dated 31 Aug. 1532 [APS, Oficio 4, Libro 1 of 1532, unfoliated]).

invested in a single cargo of general merchandise three years previously; it was about half the debt owed to him one year later by a single purchaser of his books; and it represented only a small percentage of the value of silver and gold which had already arrived from the Indies for him in the very year he sent Pablos to Mexico.[84] If he did not think of it as a venture which warranted the investment of a large amount of capital, a document written several years later indicates that he was correct in his judgement, for in 1547 Pablos maintained that the press produced no profit and provided him with little work.[85] Some ten years earlier when Pedraza had said that a printer wanted to go to the New World, he had sought royal finance on the grounds that a press would not provide enough income for the craftsman employed in it to earn a living.[86]

The surviving books from the press give an average of only just over two editions each year while it was dependent upon the Crombergers; these figures are inevitably unreliable for, if Seville editions have disappeared in large numbers, it is unlikely that those printed in Mexico have been less prone to destruction. Nevertheless, the fact that Mexico City, with its damp atmosphere and chronic shortage of paper, was unfavourable to the conservation of printed material is not, as some bibliographers would have us believe, *prima facie* evidence for a large output from Pablos's office. It is probably the case that the bulk of his work was made up of ephemeral editions of forms, decrees, and pamphlets for teaching the Indians how to read and the basic tenets of the Christian Faith.[87] We should not discount, however, the simplest explanation for the lack of surviving products of his press: namely that there were very few of them in the first place.

Juan Cromberger's motives in setting up the press were probably rather different from those implicit in the assumption of many modern commentators that he wished to open up the New World to the benefits of printed culture. An important clause of the contract signed by Pablos reads:

Item, if at any time during the said ten-year period you, the aforementioned Juan Coronberguer, send me merchandise or books to sell, I shall be under an obligation to

[84] Juan de Espinosa, the Medina book-merchant, owed him over 1,000 *ducados* in 1540 (see Gestoso, p. 86); Otte, 'Jakob und Hans Cromberger', pp. 161–2.

[85] García Icazbalceta, *Bibliografía mexicana*, p. 47. Pablos's complaint cannot be reconciled with Torre Revello's baseless assumption that the press was very active and profitable (see *Orígenes de la imprenta en España*, p. 98). Henry Wagner is probably correct to suggest that there were few literate Spaniards in Mexico at this time and the press would seldom have printed profitable editions (see *Nueva bibliografía*, pp. 11–12). Until 1550 an average of only some 300 to 400 souls travelled to the Indies each year, and many of these did not stay there (see Jean-Paul Le Flem, 'Los aspectos económicos de la España moderna', in *La frustración de un imperio (1476–1714)* [Barcelona, 1982], vol. v of Manuel Tuñón de Lara [ed.], *Historia de España* [Barcelona, 1982–], pp. 11–133 [p. 23]). Pablos's press must have worked mainly under contract for the ecclesiastical and secular authorities.

[86] Valtón, *Impresos mexicanos*, p. 10.

[87] For a description of forms printed on the early Mexican presses see Lewis Morgrage Stark, 'Bibliographical Notes: Americana', *Papers of the Bibliographical Society of America*, 34 (1940), p. 87.

sell them at the highest price possible—only for cash and never giving credit to a purchaser—and to send you the profit, duly registered according to the law, by the first vessels sailing for Spain after the sale of the goods.

He had already sent Guido de Labezaris to Mexico as his factor to deal with the sale of books there, but Guido was due to return to Seville in 1540, not long after Pablos would have arrived. From his later history we know that Guido was a restless and unreliable agent whom Cromberger may have wished to replace with somebody more strongly anchored to Mexico City. It is true that Cromberger had the use of other representatives in Mexico, but they also had to deal with the great volume of trade generated by Nuremberger and Raiser. Moreover, they were frequently absent from Mexico City dealing with family mining interests. Cromberger may well have decided to combine the roles of printer and factor in the person of Pablos and use him as a distributor of material printed in the Seville office as well as of other merchandise exported to Mexico.

It was probably this export trade, particularly of ABCs and catechisms, which was of more interest to him than the branch office itself. He may have been willing to sink money into Zumárraga's project knowing that it would not be profitable, but making it a condition of his investment that he be granted a monopoly of book exports to Mexico. All record of negotiation over this matter with the Emperor has disappeared, but two years after Juan's death the monopoly which he had acquired in 1539 on the export of books to New Spain and all printing there was renewed. The royal decree dated 6 July 1542 and confirming the privilege for his heirs contains this following passage:

Joan Conbergel . . . sent to New Spain pressmen, a press, and all the equipment necessary for printing books . . . Our bishops, realizing how beneficial it was for those lands that he should print the said books and also export to New Spain . . . works on a wide variety of subjects, agreed that he should . . . be allowed to make 100 per cent profit [from this export trade] . . . and that, if it were our royal wish, only he be allowed to export books, ABCs, and all other printed matter to New Spain, and that he alone . . . be permitted to print there.[88]

Although he was allowed to charge high prices for books actually printed in New Spain (a silver *cuartillo*, or 8.5 *mrs*, for each printed sheet and half a *real*, or 17 *mrs*, for every ABC), the profits of 100 per cent on all books exported there from Spain must have been a handsome reward for sending a single press. It is impossible to discover precisely how much of the profits would have been eroded by the cost of transport, handling, and insurance, but some indication of the potential income

[88] Document dated 6 July 1542 (AGNMex., *Mercedes*, Libro 2, Expediente 120, fols. 46^v–47^r [transcribed, with several important errors, by Millares Carlo in García Icazbalceta, *Bibliografía mexicana*, pp. 45–6, and incorrectly dated by him to 6 June 1542]).

from the monopoly is afforded by the rapid response of the other Seville printers and booksellers who, five days after the signing of this royal decree, drew up a protest pledging that they would supply the New World with printed material at a profit of only 25 per cent on condition that the Emperor annul the Crombergers' stranglehold on the trade.[89] These printers presumably thought it would be worth their while transporting their wares to the Americas at only one-quarter of the profit sanctioned for the Crombergers.

But even the earnings on this trade may have been small when compared with the family's other interests in Mexico, for not only were they involved in general trading there but they were also breaking into silver-mining. This was a natural development of their exploitation of the business opportunities offered by the new colonies, especially as Nuremberg merchants had specialized, as has been noted, in mining and the export of metal goods ever since the Middle Ages. Already in 1535 Hans Henschel had gone out to Mexico on behalf of Juan Cromberger, Lázaro Nuremberger, and Christoph Raiser to deal, among other things, with their mining activities there. Henschel frequently appears in documents concerned with their trading interests including their mines, and he also entered into financial transactions with other merchants based in Spain. Indeed, in 1537 he was a party to the first letter of exchange used in the New World trade. He bought Nuremberger's and Raiser's first mines in Sultepec (about seventy miles south-west of Mexico City), but in 1537 had got them into so much debt that they dismissed him from their employ.[90] The close co-operation between Lázaro, Raiser, and Juan Cromberger in their common mining interests is revealed in later documents. In 1548 Guido de Labezaris, who was in Seville on his return from his disastrous journey to the Islas de Poniente, was engaged by Lázaro and Raiser to deal, together with their factor Marcos Hartman, with their mines in Sultepec.[91] The contract between the Germans and Labezaris describes him and Henschel as the 'first discoverers and inventors of the machinery' used in Mexican mines, and adds that they had lost a large amount of money in this mining enterprise. This information is confirmed by a declaration made by Labezaris to the Viceroy in about 1550 in which he stated that he and Henschel and other factors of Lázaro

[89] Gestoso, pp. 103–4.

[90] Millares Carlo and Mantecón, *Índice y extractos*, ii, document No. 2316; for further evidence of Henschel's activities see documents Nos. 2517 and 2518, and Otte, 'Jakob und Hans Cromberger', pp. 152–3.

[91] Document dated 15 Sept. 1548 (APS, Oficio 15, Libro 2 of 1548, fol. 624^{r}). Hartman was a German from Ulm whom Lázaro Nuremberger sent to replace Henschel in Mexico. He had been the factor of an Antwerp company at Lisbon, but had begun to work for Lázaro in Portugal by 1534 (see Kellenbenz, 'Os mercadores', pp. 136–7). His permission to travel to the Indies was granted on 9 Nov. 1537, and his guarantors were Nuremberger and Raiser. He was accompanied by one Diego Juques from Bruges (see Cristóbal Bermúdez Plata, *Catálogo de pasajeros a Indias*, 3 vols. [Seville, 1940–6], ii, Nos. 3853, 3854).

and Raiser went to the mines of Taxco in 1536 where they introduced techniques and machinery previously unknown in the New World for extracting silver from ore. The process referred to here was probably that of grinding the ore in special mills and then smelting it with lead flux in blast furnaces. In this operation it appears that the Seville merchants lost some 10,000,000 *mrs.*[92] Lázaro later declared in his will that he had spent 5,000 *ducados* himself but had received no income from the mines.[93] The innovation introduced by Lázaro's employees seems to have been that of adding lead flux to the ores. Indeed, Juan Velázquez de Salazar later reported to the Council of the Indies that in 1542 Henschel had received reports of a German method of silver extraction which involved the addition of metallic lead, litharge, or hearth-lead to ores which were poor in natural lead (for lead was essential to the process of smelting), and that he had informed the Viceroy of this, with the result that the dwindling amount of silver being extracted from ores mined in Mexico at that time was substantially increased.[94]

Guido de Labezaris, although sent as Cromberger's factor to deal with the merchandise and books he sent to Mexico, soon abandoned this role to represent the family and Raiser in the Taxco and Sultepec mines. No doubt this accounts for the large amount of silver which reached Juan Cromberger in 1539. The first silver-mines in Mexico had not been discovered until the early 1530s and those in Taxco in 1534. The dispatch of the factors to Mexico so soon afterwards shows the speed with which the Crombergers could mobilize capital and labour for a venture in which they were really interested, and contrasts markedly with Juan's dilatoriness over the foundation of the Mexican press. The heavy involvement of the family's other factors with mining by 1539 also suggests that the sending of Juan Pablos to Mexico in that year may have had more to do with their requiring a permanent agent in Mexico City than with a belated recognition of the need for a press in the New World.

The family was, then, involved very early in Mexican silver-mining and in technical innovation in the industry. Furthermore, in 1540 Lázaro and a colleague contracted a prospector from Brabant, one Gaspar Loorman, or Lomann, to go to the New World for them and also to train two of Nuremberger's employees to prospect for precious metals there. Another document of the same date reveals that Lázaro already had a prospector working for him in Mexico, presumably in

[92] Francisco A. de Icaza, *Diccionario autobiográfico de conquistadores y pobladores de Nueva España*, 2 vols. (Madrid, 1923), ii. 258–60. The grinding equipment was probably imported from Nuremberg (see Otte, 'Jakob und Hans Cromberger', pp. 155–7).

[93] Nuremberger's will is partially transcribed in Werner's 'Zur Geschichte Tetzelscher Hammerwerke', pp. 219–20.

[94] Henry R. Wagner, 'Early Silver Mining in New Spain', *Revista de historia de América*, 14 (1942), 49–71 (pp. 69–70). I am grateful to Professor Peter Bakewell of the University of New Mexico for information on the processes used in silver-mining in New Spain.

his own mines.[95] All these mines were not, however, exclusively the property of Nuremberger or Raiser. Some of those in Sultepec must have been transferred to Juan Cromberger shortly before or after his death.[96] In July 1540 he authorized an old friend, Juan Rodríguez de Morales, to recoup money owed to him by Labezaris in Mexico as Morales was travelling to the Indies. He was probably going there to replace Guido as administrator of Cromberger's mines and by 1542 he was certainly employed in that capacity.[97] By contracting Loorman, the family was indirectly responsible for the first implementation or, at the very least, for the refinement of the revolutionary method of extracting silver by amalgamating low-yielding ores with mercury, a process which accounted for much of Mexico's prosperity, for this same Gaspar Loorman whom they had sent to the New World in 1540 had in 1556 successfully established this method in Sultepec, possibly in Nuremberger's and the Crombergers' own mines there.[98] The Crombergers are, therefore, important figures in the history of the development of silver-mining in Mexico.

Juan Cromberger's founding of the press in Mexico in compliance with the request of Zumárraga (and, possibly, of the Viceroy Antonio de Mendoza) was prudent, for his more profitable activities in the colonies could only have been enhanced by the favour of such highly placed officials. Indeed, his heirs subsequently benefited from several favours granted by the secular authorities in Mexico. It also strengthened Juan's ties with the bishop, who was one of the most important clerics in New Spain. Cromberger's interests in the Mexican Church were not limited to printing catechisms in native languages, nor to supplying the books needed for the library in Mexico City and ABCs for the missionaries. For example, Cromberger hired a mason to travel to the New World to work on Mexico cathedral and the Mexican chapter agreed to repay him in 1540.[99] The monopoly of the export of books to New Spain which Cromberger was granted was attractive, and Pablos's press itself may, in the early days, even have made some money. However, the order in which Brígida Maldonado later listed the family's interests in the Americas probably indicates its priorities: she talks of the

[95] Two documents dated 19 Jan. 1540 (APS, Oficio 5, Libro único of 1540, fols. 117^{r}–118^{v}). The unnamed prospector may have been Diego Juques.

[96] Document of 7 June 1542 reproduced in García Icazbalceta, *Bibliografía mexicana*, p. 46: 'Rodrigo Morales has informed me that he is in charge of administering the mines, ranches, and slaves which used to belong to the Germans and now are the property of Joan Converger's children.' These mines may have included that of the Niedharts, which was acquired by Hartman for Cromberger and Nuremberger (see Otte, 'Jakob und Hans Cromberger', pp. 154–5).

[97] Gestoso, p. 69. Morales had worked as Juan Cromberger's agent in Spain many years previously (see Gestoso, p. 35 and also documents dated 7 May 1529 [APS, Oficio 4, Libro 1 of 1529, unfoliated] and 29 July 1534 [APS, Oficio 4, Libro único of 1534, unfoliated]). In the Indies he seems to have been known as Rodrigo Morales.

[98] Silvio Zavala, 'La amalgama en la minería de Nueva España', *Historia mexicana*, 11 (1961–2), 416–21.

[99] García Icazbalceta, *Bibliografía mexicana*, p. 24.

'mines and ranches with all their equipment, gold, silver, the press, books, and the privilege on printing and selling books'.[100] While the founding of the first press on the American continent is the achievement for which the family is now best known, the stress laid by later historians on this aspect of their business would doubtless have surprised the Crombergers themselves.

The Death of Juan Cromberger

Pablos cannot have been printing for many months before his employer died. Cromberger made his will on his death-bed on 9 September 1540.[101] As in the case of his father, he died at the height of his activity. He had recently secured the monopoly of books exported to the Indies; only two months earlier he had been making arrangements about his silver-mines in Mexico, asking Zumárraga to deal with goods he had sent for sale in the New World; and all the time his Seville presses were printing a steady supply of fine editions. Indeed, 1540 was the press's most productive year (see below, Figure 3). Juan's last years show all the signs of prosperity. He had frequent contacts with the ecclesiastical authorities: the chapter of Mexico, the chapter of Seville, and the Franciscan Order, which entrusted him with large sums of money.[102] A leading member of the order in Seville, Fray Bernardino de Laredo, was an executor to his will. At the same time he had dealings with the secular authorities: in Mexico a grateful Viceroy gave his family grants of land, and back in Spain he used the royal chronicler of the Indies and military governor of the castle of Santo Domingo, Gonzalo Fernández de Oviedo, to collect debts for him.[103]

At the time of his death he had investments not only in Spain but in Northern Europe, while in the New World his activities were not all concentrated in Mexico, for, as has been seen, he had interests in Peru and elsewhere in South America.

[100] Document dated 17 Mar. 1545 (APS, Oficio 17, Libro 1 of 1545, fols. 420^r–421^v [*olim* 704^r–705^v]).

[101] His will was drawn up by the Seville notary, Andrés de Toledo (APS, Oficio 17), but has disappeared from the appropriate 'oficio' of the archive. On 16 Sept. 1540 his heirs said that he had died 'about eight days ago' (see Gestoso, p. 70). On 20 Sept. they said that he died 'about twelve days ago' (see Gestoso, p. 73).

[102] Document dated 12 Jan. 1540 (APS, Oficio 15, Libro único of 1540, fol. 70^v).

[103] Document dated 13 Mar. 1538 (APS, Oficio 17, Libro 1 of 1538, fol. 464^{r-v}). In 1526 Oviedo had given the privilege for the printing of his *Natural historia de las Indias* and also a work entitled *Laberinto de amor* to the Seville bookseller Rodrigo de Ayala and his brother Alonso de Alfaro (see documents dated 23 Apr. and 4 May 1526 [APS, Oficio 5, Libro 2 of 1526, fols. 204^v–205^r and 309^v–310^r]). Although Ayala and Alfaro had frequent contacts with Jacobo Cromberger, they do not appear to have given him these books to print. The only known edition of the *Natural historia* came from the presses of Ramón de Petras at Toledo in Feb. 1526, several months before Oviedo gave the privilege to the Sevillians. Nevertheless, in 1535 Oviedo had the first part of the definitive version of his history, *La historia general de las Indias*, printed in Seville by Juan Cromberger. It is one of the most important 16th-c. works on the Indies and is beautifully printed. Oviedo maintained in a letter which appeared at the end of the book that he had financed the edition himself.

Unfortunately for us, unlike his father's estate and that of Varela, Juan's possessions and investments were not valued in his inventory. It is therefore impossible to assess how much he had increased the family fortune during his management of the press. Nevertheless, the inventory of the goods remaining in the house in the Calle de Marmolejos gives a clear impression that he had done so considerably during his short life as a printer and merchant. Apart from the house itself, four printing-presses (one more than his mother had inherited from Ungut), and the slaves who worked them, about 3,000 *ducados* of credit, mainly with booksellers in Spain, and some 100,000 unsold books or broadsheets which he had in stock, he owned vineyards in Castilleja de la Cuesta and the Vega de Triana and several houses and shops in Seville itself. The contents of his own home show that, typically enough for a merchant in Spain, some of his wealth had been transformed into articles of conspicuous consumption.[104] He possessed at least ten slaves, some of whom were engaged in domestic service: one negress, Beatriz, was listed as a cook, a role frequently played by black slaves in Seville. The furniture of the house, including the money chests, was mainly imported from Flanders, while the family's extraordinary number of clothes were of velvet, silk, or fine cloths from India and were frequently embroidered with gold thread. The latest fashions such as 'pairs of new Valencian chopines' were imported for the womenfolk. Gold jewels set with precious stones, especially pearls, abound in the inventory; the family dined off silver and gold plate using silver cutlery and pouring their wine from silver jugs. They possessed religious paintings and icons set in gold and richly decorated with gems, and they even had sacred objects originating from a source as distinguished as the 'reliquary full of most authentic relics which Fray Bernardo [=Bernardino] de Laredo took from the Queen of Portugal's reliquary, all worked in gold and enamel and surmounted by a crown'.[105] Juan's connection with the New World trade is witnessed by a desk made from wood imported from the Americas, some navigational charts, the large quantity of pearls, and numerous curiosities brought back from the Indies, possibly including the 'little piece of unicorn's horn' which was listed with the family's jewels and doubtless valued for its magical properties. Perhaps the clearest emblems of ostentatious wealth are 'a man's black silk purse', 'a silk purse with gold letters on it', and 'another large purse made of fine wool which has gold buttons'.[106] This large estate was not divided among his heirs until 1 August 1547.[107]

[104] For the conspicuous consumption so characteristic of Spain at this time see Bartolomé Bennassar, *La España del Siglo de Oro*, tr. Pablo Bordonava (Barcelona, 1983), pp. 117–24.

[105] These were probably a present to Bernardino de Laredo (1482–1540) from the Portuguese royal family. As a doctor, he had tended King John III of Portugal and his consort, a sister of Charles V. On one occasion, Laredo had cured King John of a serious illness.

[106] The inventory is transcribed in part and with errors by Gestoso, pp. 73–99.

[107] The division of his estate is mentioned in a document dated 29 Jan. 1551 (APS, Oficio 10, Libro 1 of 1551, unfoliated).

Juan's printing may not have been of quite the high standard of his father's finest products, but it was still of a quality which succeeding generations of Seville printers would find hard to match. His twin interests in printing and trade were, like those of his father, closely interconnected: it was his printing which drew him to the attention of Zumárraga and won him the monopoly of the export of books to New Spain, and he used this to further his commercial interests there. On the other hand, it appears that profits from trade and mining in Mexico enabled him to invest in the press which he sent across the Atlantic. In the meantime, he did not ignore the domestic market for books, of which he continued to be the major supplier in the south of Spain.

The family's wealth is apparent not only from the goods which Juan had accumulated. Two of his daughters contracted prosperous unions: Catalina married Licenciado Luis Mexía Ponce de León, a scion of an aristocratic Seville family, bringing with her a large dowry of 4,500 *ducados*. As we have already seen, Ana married Juan's colleague, Juan de Espinosa, the book-merchant from Medina del Campo, and brought him a dowry of 3,000 *ducados*.[108] The third of the daughters to survive childhood, Francisca, entered the local convent of Nuestra Señora de Santa María de las Dueñas as a novice in 1547 and took the veil there six years later. Christ was a less promising match: her dowry was a mere 267 *ducados*.[109] The two surviving sons followed in their father's footsteps. Tomás became a merchant and traded both in Spain and the New World; in 1568 he married Juana de Llantadilla, the daughter of a merchant family from Burgos, and died a rich man six years later at Nombre de Dios, the gateway to the lucrative Peruvian markets.[110] Jácome, Juan's eldest child, was destined to carry on the family tradition of printing.

[108] Documents dated 2 May 1549 (APS, Oficio 17, Libro 1 of 1549, fols. 302^r–303^v [*olim* 1066^r–1067^v]), and 19 May 1548 (APS, Oficio 17, Libro 1 of 1548, fols. 348^v–349^v). Klaus Wagner, *Martín de Montesdoca*, p. 23, judges a dowry of 373 *ducados* given in 1548 to be 'handsome'.

[109] One document dated 8 to 10 May 1553 (APS, Oficio 9, Libro 2 of 1553, fols. 134^r–138^r).

[110] Otte, 'Jakob und Hans Cromberger', p. 158. The dispute between Tomás's widow and the Crombergers over his estate was bitter (see AGIS, *Contratación*, Legajo 215, No. 8). For evidence of remittances to Tomás from the New World see AGIS, *Contratación*, Legajo 1788, manifest of *La Magdalena* which arrived in Gibraltar in November 1572, fol. $[9]^r$, *fardo* 20. For an example of trade of other kinds in which he engaged, see the document dated 23 Jan. 1565 (APS, Oficio 16, Libro 1 of 1565, fol. 326^r).

CHAPTER 4

JÁCOME CROMBERGER AND THE DECLINE OF THE PRESS (1540–1560)

Brígida Maldonado and Jácome Cromberger

WHEN Juan Cromberger died Brígida Maldonado claimed in the inventory of his estate that she was so busy with business worries that she could not remember anything belonging to her late husband which she had failed to declare, but that she would add to the inventory any items she subsequently recalled or heard about.[1] The business which preoccupied her in her widowhood was her late husband's printing and trading activities. As her son Jácome was too young to take charge of the press, she managed it for him in the following years and became so identified with the craft that one of her servants would refer to her in 1544 as 'Brígida Maldonado the printer, widow of Juan Conberger'.[2] Jácome was only fifteen years old in 1540, but she immediately authorized him to represent Juan's heirs in business transactions, although it is unlikely that he would have taken any decisions himself at this stage.[3] It was his mother who empowered him in 1541 to seek a *merced*, or favour, from the Crown, and the result was the prolongation of the Crombergers' monopolies of printing in New Spain and of the export of books there which was contained in the royal decree of 1542 already mentioned. Juan's heirs had sought a twenty-year extension of the monopolies but, as was normal with such privileges, it was granted for only half that time. It was noted in Chapter Three that this renewal caused immediate consternation among Seville's other booksellers and printers; they claimed that the Crombergers' profits were too high and that the trade was gravely prejudiced by the family's stranglehold on the American market. Nevertheless, the Crombergers had powerful friends—their petition had been presented at court by Francisco Ramírez, Cardinal of Seville—and they were able to resist this attack on their privileges.

[1] Gestoso, pp. 74–7.

[2] Document dated 20 Mar. 1544 (APS, Oficio 20, Libro 1 of 1544, unfoliated). It was not unheard of in Spain for female members of printing families actually to work in the presses themselves: Juan de Lucena's daughter did so in 15th-c. Toledo or Montalbán (see M. Serrano y Sanz, 'Noticias biográficas de Fernando de Rojas, autor de *La Celestina*, y del impresor Juan de Lucena', *Revista de archivos, bibliotecas y museos*, 6 [1902], 245–99 [pp. 258–9]). For references to printers' widows in France who managed presses until their sons were of an age to take charge see Martin and Chartier, *Le Livre conquérant*, p. 249.

[3] Gestoso, pp. 100–1.

Brígida Maldonado thus protected the monopoly of the export trade in books for her eldest son. She was also given grants of land in Mexico connected with the family's mining interests there. In June 1542 the Viceroy, Antonio de Mendoza, gave Juan's heirs land near their mines at Sultepec for the cultivation of crops and grazing of cattle needed to support the dozen or so employees of the mines as well as their slaves and Indian servants. Twelve months later 'two sites for the location of ore-crushing machinery and smelters on the said river [at Texcaltitlán]' were granted to them.[4] Although Brígida and Lázaro Nuremberger formally liquidated their debts with each other in 1542, it is significant that she continued to be associated with him in the family's mining enterprises, doubtless because Lázaro had considerable experience in the business. This was the reason that the Tetzels took him on as a partner in 1546 for the exploitation of their copper-mines in Cuba.[5] As a consequence of Brígida's continuing interest in silver-mining in Mexico, considerable quantities of bullion were remitted to her in Seville during the latter part of the decade.[6]

The Cromberger press continued to flourish under her enterprising management and there is evidence that she not only issued books under the Cromberger imprint but also sold them in her own name.[7] Even Lázaro may have taken a part in the distribution of the press's products, for in 1542 he sold 32 volumes of St Augustine to the important book-merchant of Medina del Campo, Guillermo de Millis.[8] But the colophons of the books produced when Brígida was in charge of the press never mention her name; they were printed 'in the office of Juan Cromberger, may he be in Paradise'. Similarly, Jácome's name appears very infrequently during the early 1540s: in December 1542 Fray Juan de Ortega's *Tratado sutilísimo de aritmética y de geometría* was printed in 'Jácome Cromberger's office', and in April 1543 Pedro Mexía's *Silva de varia lección* was likewise finished 'in Jácome Cromberger's house', but it is not until December 1545 that we find a book printed '*by* Jácome Cromberger' (*Los cuatro libros del valeroso caballero don Cirongilio de Tracia*). From August of the following year onwards, editions issued

[4] Documents dated 7 June 1542 and 8 June 1543 (AGNMex., *Mercedes*, Libro 1, Expediente 144, fols. 68ᵛ–69ʳ, and Libro 2, Expediente 234, fol. 91ᵛ, both transcribed by Millares Carlo in García Icazbalceta, *Bibliografía mexicana*, p. 46; he dates the latter document wrongly and gives an incorrect archival reference to it). There was a burst of mining activity in the Sultepec region in June 1543. Sites for smelters were granted to several individuals in that month; for example: AGNMex., *Mercedes*, Libro 2, Expedientes 222, 225, 226, 227, 233, fols. 87ᵛ, 89ʳ–90ʳ, 91ʳ⁻ᵛ.

[5] Gestoso, pp. 102–3; Otte, 'Jakob und Hans Cromberger', p. 156; Werner, 'Zur Geschichte Tetzelscher Hammerwerke', p. 219.

[6] In 1546 she received 150 *marcos* of silver and 300 *ducados* or *pesos* of gold; in 1548, 1,822 or 2,441 *marcos* of silver; and in 1549, 491 *marcos* of silver (see Otte, 'Jakob und Hans Cromberger', p. 162).

[7] Gestoso, pp. 102, 125. Documents dated 13 May 1546 (APS, Oficio 17, Libro 1 of 1546, fol. 302ʳ) and 22 Feb. 1555 (APS, Oficio 6, Libro 1 of 1555, unfoliated).

[8] Document dated 10 Mar. 1542 (APS, Oficio 5, Libro 2 of 1542, fols. 795ᵛ–796ʳ). For more information about Millis see Pérez Pastor, *La imprenta en Medina del Campo*, pp. 484–6.

by the Cromberger press were regularly said to be printed by Jácome.[9] Although too much faith should not be placed in the formulae used in colophons, the disappearance of Juan Cromberger's name at this date suggests that in 1545 or 1546 Jácome did indeed take over responsibility for the press from his mother. He was at that time of a suitable age.

His future was bright: he was young, but could already have spent several years training to be a master-printer and may even have been introduced to the craft by his father. Apprentices in most trades in Seville began work at the age of thirteen, and Jácome was already fifteen when his father died. He also had considerable experience of business by 1546, especially in trade with the New World.[10] By the early 1550s there is evidence that he owned land near Seville and also carried on trade with Lisbon.[11] He was well connected with the German and Italian business communities of Seville, as is witnessed by his being invited to become the godfather to the son of Catalina Alemán and Federigo Alborgo in 1546. The seal was set on his good fortune by his brilliant marriage in late 1550 or early 1551 to Inés de Alfaro, the daughter of Juan Varela de Salamanca.[12] Varela had three sons, Andrés, Pedro, and Juan, but they had not followed their father into the world of printing; Andrés and Juan had gone into the Church and at a very young age had become dignitaries of Seville cathedral, while Pedro had emigrated to New Spain to represent his father there in business. By 1551 all three were dead. Varela's sons-in-law, Pedro de Ávila and Pedro Farfán, were not interested in printing, and so it may be for this reason that Varela married Inés to the grandson of his old colleague, for in Jácome Cromberger could be combined the prestige of the two presses which had virtually monopolized printing in Seville for almost half a century.

It appears to have been a prosperous union on both sides. Inés brought with her a dowry of over 5,000 *ducados*, a considerably larger sum than that given to either of her sisters.[13] However, none of this was in cash; most was in the form of a house in the parish of Santa María Magdalena which Jácome immediately mortgaged to raise cash and later claimed had been overvalued in the dowry agreement.[14] For her part, Brígida Maldonado promised Jácome his share of his

[9] An exception is the 1553/4 edition of Diego de Valera's *Crónica de España abreviada*, possibly the last surviving book from the press. This was a reprint of an edition of 1542/3 and copied the colophon of that earlier edition (see below in App. I).

[10] Gestoso, p. 101.

[11] Documents dated 23 Nov. 1552 (APS, Oficio 20, Libro 4 [vol. 2] of 1552, fols. 1417^v–1418^r) and 13 Mar. 1553 (APS, Oficio 20, Libro 1 of 1553, fols 727^v–728^r).

[12] Hazañas, i. 175–6. The date of the wedding is not known, but on 29 Jan. 1551 Varela arranged his daughter's dowry saying she was already Jácome's wife (APS, Oficio 10, Libro 1 of 1551, unfoliated).

[13] Ana de Alfaro had received 900 *ducados* and Isabel 2,000 *ducados* (see document dated 7 Aug. 1556 [APS, Oficio 5, Libro 2 of 1556, unfoliated]).

[14] Documents dated 29 Jan. 1551 (APS, Oficio 10, Libro 1 of 1551, unfoliated) and 26 May 1551 (APS, Oficio 15, Libro 1 of 1551, fols. 448^r–451^v).

father's estate in Seville and the Indies as well as providing Inés with jewels and a trousseau, and agreeing to the couple's living with her in the family home in San Isidoro. Jácome was not without his own independent income at this time, for he stated that the 1,000 *ducados* which he gave his bride as her *arras* (the customary gift from the bridegroom to his bride) had come from his own possessions.[15]

It has not been possible to discover whether this apparent wealth was illusory, the dowry and income from the New World being swallowed up by debts or unwisely invested in unprofitable projects of which no record has been found, or whether, on the other hand, it was growing prosperity which led Jácome to neglect his printing-office for quicker and larger returns in business, especially after 1556 when the amalgam method of silver extraction made mining far more profitable in Sultepec. What is, however, certain is that after the demise of Juan Cromberger there were several signs of decline in the press and that these became much more apparent when Jácome took over from his mother.

This decline can be appreciated if one compares the books printed in the early 1540s with those produced in the period from 1546 onwards, and particularly in the 1550s. The number of surviving editions printed from 1541 to 1545 averaged about eleven per year, a slight reduction from the days of Juan, but still considerably larger than the figure of about five during the period when Jácome was in charge. Although such statistics are not altogether reliable, other observations support the conclusion that all was not well with the press once Jácome took control of it. Despite some deterioration in the standard of typography under Brígida, the presswork becoming more careless and the material muddled, she did at least attempt to maintain her equipment by having types recast.[16] This all changed after 1545. No new matrices and few, if any, new woodcuts were acquired by Jácome. Even the existing types appear not to have been recast, for they show signs of severe wear. It is no exaggeration to say that the books he printed were shoddy, and it is usually easy to tell at a glance whether a Cromberger edition dates from this period.[17] The larger heading types (Types 2 and 4) became hopelessly hybrid, producing an ugly effect in books in which they were employed and revealing either that pieces of type had been broken and, as they were not being recast, had to be replaced from founts of another size, or that the cases were muddled and the compositors could not be bothered to sort them out. Likewise, the ornamental initials were now seldom used in their sets but, rather, chosen at

[15] Two documents dated 29 Jan. 1551 (APS, Oficio 10, Libro 1 of 1551, unfoliated).

[16] See App. II for the notes to Type 10.

[17] Domínguez Guzmán's comment about Jácome's 1552 edition of Guido's history of Troy—'an excellent and careful piece of printing'—is inexplicable, for all the copies I have seen are very poorly printed (see her 'Veinte años de impresiones sevillanas (1551–1570)', *Cuadernos bibliográficos*, 37 [1978], 1–57 [p. 10]).

random. The result was books which had little uniformity and showed no evidence that the compositors had attempted to meet the most modest criteria of typographical elegance. An explanation for the motley collection of ornamental initials could be that the sets were no longer complete, some letters having been broken or lost, but only careless workmanship can account for the excessive number of misprints which are found in Jácome's editions. The compositor's and proof-reader's apology that misprints in Jácome's edition of the Spanish translation of St Jerome's letters printed in 1548 were 'as inevitable as a shadow is to a body or weight to a stone' becomes something of an understatement. Poor and hurried presswork resulted in bent lines of type, letters falling out of their proper place and not being realigned, misleading pagination, and occasional incomprehensibility of the text.

Similarly, books printed by Jácome frequently contain a variety of paper with different watermarks and even of different thickness and quality. The reason for this is unclear, but it could be that adequate supplies of paper from a single source became difficult to obtain in Seville in the late 1540s and early 1550s. This would explain the sudden reappearance of contracts signed by Jácome for the purchase of paper after so many years' silence, for the office's former arrangements for a steady supply may have now ceased. The variety of paper found in books dating from this period could also indicate that Jácome was buying small quantities as the printing of an edition progressed, which would suggest that he was in financial difficulties. On the other hand, such a mixture could merely show that he was no more concerned to produce his editions on paper of uniform quality than he was to print elegantly and accurately. All this evidence leads one to suspect that the press was suffering at this time from a lack of investment, a shortage of skilled craftsmen, and an absence of discrimination in the master-printer.[18]

The titles which Jácome printed also indicate either that he had lost interest in the press or that it was short of money. As has been seen, the Crombergers had always issued many reprints of popular works, in particular of literature of

[18] Jácome was not alone in producing shoddy books. In 1562 or 1572, the king ordered the University of Salamanca to investigate the poor quality of Spanish editions, attributing their shortcomings in comparison with foreign imports not only to inferior materials, but also to insufficiently skilled personnel in the presses (see letter dated Madrid, 12 Nov. 1572 [dated to 1562 by Gil Fernández, *Panorama social*, p. 565], reproduced in C. Mª Ajo G. y Sáinz de Zúñiga, *Historia de las universidades hispánicas: orígenes y desarrollo desde su aparición a nuestros días*, 2 vols. [Madrid, 1957; Ávila, 1958], ii. 587–8). Inferior workmanship was, however, evident in Spanish presses at a much earlier date (see the Oviedo Chapter's letter of 1535 to Valdés reproduced in Novalín, *El inquisidor general Fernando de Valdés*, ii. 27–9). One reason for the large number of misprints in Spanish books was that printers in 16th-c. Spain normally corrected proofs themselves rather than employing a *corrector* (see Jaime Moll, 'Problemas bibliográficos del libro del Siglo de Oro', *Boletín de la Real Academia Española*, 59 [1979], 49–107 [p. 94]) and Spanish printers were generally no scholars (see Juan Maldonado's comment made in 1531: 'You must remember how deep is the slumber of ignorance in which printers in our land are sunk', cited by Gil Fernández, *Panorama social*, p. 565).

entertainment and of devotional guides. However, Jacobo had balanced a large output of such editions with a healthy proportion of more ambitious or untried works; his knowledge of the market meant that many books of which he printed the first edition subsequently became best-sellers. This policy had been followed by his son. Likewise, Brígida Maldonado was remarkably innovative: many of the editions she issued were of books which the press had not previously printed, among which were important works several of which were first editions.[19] It is, however, indicative that over one-third of Jácome's editions were reprints of well-tried romances of chivalry and only one of the remainder was definitely of a book which had not previously been issued by the press. This single edition was an opportunistic piece of printing for it was Bartolomé de Las Casas's *Entre los remedios*, which Jácome issued in the aftermath of the great Valladolid debate of 1550 and 1551 between the bishop and Sepúlveda. As the Protestant printers of Geneva discovered, cheaply produced controversial pamphlets were good money-spinners and Jácome no doubt issued a large edition of the *Entre los remedios* in an attempt to make a quick killing.[20] So popular were these treatises on the 'Affair of the Indies' that Sebastián Trugillo issued no fewer than six different works by Las Casas in 1552 and one, *Aquí se contiene una disputa . . . entre . . . Las Casas . . . y el doctor Ginés de Sepúlveda*, was not only printed without a licence, but was immediately imitated by an anonymous printer despite the Crown's attempt to suppress it.[21] Even in this case of a virtually guaranteed market, Jácome only risked printing one of Las Casas's works.

Seville Presses and their Problems

The extraordinary conservatism of Jácome's output was probably due to many causes: the undercapitalization of his office suggested by the state of his material would have led to his being unable to buy privileges on new books, while the declining quality of his products may in turn have meant that he was not sought out by the holders of such privileges to print editions for them. In any case, it was a less skilful and a cheaper operation to set up a book page for page from one which had already been printed than from a manuscript. Jácome may have found himself in such a parlous financial position that he could not afford to take

[19] See below, pp. 149–50, 155, 157, 164.

[20] 43 copies are listed in App. I; this is a greater number than for any other Cromberger edition recorded. It is difficult to judge whether this was an exceptionally large edition or whether many copies have been found because it is a controversial work in which bibliographers and librarians, particularly in the Americas, have taken a special interest. One of the distorting factors in any list of a printer's works is the predominance of books which later bibliographers have bothered to search for; other editions which are of little interest to historians or literary critics remain unknown although copies may still survive.

[21] Anthony R. D. Pagden, 'The "School of Salamanca" and the "Affair of the Indies" ', *History of Universities*, 1 (1981), 71–112 (p. 94).

the risk of being left with a single unsold edition on his hands and therefore relied on titles which he knew were popular. On the other hand, as was mentioned previously, there was a continuing increase in the penetration of the Spanish market by foreign manufactured goods both because of the weakness of Spanish industry and because the 'price revolution' caused the rise in price of those produced in Spain, thus reducing their competitiveness.[22] As far as books were concerned, this penetration now included even editions of popular works as the powerful merchant-publishers of Lyons, Venice, and the Low Countries sent large shipments of books to Spain aimed at this market.[23] It is significant that printers in Antwerp took advantage of their relatively low prices and the long-established trade links with Spain to make devastating inroads into the Spanish market. This trade was nothing new. Illuminated manuscript books of hours from Flanders had poured into Spain ever since the mid fifteenth century as part of the commerce in luxury goods. They would have affected Seville in particular because it was a major port of entry for products of the Low Countries. Antwerp printers therefore just continued this established trade and produced editions intended for export. Spain was not the only export market they supplied; for example, the Antwerp presses also printed books in English. However, the law of 1534 restricting the activity of foreigners in the English book-trade helped English printers to survive this onslaught.[24] Their Spanish colleagues enjoyed no such protection. The printer Martin Nutius (Nucio) and the publisher Jean Steelsius (Steelsio), both of Antwerp, exported large numbers of Spanish books to Seville in the 1540s and 1550s. Nutius had first-hand experience of working in Spain; the list of Spanish titles he printed and those for which he sought a privilege in 1544 show that he was issuing the very works of literature of entertainment and popular devotion which were the staple business of Seville printers.[25] In the mercantilist *Memorial* which Luis Ortiz addressed to Philip II in 1558 one of the measures proposed for preventing the flight of Spanish specie abroad was to

[22] Péligry, 'Les Difficultés', p. 269; id., 'Les Éditeurs lyonnais et le marché espagnol', p. 93; Marjorie Grice-Hutchinson, *Early Economic Thought in Spain 1177–1740* (London, 1978), pp. 122–37.

[23] The 'international book' also continued, of course, to be imported into Spain in large numbers (see Pérez Pastor, *La imprenta en Medina del Campo*, pp. 430–1).

[24] H. S. Bennett, *English Books and Readers 1475–1557* (Cambridge, 1969), pp. 31–2.

[25] J. F. Peeters-Fontainas, 'L'Officine espagnole de Martin Nutius à Anvers', *De Gulden Passer*, 35 (1957), 1–106 (p. 13); B. A. Vermaseren, 'De Spaanse Uitgaven op Godsdienstig Gebied van M. Nutius en J. Steelsius: een voorlopige oriëntatie', ibid., 50 (1972), 26–99. Clara Louisa Penney was preparing a book on Nutius when she died. I have not been permitted to see this unpublished work, but Theodore S. Beardsley Jr. kindly informs me that she suggests that Nutius had served as an apprentice in the Crombergers' office at Seville; I have been unable to discover on what evidence, if any, this suggestion was based. Beardsley also notes in his 'Spanish Printers and the Classics', pp. 30–1, that she believed that Nutius and Steelsius produced books for Spain and the New World through an arrangement with the Crombergers; this is unlikely to be true, although my researches in Mexican libraries do confirm that large numbers of Steelsius' and Nutius' books were exported to the colonies.

prohibit the importation of foreign editions of profane and liturgical books. It is noteworthy that Ortiz should have considered the purchase of these items to be a significant factor in increasing the kingdom's deficit in its balance of payments; he estimated that over 200,000 *ducados* went abroad every year in this trade.[26] Two years earlier, Juan Páez de Castro had explained to the king that Spanish presses were inactive because all the money from the book-trade was ending up in the hands of foreigners.[27]

If the market was under attack from foreign imports, it was also threatened by excessive competition within Seville itself. The crash suffered by Venetian printers in the early 1470s had revealed that competition was often unhealthy in the printing industry.[28] It certainly seems to have been so in Seville in the 1550s, where, just as in London in the 1580s or Seville again in the 1620s, too many printers were chasing too small a slice of the market. The result in Seville was that printers produced editions of increasingly poor quality and in order to survive may, like their counterparts in the following century, have had to rely on the printing of chap-books and other ephemera to obtain a rapid return on their investment. An indication of the poor state of printing in Seville—and, indeed, elsewhere in Spain—in the mid century is afforded by a study of editions of works by Francisco Guerrero, a musician at Seville cathedral. His books had been printed at Seville by Montesdoca, but when that printer went out of business, Guerrero turned to printers in Rome and Venice.[29] One reason why no local printer could be trusted to print a complex book of music was probably the decline in the quality of craftsmanship resulting from this competition.

While Jacobo and Juan Cromberger had virtually no rivals in Seville, at least five new printers set up shop there while Brígida Maldonado was in charge of the family office and at least another nine when it was under Jácome's management. Nevertheless, Brígida's share of the market held up, over half the surviving editions printed in Seville from 1541 to 1545 coming from the Cromberger press, but it fell to one-quarter in the remainder of the decade. From 1550 until the appearance of the last surviving book printed by Jácome it had slumped to only one-sixth. As the total volume of editions printed in Seville appears not to have risen during

[26] Ortiz, *Memorial*, pp. 45–51, 105–6. Again in 1558, Pedro de Mercado complained of the weakness of Spanish industry, which he attributed to Spaniards' belief that involvement in such activities was dishonourable. The result, according to this writer, was the flight of Spanish bullion to Flanders to pay for tapestries, to France for cloth and books, and to Turkey for stuff. He inveighs against the arid study of law, claiming that it serves merely to enrich French booksellers and impoverish Spain (see his *Dialogos de philosophia natural y moral*, 2nd edn. [Granada, 1574], fols. 105^r, 131^r).

[27] Gil Fernández, *Panorama social*, p. 711.

[28] Martin Lowry, 'The Social World of Nicholas Jenson and John of Cologne', *La Bibliofilia*, 83 (1981), 193–218 (p. 214).

[29] Klaus Wagner, *Martín de Montesdoca*, p. 36. It was, however, quite common for Spanish authors to send their manuscripts abroad for printing and then to import the printed books back into Spain (see, for example, Martin and Chartier, *Le Livre conquérant*, p. 241).

this period and the number of printers increased dramatically, the result was inevitable. Some printers would go bankrupt.

The careers of very few Seville printers have received attention from scholars. One exception is Martín de Montesdoca, who began his association with the book-trade as a paper merchant in 1550 and set up as a printer the following year.[30] If the number of books printed by him which survive is an accurate reflection of his output, he printed an average of only some four editions each year; in 1559 he found himself so deeply sunk in debt that he stopped printing, sold his equipment, and later fled to Honduras and Guatemala, where he was eventually run to ground by his unpaid creditors.[31] His case was not untypical. Many Seville printers were beset by financial problems at this time. In 1548 Andrés de Burgos rented a press to his brother-in-law, Diego de Rendón, for two years; Rendón returned it, however, after only two months had elapsed. It emerges from the litigation which followed this apparently innocent transaction that it had all been a desperate bluff to extract money from the unfortunate third party who had guaranteed the original contract for the hire of the press and who later complained that 'the said Andrés de Burgos is swamped by his debts and other excesses. His possessions, if he has any left, are hidden away out of sight, and those which he has not managed to conceal have been confiscated by his creditors'.[32] Not surprisingly, the last book we know which came from Burgos's Seville presses is dated 1548, although he appears to have decided to try his luck where there was no competition, and in 1553 set up shop in Portugal in the city of Évora.[33] Pedro de Luján, the minor Seville humanist, author of Book Twelve of the Amadís cycle, *Silves de la selva*, and son-in-law of the famous printer, Domenico de Robertis, fared no better. In 1554 he was in prison, doubtless for debt; three years later, now at large, he had to mortgage some of the books he had printed to Jácome Cromberger in order to secure a loan which he needed to pay a creditor. Even then his struggle to remain afloat was unsuccessful, for in 1559 another creditor had him incarcerated once again.[34]

Jácome may have faced the same sort of difficulties, for his output dwindled

[30] Klaus Wagner, *Martín de Montesdoca*, p. 24.

[31] Klaus Wagner, *Martín de Montesdoca*, pp. 24–7. Evidence, unknown to Wagner, of Montesdoca's activities as a judge in the town of Trinidad, Guatemala (1564–5) can be found in AGNMex., *Inquisición*, vol. 5, Expediente 13, fols. 296^{r}–302^{v}, 305^{r}–314^{v}.

[32] Gestoso, pp. 111–12.

[33] If the Évora printer of this name is really the same Andrés de Burgos, he did not change in character; he was involved in some shady business in Portugal (see António Joaquim Anselmo, *Bibliografia das obras impressas em Portugal no século XVI* [Lisbon, 1926], p. 102). See also Norton, *Printing in Spain*, p. 64.

[34] Gestoso, pp. 123–4, 126, 133. A printer called Domenico de Robertis was also imprisoned in 1560, but this was probably only a relation of the famous printer of that name (see Gestoso, p. 134). Hazañas, ii. 146, claims that Robertis died in 1548 or 1549; the colophons of editions coming from his press in 1549 indicate as much.

in the 1550s and, if we were to judge from his surviving editions, it would appear to have ceased altogether in 1553–4. Indeed, Domínguez Guzmán, basing her observations on slender bibliographical evidence, claims that Jácome stopped printing in 1552, passing his material to Sebastián Trugillo, who completed the series of Las Casas's works which Jácome had begun in August of that year.[35] Her conclusions illustrate the dangers of disregarding archival sources in a study of early printing. It is true that Trugillo's September 1552 edition of Las Casas's dispute with Sepúlveda employed four blocks on the title-page which had frequently been seen in earlier books printed by the Crombergers, the last being Jácome's edition of Las Casas's *Entre los remedios* issued only one month earlier, but this does not mean that Jácome had stopped printing nor that all his material had passed to Trugillo, for the blocks were probably hired or lent, and Trugillo may have acquired no more than these four pieces.[36] Indeed, his use in an edition of Valtanás's *Flos sanctorum* of 1558 of a title border which was clearly only an imitation of one frequently used at the Cromberger press suggests that he had not gained possession of Jácome's material by this date. Nevertheless, by 1562, Trugillo does seem to have owned a large number of blocks that had previously belonged to the Crombergers: in his edition of Valera's *Crónica de España abreviada* of that year he used several sets of the Crombergers' ornamental initials (see Appendix Four, sets 01:1, 01:35, 01:36, 01:56, and 01:65), at least one set of which (01:35) subsequently passed to his son, Alonso de la Barrera, who used it in his 1569 edition of Francisco Franco's *Libro de enfermedades contagiosas*. As we shall see, it is not surprising that Trugillo should have possessed Cromberger's material in 1562; in 1552, on the other hand, he had probably borrowed or purchased only a few blocks from Jácome.[37]

The Press in the 1550s and the Death of Jácome

The archives reveal that Jácome was still active in the book-trade well on into the 1550s; in 1556 he sold books to Alonso de Huete, a book-merchant from Medina del Campo, and in the following year he received books as a surety against

[35] Domínguez Guzmán, *El libro sevillano*, p. 33.

[36] Ornamental material and types frequently passed temporarily from one Spanish printer to another. For example, in the same year Trugillo borrowed blocks from another Seville printer, Juan de León (see Klaus Wagner, *Martín de Montesdoca*, pp. 26, 53–4; and Norton, *Printing in Spain*, pp. 28, 109).

[37] It has been claimed that some of the Crombergers' ornamental material was employed by Montesdoca at Seville in 1554 (see Klaus Wagner, *Martín de Montesdoca*, p. 54). A careful examination of Montesdoca's books reveals, however, that he was using only a close imitation of those blocks. After Jácome's death, material from his office was dispersed among many Seville printers. In 1563, for example, Alonso Coca used the Crombergers' blocks OM:1, 2, and 3 (see Francisco Vindel, *Manual gráfico-descriptivo del bibliófilo hispano-americano (1475–1850)*, 12 vols. [Madrid, 1930–4], x, No. 3168); in 1559 Juan Gómez employed the Crombergers' WC:1067, but this may only have been borrowed.

a debt.[38] Moreover, although there is no bibliographical evidence to show that he was still a printer after 1553, documents indicate that he was active in this role for several years after this date.

In 1546 the Inquisitor-General, Fernando de Valdés, became Archbishop of Seville. Five years later, in the thick of his battle with the cathedral chapter, he appointed to the post of *provisor*, or Vicar-General, Gaspar Cervantes de Gaete, who was close to the Jesuits in Seville, an Inquisitor, and no friend of the Erasmian canons.[39] One of the measures Valdés and Gaete took to further the Catholic Reform and to discourage the notorious abuses of some of the Seville clergy was to publish revised editions of liturgical books. Valdés's predecessor-but-one, the Erasmian Alonso Manrique, had commissioned a humanist scholar to revise the local missal, which was then printed at Seville by Varela in 1534 and again in 1537. Valdés ordered the preparation of an emended edition based upon this earlier one. Likewise in 1549 work began on a new breviary which was ready for printing in 1554.[40] The archbishop considered editions of liturgical books circulating in Seville and nearby towns to be potentially dangerous, so when all Bibles were called in for examination in 1552, he took the opportunity to have liturgical editions collected as well.[41]

We have seen that Jacobo Cromberger's monopoly of the printing of service books for the diocese of Seville had on his death passed to Varela (although it seems likely that Juan Cromberger printed an edition of the Seville breviary in 1531, possibly by agreement with Varela). By the time that Gaete was looking for a printer for the new editions, Varela had retired and so he turned to Jácome Cromberger to produce the books. Fortunately the contracts agreed between them survive. Jácome was summoned to the Archbishop's Palace on 4 July 1554 and there agreed to print one thousand copies of each of three revised liturgical books: a breviary, a diurnal, and a missal.[42] For the breviaries he promised to use the same type, or one of similar design, that Varela had employed in a book of hours printed at Seville in 1531 (presumably Palau 116150[*c*]). As for the missals, a sample page of one of Varela's editions of what was by then called the 'misal antiguo' had been torn out of a copy and signed by the parties as a model on which the typography of the new edition was to be based. Other instructions were given about the sort of paper to be used: 'papel de buitre' (paper with a watermark of a vulture) for the breviaries and diurnals, and 'papel de veta verde'

[38] Document dated 9 May 1556 (APS, Oficio 1, Libro 1 of 1556, fol. 964r); Gestoso, p. 126.

[39] Novalín, *El inquisidor general Fernando de Valdés*, i. 167, 188–9. Gaete subsequently became Bishop of Messina, Salerno, and finally, Tarragona. He represented Valdés at Trent (see Constantino Gutiérrez, SJ, *Españoles en Trento* [Valladolid, 1951], pp. 523–7).

[40] Novalín, *El inquisidor general Fernando de Valdés*, i. 175.

[41] J. Ignacio Tellechea Idígoras, 'Biblias publicadas fuera de España, secuestradas por la Inquisición de Sevilla en 1552', *Bulletin hispanique*, 64 (1962), 236–47.

[42] APS, Oficio 10, Libro 3 of 1554, unfoliated.

('green-stripe' paper [?])—presumably of a higher quality—for the missals. The *provisor* was to pay Cromberger half a *ducado* for each breviary, three *reales* for each diurnal, and 14.5 *reales* for each missal: a total of over 2,000 *ducados*. These were sizeable editions and required careful setting-up, because it was Valdés's intention that the errors of previous editions be eliminated, and a good deal of energy had gone into the preparation of the copy. In addition, the books were to be printed, as was usual for such works, in two inks, and so the amount of labour involved was considerable. In view of this Jácome was to be paid half the sum when printing commenced and to receive the remainder when the completed books were delivered.

Jácome had landed a good order; sales were guaranteed and so his investment would not be uselessly tied up in unsold copies. Gaete and Valdés had laid a good deal of emphasis on the preparation and printing of new liturgical works for the diocese, and so they would not have given the contract to somebody who was no longer working as a printer, even though it must be admitted that, in view of what we know about the poor quality of Jácome's presswork, it is slightly surprising that they chose him rather than one of his many rivals, some of whom, like Montesdoca, were excellent craftsmen. However, such a large order may have been beyond the capacity of other Seville presses, and the formidable reputation of the two presses which had traditionally printed liturgical works in the city, those of the Crombergers and Varela, had been inherited by Jácome. Doubtless it was the prestige of his long-established office, rather than his own more recent work, which secured him the contract. Jácome was given generous deadlines for the printing: the breviaries and diurnals were to be delivered at the beginning of Lent 1555, but the missals were not due until the end of December of that year, eighteen months after the contract was signed. This might indicate that both parties to the contract were aware that printing in Seville at this time was not without its problems.[43]

Later in July the cathedral chapter confirmed the contract with 'Xacome Conbergel, the printer' for the printing of the three editions. It is some indication of the tensions among the ecclesiastical authorities in Seville that Jácome was concerned lest a contract signed by Gaete should not be honoured by the chapter; the canons accordingly guaranteed that if those who held the purse-strings of the cathedral refused to pay for books printed on the authority of the *provisor*, the dean and chapter would honour the debt.[44] In the following month, and again in

[43] Compare with Plantin's promise to Philip II in 1571 that he would send 2,000 breviaries, 2,000 diurnals, and 1,000 missals to Spain every four months (see Gil Fernández, *Panorama social*, p. 619). Admittedly, Plantin ran a very large business. An example concerning a humbler printer than Plantin is that of the Parisian Jean Kerbriant. In only about eight months during 1539 and 1540 he printed 650 copies of a Liège missal of a size comparable to that of the Seville missal (see Martin and Chartier, *Le Livre conquérant*, p. 299).

[44] ACS, *Actas Capitulares*, 1553–4, fols. 167^{v}–168^{r} (13 July 1554).

1555, the chapter minutes record that certain canons of the cathedral were to attend to the final revision of the breviary and missal, supervise the printing, and correct the proofs.[45]

It comes as no surprise to learn that Jácome purchased paper in 1555 and 1556, but some of it was probably not intended for these editions because it was not of the sort stipulated in the contract with Gaete. It is therefore likely that he was still printing other works at this time. He may even have continued to do so until at least 1557, when he is said by some bibliographers to have issued an edition of Guevara's *Libro áureo de Marco Aurelio con el Relox de príncipes.*[46]

The printing of the missals was beset with problems, either through Jácome's own fault or because of the general decline suffered by Seville's presses at this time. He probably managed to deliver the breviaries and diurnals on time for, although no copies survive, they are not said to be still unprinted in subsequent contracts. The missals were another story. In October 1555, a mere two months before they were due, Jácome went with his brother, Tomás, and brother-in-law, Pedro de Ávila, to Gaete's residence and there drew up a fresh agreement with him.[47] It seems that Cromberger had printed part of the missal, and that it had been rejected by his client on the grounds that it was riddled with misprints. Jácome blamed the quality of his product on factors outside his control:

> I, the said Jácome Conberjel, have been rather negligent in the correction of the said missal and, as I fear that it cannot be printed to a sufficiently high standard in Seville because of the shortage of paper and personnel here, I, the said Jácome Conberjel, have agreed with the said *provisor* to print the said missal . . . at Lyons or Paris.

He now promised to deliver the edition by May 1556, but asked for a larger advance to cover the expenses he would incur by having to travel to France to print it. It is difficult to decide whether Jácome was wriggling or whether his excuses were valid. Printing was indeed in decline in the city, and yet fine books were still being produced there, as is witnessed by Montesdoca's splendid 1554

[45] ACS, *Actas Capitulares,* 1553–4, fols. 171^v, 177^r (3 and 31 Aug. 1554); ACS, *Actas Capitulares,* 1555–6, fol. 24^v (21 Mar. 1555).

[46] Documents dated 11 June 1555 (APS, Oficio 23, Libro 2 of 1555, fol. 346^v) and 18 Mar. 1556 (APS, Oficio 23, Libro 1 of 1556, fol. 921^{r-v}). Several bibliographers list the 1557 edition of Guevara (e.g. Haebler, *The Early Printers,* p. 67), but Domínguez Guzmán, 'Veinte años', p. 54, dismisses it as an impossibility because the colophon is said to refer to Juan and not Jácome. However, the 1553/4 edition of Valera's *Crónica de España abreviada* shows that Juan's name did appear in books printed by his son, especially when they were reprints. Dr Jack Gibbs informs me that Bonsoms once owned a copy of the 1557 Guevara, but I have been unable to find any evidence to substantiate this claim. Bonsoms's collection is now in the Biblioteca de Catalunya, but the Guevara is not among it. Sr Amadeu-J. Soberanas i Lleó, the Keeper of Manuscripts in that library, kindly allowed me to consult Bonsoms's own catalogue of his collection, but no reference to this edition appears. It is, nevertheless, possible that Bonsoms once owned a copy but had sold it before his books passed to the library. The Marqués de Morante claimed to have a copy, but probably confused it with the 1537 edition.

[47] Document dated 17 Oct. 1555 (APS, Oficio 10, Libro 3 of 1555, fols. 74^r–75^r).

edition of Fuenllana's *Libro de música para vihuela*. The claim that pressmen were in short supply may have been justified, because there could well have been too many presses in the city for the skilled workforce available. Indeed, when Juan Pablos was looking for staff for his Mexican press in 1550 he had authorized an agent to search

for one, two, or even three craftsmen skilled in the said art of printing. You shall look in the said city of Seville or in any other city, town, or village of the kingdom of Castile, if any such men are to be found there, or in the city of Lyons in France, or anywhere else you can recruit them.[48]

This suggests that he did not expect to find enough skilled craftsmen in Seville itself at that time. But there is, on the other hand, evidence that pressmen could not find employment in printing-shops (or their wages were too low) and so were drifting, like Juan Zillo and Claudio Borgoñón, into other activities from the late 1540s onwards.[49]

There may also, as Jácome claims, have been paper shortages, which would explain the variety of paper he used in his later productions if he had to snap up whatever paper became available in the city regardless of its quality. War and the rise in production of foreign presses could have substantially reduced shipments to Spain.[50] We find the Cromberger press buying paper on the open market in 1555 for the first time since 1520 and, as I have suggested, one possible reason for this was that their steady supply had dried up. Certainly the Mexican press was starved of paper at this time, a clear indication that there was a shortage in Seville. This interpretation is supported by a later document concerning the missals in which special provision was made for extra payment to Jácome if he were obliged to use paper of a better quality than the one stipulated in the contract, presumably because large enough quantities of 'papel de veta verde' could not be found.[51] Nevertheless, the situation is not altogether clear, for Martín de Montesdoca was able to buy substantial quantities, even including 'papel de veta verde', from the usual Genoese suppliers at this very time.[52]

Whatever the causes of Jácome's failure to produce the missal on time, Cervantes de Gaete must have realized that his choice of printer had been unfortunate, for when the cathedral decided to commission an edition of Fray Felipe de Meneses's

[48] García Icazbalceta, *Bibliografía mexicana*, pp. 47–8.

[49] See above, Ch. 3, n. 76. Montesdoca's financial position obliged him to sack a faithful employee in 1557 (see Klaus Wagner, *Martín de Montesdoca*, pp. 25, 120).

[50] Febvre and Martin, *L'Apparition*, p. 94. For the drastic reduction in shipments of paper from France to Spain by the Ruiz between 1552 and 1564 see Lapeyre, *Une Famille*, pp. 564–5.

[51] Document dated 10 Feb. 1556 (APS, Oficio 10, Libro 1 of 1556, fols. 665r–657r). I have been unable to discover what this 'green-stripe' paper was.

[52] Klaus Wagner, *Martín de Montesdoca*, p. 114. Montesdoca may have been at some advantage because of his experience as a paper-merchant.

Luz del alma cristiana, he turned to the Franciscan, Gregorio de la Torre, rather than to Jácome to print it. However, Torre proved no more reliable and, when he failed to fulfil the contract, Gaete asked Martín de Montesdoca to take over. The *Luz del alma* duly appeared in June 1555.[53] In future Gaete would rely on Montesdoca; in August of the same year he commissioned him to print an edition of the *Constituciones* of the archbishopric of Seville, and three months later placed an order with him for a *Calendario para rezar*.[54]

By January 1556, Gaete had become suspicious about Jácome's good faith, and he demanded a surety for the advance which had been made. Jácome did not give this personally, but one Martín de Balcázar guaranteed the money on his behalf by mortgaging some houses worth 400 *ducados* which he owned in the parish of Santa María la Mayor. Balcázar claimed that the advance had been of 30,000 *mrs* and that half the remaining cost of the missals would be paid by the *provisor* when the edition was half printed and the other half when it was delivered.[55] The implication is that Jácome was desperately short of capital to finance the printing. Gaete was wise to have his doubts, for Cromberger realized that he was not going to meet even the new deadline. In February 1556, as Gaete had by then left Seville, he drew up a fresh contract with the new *provisor*, Toribio Rojo, to deliver the missals in July of the same year.[56] This time, however, he contracted to deliver them at the slightly lower price of 14 *reales* each and to print them in company with Gregorio de la Torre, to whom he had sold two of Varela's old presses in November 1555.[57] Again the printers were working on a shoestring, for the new agreement provided for payment by regular instalments as the work progressed. These payments were no doubt to cover wages and the purchase of more paper. Jácome then authorized Torre to receive the instalments from Rojo, a fact which suggests that he may have been expecting Torre to carry out the printing or at least take charge of the financial arrangements.[58] If this were the case, Cromberger had withdrawn from the production of the edition and was acting merely as an agent. Predictably enough, the missals did not appear on time and Valdés had to wait for delivery until June 1558. There is no indication that Cromberger had a hand in the printing, for the colophon reads : 'Printed at Seville by Master Gregorio de la Torre' and only Torre's printer's-mark appears in the books.[59] The printing

[53] Klaus Wagner, *Martín de Montesdoca*, pp. 37–8, 67.

[54] Klaus Wagner, *Martín de Montesdoca*, pp. 112-16.

[55] Document dated 21 Jan. 1556 (APS, Oficio 10, Libro 1 of 1556, fol. 432^{r-v}).

[56] Document dated 10 Feb. 1556 (APS, Oficio 10, Libro 1 of 1556, fols. 655^{r}–657^{r}). Valdés had transferred Gaete to the post of Inquisitor at Zaragoza (see Novalín, *El inquisidor general Fernando de Valdés*, ii. 165).

[57] Document dated 29 Nov. 1555 (APS, Oficio 5, Libro 4 of 1555, unfoliated); also see below, p. 172.

[58] Document dated 19 Feb. 1556 (APS, Oficio 10, Libro 1 of 1556, fol. 685^{r-v}).

[59] A copy is described by Domínguez Guzmán, 'Veinte años', p. 29. On Torre and the long-awaited Seville missal see also Klaus Wagner, 'La Giralda como marca de impresor (a propósito del *Missale*

of the missals probably took many months, but the four-year delay between the original contract and eventual delivery was doubtless due more to Jácome's shaky finances or lack of interest in his craft than to the mitigating factors he cited in his defence.

There is no conclusive evidence about the state of his finances in the mid and late 1550s. If the monopoly on the export of books to New Spain had provided any substantial income for the family in the early 1540s, by the 1550s it had expired and there is no indication that the Crown had renewed it. Long before that date the *Audiencia* of Mexico and the Viceroy had complained to the Emperor that the Crombergers were not supplying the colony with much-needed books as they were bound to do by the terms of their monopoly.[60] It is therefore unlikely that Jácome became rich through the export of books. Ever since his marriage he and his wife had lived with Brígida Maldonado in the family home, but when she had originally agreed to accommodate them, Brígida had made it clear that she could ask them to leave whenever she wished. It may be no coincidence that it was in 1554, just when the press was experiencing serious difficulties, that Jácome moved from the parish of San Isidoro to that of Omnium Sanctorum.[61] Then as now, this parish was very different from Santa María and San Isidoro; it was a run-down working-class quarter which had long been the centre of popular unrest in Seville.[62] If the presses remained in the workshop in San Isidoro, the move may indicate a split between mother and son because it is known that Brígida retained possession of them. Nevertheless, in April of the following year Jácome sold Brígida that share of the family home which he had inherited from his father, and she paid for it with three printing presses (one fewer than her husband had owned in his heyday), their equipment, and a sum of money. This transaction suggests that Jácome was short of funds to buy the presses and therefore had to exchange property for them. It also suggests that he wished to install them in his new house.[63]

In 1555, however, his father-in-law, Juan Varela de Salamanca died. A long document detailing the division of his estate between his three daughters survives.[64] It reveals that Jácome was appointed to supervise the lavish funeral arrangements,

Hispalense de 1558)', in T. Falcón Márquez (ed.), *Homenaje al profesor Dr Hernández Díaz* (Seville, 1982), i. 269–73.

60 García Icazbalceta, *Bibliografía mexicana*, pp. 30, 47.

61 On 25 Jan. 1554 he was still in San Isidoro (APS, Oficio 6, Libro [2] of 1554, unfoliated), but by 26 June he was living in Omnium Sanctorum (APS, Oficio 23, Libro 2 of 1554, fol. 497^{r-v}).

62 Pike, *Aristocrats and Traders*, pp. 16, 149–50.

63 Document dated 20 Apr. 1555 (APS, Oficio 15, Libro 1 of 1555, fols. 781^{r}–784^{v}). On 29 Nov. 1555 Jácome is said to be resident in the neighbouring parish of San Martín (APS, Oficio 5, Libro 4 of 1555, fol. 1190^{r-v}) but, if he had really moved there, he was back in Omnium Sanctorum again by 10 Feb. 1556, (APS, Oficio 10, Libro 1 of 1556, fols. 655^{r}–657^{r}).

64 Document dated 7 Aug. 1556 (APS, Oficio 5, Libro 2 of 1556, unfoliated).

pay the immediate expenses, and collect and honour outstanding debts. The estate was so large that the share received by Inés and Jácome was valued at more than 12,000 *ducados* made up of agricultural land, houses, and shops in Seville, and money owed to Varela from rents and debts. The greater part of the sum was not, however, in cash: it consisted of investments and *tributos* (mortgage loans), on which annual payments were made by the mortgagees at about 10 per cent of the value of the mortgage. Jácome was therefore assured a steady income if not a large sum of disposable money.

It is no coincidence that Jácome and Inés left Omnium Sanctorum shortly after receiving their share of Varela's estate and moved to the fashionable quarter of El Salvador, where they took up residence in a house in the Calle Sierpes, to which they carried out extensive alterations, possibly to accommodate the presses.[65] On the other hand, the inheritance may have enabled Jácome to abandon the mechanical art of printing altogether so that subsequently he would act as a merchant-publisher or subcontract work out to other printers like Gregorio de la Torre; such a move could explain his arrangement with Torre for the printing of the Seville missal. Whether Varela's death had really brought him considerable wealth or not, by 1557 there were signs that all was not well, for his brother-in-law, Pedro de Ávila, made him a large short-term loan probably to relieve him of immediate financial embarrassment.[66] In the following year there is evidence that something serious had gone wrong: in August Jácome defaulted on the payment of a debt and he was threatened with having his home summarily sold over his head in order to meet it (this was the customary method employed in Seville at the time by creditors to extract payment from their debtors).[67] Then on 13 September Inés de Alfaro promised to pay a creditor out of her own dowry in order to have her husband released from the Archbishop's Prison, where he had been incarcerated for debt. Even her promised payment was late.[68] Finally, some ten days later, his brother, Tomás, had to bail him out of the Royal Prison, where he was once more being held for defaulting on a financial obligation; yet again the debt was not honoured on time, the creditor having to wait for over a year to receive his money.[69] Nevertheless, Jácome's predicament is unclear; he still owned property and slaves so his imprisonment could have been due to cash-flow problems rather than to destitution.[70]

[65] Gestoso, p. 126, and document dated 8 Oct. 1557 (APS, Oficio 4, Libro 2 of 1557, fol. 2329^{r-v}).

[66] Document dated 11 Dec. 1557 (APS, Oficio 1, Libro 2 of 1557, fols. 1328^{r}–1329^{v}).

[67] Gestoso, p. 130.

[68] Documents dated 13 Sept. 1558 (APS, Oficio 19, Libro 2 of 1558, fols. 1382^{r}–1384^{v}) and 18 Jan. 1559 (APS, Oficio 19, Libro 1 of 1559, fol. 50^{r-v}).

[69] Document dated 22 Sept. 1558 (APS, Oficio 4, Libro 2 of 1558, fols. 1841^{r}–1842^{r}).

[70] Documents dated 24 Mar. 1559 (APS, Oficio 1, Libro 1 of 1559, fols. 638^{r}–639^{v}); 9 Aug. 1559 (APS, Oficio 1, Libro 2 of 1559, fol. 242^{r}); 29 Nov. 1559 (APS, Oficio 1, Libro 2 of 1559, fols. 1066^{r}–1067^{v}); and 20 Dec. 1559 (APS, Oficio 1, Libro 2 of 1559, fol. 1204^{r-v}). In 1561 Inés de Alfaro was

His career as a printer had now come to an end. He tried to raise capital by selling *tributos* and then optimistically empowered his wife to take possession at the Seville docks of all the 'consignments of gold, silver, pearls, precious stones, hides, sugar, and cochineal which I shall send from the Indies of the Ocean Sea'.[71] Like many of his contemporaries in Seville, he had decided to go to the New World, either succumbing to the lure of the riches to be gained there or merely hoping to escape from debts at home. He may have accompanied his brother, Tomás, who had obtained permission to travel to the Isthmus of Panama or the mainland of South America ('Tierra Firme' rather than 'Nueva España') two days before Jácome authorized Inés to act on his behalf during his absence. Characteristically, Jácome himself did not obtain an official licence to make the voyage.[72] He set sail from Seville in the same year, for in November his wife was acting on behalf of 'Jácome Conberger who is away from this city of Seville and is in the Indies'.[73] Some of his property was compulsorily auctioned to meet debts which he had left behind him, while court cases brought against him by other creditors mounted up in his absence.[74] In August 1560 Inés still maintained that her husband was in the Indies, but by the beginning of the next year she was calling herself a widow.[75]

No will drawn up by Jácome before leaving Spain, nor any inventory of his estate after his death, has come to light in the Seville or Mexican archives, although it would be surprising if even such an improvident man as Jácome had not drawn up a testament before setting out on the hazardous journey to the Americas. It is therefore impossible to discover whether he still owned his presses and books when he departed for the New World. He left behind him a widow who was still alive over twenty-two years later; a daughter, Brígida Agustina, who had been

prosperous enough to give 60,000 *mrs* to one Domingo de Portonaris to enable him to continue his studies at Salamanca University (see document dated 6 January 1561 [APS, Oficio 1, Libro 1 of 1561, fol. 99ᵛ]). He may well have been the same man who styled himself eighteen years later 'Domingo de Portonaris de Ursino, Royal Printer' and who became a famous printer at Salamanca. If this is the case, he was no better at avoiding gaol than his benefactress's husband (see Abizanda y Broto, *Documentos*, i. 355).

71 Documents dated 10 Mar. 1559 (APS, Oficio 1, Libro 1 of 1559, fols. 533ʳ–535ʳ and one unfoliated sheet) and 24 Mar. 1559 (APS, Oficio 1, Libro 1 of 1559, fols. 640ʳ–643ʳ).

72 Bermúdez Plata, *Catálogo de pasajeros a Indias*, iii. 316. Jácome's name does not appear in AGIS, *Contratación*, Legajo 5537, which contains a list of all licences granted for emigration in 1559. There were severe penalties for those going to the Indies without a licence but, predictably enough, this control was not effective (see C. H. Haring, *Trade and Navigation*, pp. 103, 109–10).

73 Document dated 24 Nov. 1559 (APS, Oficio 1, Libro 2 of 1559, fol. 1044ʳ).

74 Documents dated 23 Jan. 1560 (APS, Oficio 21, Libro 1 of 1560 [wrongly labelled '1550'], fol. 201ʳ⁻ᵛ); 30 Mar. 1560 and 4 May 1560 (APS, Oficio 1, Libro 1 of 1559 [*sic*], loose sheets between fols. 642ᵛ and 643ʳ); 18 May 1560 (APS, Oficio 1, Libro 2 of 1559 [*sic*], loose sheet between fols. 1043ᵛ and 1044ʳ); and 2 Aug. 1560 (APS, Oficio 5, Libro 2 of 1560, fols. 668ʳ–673ᵛ).

75 Documents dated 2 Aug. 1560 (APS, Oficio 1, Libro 2 of 1560, fols. 198ʳ–199ᵛ) and 6 Jan. 1561 (APS, Oficio 1, Libro 1 of 1561, fol. 99ᵛ).

born in 1552; and a son, Juan Cromberger Maldonado, who in 1577 went to the Indies as an administrator of the valuable ecclesiastical revenue, the Santa Cruzada, thus following the normal pattern among the descendants of immigrant businessmen in Seville: the abandonment of trade for a post in the civil service.[76] Juan Cromberger Maldonado had no known connections with printing; his father was the last of the Crombergers to practise the craft which the family had followed for over half a century.

Jácome was an unremarkable and unadventurous printer: he made no innovations in the sort of books the press produced nor in the material with which they were printed; his typography was poor; and his business activities undistinguished. Only his marriage had been brilliant, and he may soon have managed to dispose of whatever wealth it had brought him. However, he was working at an unhappy time for Seville printers. It has been shown that they found themselves beset by financial problems, and new offices came and went rapidly. The influx of foreign books and competition at home must have contributed to this decline.

Printing in Seville, the Economic Crisis of the 1550s, and the Inquisition

It is unsatisfactory to consider the problems experienced by Seville printers in the mid-sixteenth century solely in terms of their incompetence or of problems peculiar to the printing industry. The 1550s were a period of financial crisis in Seville. Trade was depressed and shipping on the New World routes, which had been increasing steadily ever since the beginning of the century, now slumped.[77] It is true that during this decade the greatest amount of gold ever to be sent from the Indies reached Seville, while silver came in larger quantities than in previous years. Indeed, so great was the amount of bullion arriving in Seville in 1556 that it had to be stored in stables from which, inevitably, some was stolen.[78] Even these records would be broken in the 1560s and, by the end of the century remittances

[76] Gestoso, pp. 122–3. There appear to be two people of this name: one was a priest, Licenciado Juan Cromberger Maldonado, the son of Ana de Maldonado (Jácome Cromberger's sister) and Juan de Espinosa (see document dated 13 May 1577 [AGIS, *Contratación*, Legajo 215, unfoliated]) and his cousin, Licenciado Juan Cromberger Maldonado 'son of Jácome Comberger and Doña Ynés de Maldonado' (presumably a mistake for 'Inés de Alfaro') who emigrated to the Indies (see his licence to travel to New Spain dated 25 May 1577 [AGIS, *Contratación*, Legajo 5538, Libro 1, fol. 429^r]). One of them was prosecuted for smuggling a silver ingot into Spain from the Indies in 1596 (see AGIS, *Contratación*, Legajo 63B, Ramo 24). For the abandonment of trade by the descendants of immigrant families see Pike, *Enterprise and Adventure*, p. 3, and *Aristocrats and Traders*, p. 110.

[77] J. H. Elliott, *Imperial Spain 1469–1716*, 2nd edn. (Harmondsworth, 1970), p. 231. Pierre and Huguette Chaunu, *Séville et l'Atlantique 1504–1650*, 8 vols. (Paris, 1955–9), vi. 160–7.

[78] Le Flem, 'Los aspectos económicos', pp. 70–2. By the late 1560s the amalgam method for the extraction of silver was well established in Mexico, and the Peruvian silver-mines of Potosí (discovered in 1545) began to be worked on a large scale. For the theft of silver see Icaza, 'Miscelánea histórica', document No. 76.

of silver had increased ninefold. The influx of so much bullion does not appear, however, to have benefited Seville's economy. Much of it was destined to go abroad to service Charles V's enormous debts, while Spain, and especially Seville, were affected particularly early by the 'price revolution'. The reasons for this inflation were complex and have been much debated by economic historians, but among the causes were doubtless a rise in population leading to increased demand; probably less importantly, the arrival of precious metals from the Americas; and the demand for exports to the colonies coupled with the high prices which both shippers and purchasers in the New World were prepared to pay for European goods.[79]

More immediately crippling were the actions of the new king. When Philip came to the throne in 1556, the Empire's economy was reeling from the strain put upon it by his father's imperial policy, by the war with France, the struggle against the Turk and the Protestants of Germany, and by the natural disasters which had virtually destroyed Spanish agriculture.[80] Philip had to finance his own campaigns against the French, while the victories won at Saint-Quentin in August 1557 and at Gravelines in July of the following year had to be paid for. In 1557 reality intruded: the Crown's finances could bear the burden no longer, payments to bankers were suspended and what amounted to a public bankruptcy was declared.[81] One immediate and well-tried expedient was to sequester the bullion which was brought back to Spain from the New World by individuals or remitted to merchants, either as the income on cargoes sent from Seville or as a return on their investments in the colonies. In 1556 and again in each of the next two years the royal authorities confiscated the bullion which arrived with the Indies fleets, taking 1,800,000 *ducados* in the winter of 1556/1557 alone.[82] The effect on merchants who had already suffered Charles V's confiscation of 600,000 *ducados* in 1553, was devastating, for they relied on the arrival of their silver and gold to repay the principal and interest on loans taken out to purchase the cargoes they had originally sent across the Atlantic. Although Philip later went some way to compensating the unfortunate merchants by giving them *juros* which produced an annual income, these consolidated fixed-interest bonds neither represented the true value of their loss nor provided them with the disposable cash which they required if they were to stay in business.[83] The result was a spate of bankruptcies in Seville. The attendant shortage of money in the

[79] For a succinct account of the 'price revolution' see Pierre Vilar, *A History of Gold and Money 1450–1920*, tr. Judith White, 2nd English edn. (London, 1984), pp. 76–89.

[80] Geoffrey Parker, *Philip II* (London, 1979), p. 62.

[81] Fernand Braudel, *The Mediterranean and the Mediterranean World in the Age of Philip II*, tr. Siân Reynolds, 2nd English edn., 2 vols. (London, 1972), ii. 960–2.

[82] Haring, *The Spanish Empire in America*, pp. 277–8.

[83] Already in 1555 complaints had been made to the Crown about the deleterious effect of such action on trade (see Haring, *Trade and Navigation*, p. 173).

city explains to some extent the printers' financial problems and the undercapitalization of the presses which led to unambitious printing with worn material and the need to issue the sort of items which gave quick returns on limited investment. The researches of Febvre and Martin in France indicate that demand for books was sensitive: when there was an economic crisis, it dried up almost completely.[84]

If the Crown's fiscal policies touched other Seville printers, they affected the Crombergers more directly because of their interests in trade with the New World.[85] In 1557 two large shipments of bullion which had reached Seville for Brígida Maldonado were confiscated; she was given *juros* in exchange, but in 1559 she sold these to an Italian dealer in such bonds, doubtless because she needed ready cash.[86] There is other evidence of her owning *juros* which she was probably obliged to accept as compensation for sequestered bullion. In 1571 she attempted to redeem one of these bonds which she had held for at least ten years, declaring that she was in desperate need of the money. Eventually she received only half its face value, because the coffers of the Casa de Contratación against which it was drawn were said to be bare.[87] Brígida was not the only member of the family to suffer such losses. On his death-bed Tomás Ungut gave a share of his estate to his servant, María de Escobar, for the many years of faithful service she had given him. It transpires that she had indeed ministered to his every need, for she had borne him an illegitimate daughter, Catalina, who married Diego Herver, the grandson of an eminent *converso* merchant of Seville.[88] In 1561 the king recognized that three years earlier he had confiscated bullion belonging to the Hervers.[89]

Such government policy in the 1550s also led to the collapse of several of the most important of the city's banks. One was that of Domingo de Lizárrazas, a Basque who was closely associated with the Genoese banking community. The Crombergers had family and business connections with him: he was married to Inés de Alfaro's niece, Isabel de Sandoval; he and Lázaro Nuremberger appear together in several documents; and Jácome was associated with Isabel after her husband's death. Lizárrazas's bank collapsed in 1553, causing a local scandal and doing the Crombergers no good whatever.[90] It appears that the whole family was

[84] Febvre and Martin, *L'Apparition*, p. 173.

[85] No other Seville printer living at this time was so much involved with New World trade. There is evidence, however, that Simón Carpintero had sent merchandise there before 1555 (see Gestoso, p. 125).

[86] Document dated 14 Mar. 1559 (APS, Oficio 15, Libro 1 of 1559, fols. 574^r–575^r).

[87] Document dated 5 Oct. 1571 (AGIS, *Contratación*, Legajo 5767, No. 6, Ramo 2). See also Vilar, *A History of Gold and Money*, p. 154.

[88] Document dated 20 June 1548 (APS, Oficio 10, Libro 3 of 1548, fols. 1029^v–1031^r).

[89] Hazañas, i. 159–60; and document dated 1 Mar. 1561 (AGIS, *Contratación*, Legajo 1059, No. 18, unfoliated).

[90] Pike, *Enterprise and Adventure*, pp. 87–93; Hazañas, i. 163; document dated 18 Feb. 1558 (APS, Oficio 15, Libro único of 1558, fols. 313^v–314^r).

affected by the Crown's policy of confiscations during the 1550s, for Lázaro had bullion coming back from the Indies, and Juan Cromberger's heirs, including Jácome, would have counted on recouping money from their father's estate in the New World.

As if this were not enough, ideological upheavals affected booksellers and printers at the same time. Control of printing by the authorities had been slack in Spain during the first half of the sixteenth century.[91] In the 1550s there was an attempt to change all this. In 1550 or 1551 Valdés, who was, as has been noted, Inquisitor-General as well as Archbishop of Seville, had received a copy of the 1550 Louvain *Index*. It was decided that this *Index* should be reprinted in Spain with an additional section containing a list of books previously prohibited by the Holy Office in the Spanish Kingdoms. The result was the publication of the first Spanish *Index* (the 1547 *Index* referred to by many historians is a fiction), of which editions appeared in 1551 at Valladolid, Seville, Valencia, and Toledo. In 1554 Valdés authorized the printing of a *censura* of editions of the Bible. Again in 1554 the power to license works for printing—which had previously been in the hands of a motley collection of secular and ecclesiastical authorities depending on the region in question—was secularized and centralized in the Royal Council, and two years later it was decreed that special licences were required to print any book concerning the Indies.[92] But it was from 1557—possibly the last year in which Jácome Cromberger issued books—onwards that the authorities made a serious bid to control the trade, especially in Seville, where bitter ideological struggles had long been raging. In 1557 Julián Hernández, the hapless 'Julianillo', was caught red-handed attempting to smuggle into Seville Protestant books mainly printed at Geneva and, in the aftermath of his arrest, the infamous 'Lutheran' circle of the Hieronymite house of San Isidoro del Campo just outside the city was discovered.[93] A few months later Protestant cells were smoked out all over Castile. A veritable witch-hunt was unleashed in Spain and printed books figured at the centre of it. One result of this was the hurriedly composed *Index* of 1559 and the burning of printed books as well as heretics. The possession of certain works was subsequently used as *prima facie* evidence against those suspected of heresy. Some measure of the atmosphere of those times as well as of the Inquisition's extraordinary powers was the arrest on Valdés's orders, in the very same week of August 1559 in which the Inquisitor-General authorized the

[91] Edicts had been issued from 1502 onwards in an attempt to control the sort of books published in, or imported into Spain. Of particular interest is that of 1521 calling in copies of Luther's works. These regulations appear to have been ineffectual (see José Simón Díaz, *El libro español antiguo: análisis de su estructura*, Teatro del Siglo de Oro, bibliografías y catálogos, 1 [Kassel, 1983], pp. 21–8).

[92] Simón Díaz, *El libro español antiguo*, p. 7.

[93] Novalín, *El inquisidor general Fernando de Valdés*, ii. 181–3; E. Droz, 'Notes sur les impressions genevoises transportées par Hernández', *Bibliothèque d'humanisme et Renaissance*, 22 (1960), 119–32.

publication of the *Index*, of Bartolomé de Carranza, Archbishop of Toledo and primate of all Spain. Carranza was the victim of Valdés's personal hatred, but it is important to note that the archbishop was arrested on the grounds that he had written and published a spiritual guide which contained heretical opinions, even though Philip himself had given permission for its publication in Antwerp the previous year.

Spiritual works were, indeed, at the heart of the trouble, and Valdés's *Index* caused Spanish publishers to be wary of printing such books for several years.[94] The Inquisition was even more terrified of Protestant editions printed in Spanish at Geneva and Lyons, and then sent direct from there or via Antwerp. Many of these heretical books were destined for Seville, which was doubly worrying for the Holy Office because the port was the gateway to the Indies.[95] Later in the century thousands of Protestant tracts were smuggled into Spain from England and Holland.[96] When it is remembered that some 30,000 copies of Calvin's *Institutes* alone were dispatched to Spain, these fears can be appreciated.[97] But it was not just foreign editions which caused concern: works such as those written by Constantino Ponce de la Fuente, who was arrested at Seville in 1558, were also called in. Likewise the inspection of bookshops was ordered. Such inspections were not new, but a letter sent to a higher authority in the Inquisition at this time indicates why readers and those connected with the book-trade were thrown into a panic: among other regulations there were to be monthly inspections of printing-shops, all books printed in Spain were to be examined by the Inquisition, all editions imported for sale were to be carefully scrutinized, all works in the vernacular were to be called in for inspection, it was to be a capital offence to import books in the vernacular without a licence, printers were to sign every edition they printed, and booksellers were to be obliged to write their names in every book they sold.[98]

[94] Melquíades Andrés, *La teología española en el siglo XVI*, 2 vols. (Madrid, 1977), ii. 625. Secular literature had also come in for much criticism. In 1555 the Cortes of Valladolid petitioned the Crown to prohibit a wide range of secular literature of entertainment, but the ban was not put into effect (see Barry Ife, *Reading and Fiction in Golden-Age Spain: A Platonist Critique and Some Picaresque Replies* [Cambridge, 1985], pp. 12–17).

[95] The Inquisition in the New World was particularly concerned about threats to orthodoxy there posed by imported books (see Richard E. Greenleaf, *The Mexican Inquisition of the Sixteenth Century* [Albuquerque, 1969], pp. 182–6).

[96] Georges Bonnant, 'Note sur quelques ouvrages en langue espagnole imprimés à Genève par Jean Crespin (1557–1560)', *Bibliothèque d'humanisme et Renaissance*, 24 (1962), 50–7 (p. 51).

[97] A. Gordon Kinder, *Casiodoro de Reina: Spanish Reformer of the Sixteenth Century* (London, 1975), pp. 12–13, 39. The Seville Inquisition was severely understaffed during this period (see Novalín, *El inquisidor general Fernando de Valdés*, ii. 162–5). See also Valentín Vázquez de Prada, 'La Inquisición y los libros sospechosos en la época de Valdés-Salas (1547–1566)', in J. Ignacio Ruiz de la Peña *et al.*, *Simposio "Valdés-Salas", Oviedo, 8–11 diciembre 1968* (Oviedo, 1970), pp. 147–55 (pp. 150-5).

[98] Novalín, *El inquisidor general Fernando de Valdés*, i. 267–9, ii. 199–201. Liturgical books, books of hours, grammars, ABCs, etc. were not covered by the regulations laid down in the pragmatic issued

The implications for the trade were serious even if the Draconian measures contained in the royal pragmatic of 7 September 1558, which set out the new regulations, were not always fully implemented. Indeed, subsequent edicts promulgated later in the sixteenth and even in the seventeenth centuries clearly indicate that a tight control over reading matter could not be exercised in Golden Age Spain.[99] Nevertheless, in theory at least, manuscripts had now to be closely examined and each page signed by the censor before a licence to print could be issued, so there would inevitably be long delays before a book could be set up. An author who was trying to obtain a licence for publication may have been affected more than his printer, but in many cases the printer too would have encountered problems. The fixing of a *tasa*, or official retail price, on books was, on the other hand, bound to have a deleterious effect on the trade, because it reduced publishers' and booksellers' ability to respond flexibly to the market, while the obligatory comparison of the printed book with the licensed manuscript or previous edition from which it was set up would inevitably cause serious delays in publication which printers short of funds could ill afford.[100] Moreover, as it was impossible, given the way in which a book was printed at the time, to produce the two required inspection copies without having the whole edition printed, any major alterations required by the censors would mean that the entire sheet or, at the very least, the offending folio had to be destroyed, reset, and as many copies run off as there were books in the edition. This is not to mention the cost of the examination copies, which often exceeded the two stipulated in the decree.[101] Even more disturbing was the threat that an increasingly zealous Inquisition, which exercised censorship only after the book had been printed, would confiscate and destroy entire editions, thus dealing the hard-pressed publisher or bookseller a mortal blow. In the 1550s Valencian booksellers protested that they and their families would be thrown into penury if their stocks of books edited by heretical writers like Melanchthon were confiscated. Likewise, in 1560 a group of Salmantine booksellers and printers sought compensation for editions confiscated as a result of the Valdés *Index*.[102]

Controls on imported editions might have been as much a response to the complaints of the *arbitristas* (the authors of those innumerable memoranda suggesting cures for Spain's economic crisis, some of which have been mentioned

by Princess Juana on 7 Sept. 1558, but were included in a later one (see Simón Díaz, *El libro español antiguo*, pp. 8–10).

[99] Antonio Sierra Corella, *La censura de libros y papeles en España y los índices y catálogos españoles de los prohibidos y expurgados* (Madrid, 1947), pp. 96–9.

[100] *Tasas* had been fixed for certain books before this date, but only sporadically (see Simón Díaz, *El libro español antiguo*, p. 91).

[101] Péligry, 'Les Difficultés', p. 260.

[102] Gil Fernández, *Panorama social*, pp. 509–10. Also AHNM, *Inquisición*, Legajo 4442(1), Expediente 36. I am grateful to my colleague, Dr R. W. Truman, for the latter reference and the one contained in n. 105 below.

on pp. 106–7) as an attempt to build a spiritual *cordon sanitaire* around Spain; it has recently been argued, for instance, that Philip II's prohibition on study abroad was as much an economic as an ideological measure.[103] Whatever the motives for the control on foreign books it was not as beneficial to printers in Castile as might be assumed, for booksellers continued to import editions in large numbers, merely seeking official permission to take possession of the copies once they arrived in Spain.[104]

Seville printers may have been affected more than most by the atmosphere of the late 1550s. The popular nature of even their secular editions made them particularly vulnerable to Valdés's ban on anonymous works in the vernacular. In 1560 a group of Seville printers and booksellers protested that 'there are some good books in Castilian, like the stories of the Cid, the Infante don Pedro, Abad don Juan, and other similar tales, which children are accustomed to reading. They have always been anonymous, and now we do not dare to print them.'[105] But spiritual works, especially in Castilian, were the main target of Valdés's prohibitions. The fact that he was harsher on works in the vernacular than was the Roman *Index* could again have affected Seville printers particularly badly.[106]

Among the books burnt at Valladolid in 1558 a large number were titles which had appeared at Seville; some were even the very editions printed there. Indeed, Seville was the only Spanish printing centre specified in the Inquisition's list of books committed to the flames. Among the books burnt were 'Gamaliel, newly printed'; the Cromberger press had issued an edition of *Gamaliel* rendered into Spanish by Erasmus' translator, Juan de Molina. One of Constantino's works, which loom large in the list, had first been printed by the Crombergers at Seville, and a version of his *Suma* had also been commissioned by Bishop Zumárraga, who had it printed in their Mexican office.[107] When the prohibition of books containing excerpts from the Scriptures was being discussed in 1558 prior to the publication of the *Index*, the provisional list which was drawn up could well have been intended to describe the Crombergers' output of devotional editions: it contained books of hours in the vernacular, compilations of passages from the Epistles with accompanying sermons, Ludolph of Saxony's *Vita Christi*, and the hagiographical *Flos sanctorum*.[108] Indeed, many of the works eventually listed in the 1559 *Index* had been printed at Seville, a good number of them by the

[103] Information from current unpublished research by Dr R. W. Truman.

[104] Novalín, *El inquisidor general Fernando de Valdés*, i. 285.

[105] AHNM, *Inquisición*, Legajo 4442(1), Expediente 40.

[106] J. M. de Bujanda, 'La Censure littéraire en Espagne au XVIe siècle', *Canadian Journal of History*, 7 (1972), 1–15.

[107] AHNM, *Inquisición*, Libro 323, fol. 146r. I am indebted to Dr A. Gordon Kinder for a photocopy of this document and of the one referred to in n. 111 below.

[108] Novalín, *El inquisidor general Fernando de Valdés*, ii. 206. The Inquisitional ban on this sort of work was applied remarkably rapidly even in far-flung parts of the New World (see the document dated

Cromberger press which was referred to by name in the list of the proscribed editions.[109]

At least one Seville printer, Gaspar Zapata—probably the same man who in happier days had shared with the Crombergers the poet Juan de Vilche's praise—had reason to fear the stake.[110] Fortunately for him it was only his effigy which was reduced to ashes at the *auto de fe* celebrated in Seville in October 1562, for he had fled the city.[111] Any heretic associated with the book-trade was considered by the Inquisition to be particularly dangerous because of his power to disseminate heterodox ideas. Thus, one of the victims of the 1559 *auto* held in Seville's Plaza de San Francisco was 'Luis de Ábrego, a copyist of service books, native of Niebla, *vecino* of Seville, burnt at the stake for Lutheranism, for being in receipt of many

Zacatecas, Mexico, 15 Feb. 1561 [AGNMex., *Inquisición*, vol. 43, Expediente 10, fols. 255^{r}–256^{v}] which deals with an unfortunate shopkeeper who accepted a book of *Epístolas y evangelios* from an Indian who hocked it for sustenance of a more intoxicating nature; or the case of a hapless silversmith in Santiago, Guatemala, who in 1560 was tried for saying that those who prohibited such works needed their heads tested [AGNMex., *Inquisición*, vol. 16, Expediente 4, fols. 76^{r}–126^{r}]).

[109] Fr. Heinrich Reusch, *Die Indices Librorum Prohibitorum des sechzehnten Jahrhunderts*, Bibliothek des Litterarischen Vereins in Stuttgart, 176 (Tübingen, 1886), pp. 231–41. Books printed at Seville 1500–59 and listed in the Valdés *Index* were Cazalla, *Lumbre del alma* (heirs of Juan Cromberger, 1542 [Cazalla is not mentioned by name, but this work is the 'Work in Spanish printed at Valladolid by Master Nicolás Tierry in the year 1528'; see Bujanda, 'La Censure', p. 10]); Constantino, *Suma de doctrina cristiana* (heirs of Juan Cromberger, 1543 and 1544; Juan de León, 1545; Cristóbal Álvarez, 1551), *Exposición del primer salmo de David: Beatus vir* (Juan de León?, 1546; printer unknown, 1547; printer unknown, 1548), *Catecismo cristiano* (printer unknown, 1547), *Doctrina cristiana* (Juan Canalla, 1548); Erasmus, *Querella de la paz* (Jacobo Cromberger, 1520), *Coloquios* (Juan Cromberger, 1529), *Lengua* (Juan Cromberger, 1533 and 1535; heirs of Juan Cromberger, 1542 and 1543/4), *Enchiridion* (Jacobo Cromberger, 1528?; Gregorio de la Torre and Juan Canalla, 1550), and *Tratado de la oración* (Andrés de Burgos, 1546); Osuna, *Gracioso convite de las gracias del santo sacramento* (Juan Cromberger, 1530; his heirs, 1544; Martín de Montesdoca, 1554); Torres Naharro, *Propalladia* (Jacobo Cromberger?, 1520; Jacobo and Juan Cromberger, 1526; Juan Cromberger, 1533–4?; Andrés de Burgos, 1545); *Fasciculus mirrhe* (Jacobo Cromberger, 1517; Juan Varela de Salamanca, 1518 and 1524; Juan Cromberger, 1531; Domenico de Robertis, 1536; Jácome Cromberger, 1550); *Gamaliel* (Juan Cromberger, 1534; Domenico de Robertis, 1536); Roman Hours in Spanish (Juan Cromberger, 1528, 1537, and 1538; Juan Varela de Salamanca, 1531 and 1539; heirs of Juan Cromberger, 1542; Domenico de Robertis and his heirs, 1541 and 1550; Jácome Cromberger, 1550; Gregorio de la Torre, 1556); *Las lecciones de Job trobadas* (Estacio Carpintero, 1545); Jiménez de Prejano, *Lucero de la vida cristiana* (Juan Varela de Salamanca, 1515; Jacobo Cromberger, 1524 and 1528; heirs of Juan Cromberger, 1543); Caviceo, *Libro del Peregrino* (Jacobo Cromberger, *c.* 1515; Jacobo Cromberger, 1527; heirs of Juan Cromberger, 1544; Jácome Cromberger, 1548); *Las preguntas que el emperador Adriano hizo al infante Epitus* (Juan Cromberger?, *c.* 1535?); *Rosario de Nuestra Señora* (so many copies of a work of this name were listed in the inventory of Jacobo Cromberger's shop that it is almost certain that he had printed an edition); Sánchez de Vercial, *Sacramental en romance* (heirs of Juan Cromberger, 1544; Jácome Cromberger, 1549; printer unknown, 1551?); Pereç, *Vergel de Nuestra Señora* (Domenico de Robertis, 1542); *Vida de Nuestra Señora* (so many copies of a work of this name were listed in the inventory of Juan Cromberger's shop that it is almost certain that he had printed an edition).

[110] For Vilche's praise of printers in Seville see Domínguez Guzmán, *El libro sevillano*, pp. 282–3.

[111] AHNM, *Inquisición*, Legajo 2072 (1), fol. 18^{v}: 'found guilty of Lutheranism, condemned to death in their absence, and burnt in effigy forfeiting all their possessions: . . . Gaspar Çapata, printer, *vecino* of Seville . . . '.

forbidden heretical books, and for disseminating Luther's heresies'.[112] This same concern about the orthodoxy of printers and members of the book-trade is reflected in the Mexican branch of the Seville Inquisition. In the 1570s both Juan Pablos's son-in-law, Pedro Ocharte, and of one of the employees in Ocharte's printing-office, Juan Ortiz, were tried in Mexico on charges of heresy.[113] Even Martín de Montesdoca may have fled to the Indies as much for fear of the Inquisition as a result of financial embarrassment: he had been a major printer and publisher of the works of Fray Domingo de Valtanás, who was arrested shortly after Montesdoca had left Seville.[114]

There is, however, no evidence that Jácome feared for his own safety in the Inquisitional clamp-down on the printed word. It is true, as will be seen in Chapter Six, that the Cromberger press had earlier been well connected with members of the Erasmian circle of Seville and had printed works by Erasmus and his devotees in the city. Jacobo's special arrangement for exchanging books with Miguel de Eguía involved him with one of the Spanish printers most closely associated with the Erasmians and one who was imprisoned by the Inquisition in the 1530s.[115] He would have handled Erasmian editions from Eguía's office as well as those he printed himself at Seville. Brígida Maldonado's editions of the early 1540s also reflect the recrudescence in those years of an atmosphere of spiritual reform at Seville. Nevertheless, it should be remembered that the Crombergers had printed such works only at a time when they were permitted to circulate; the family does not appear to have been involved in any way with the printing of prohibited works or with the clandestine importation of banned books. Jácome's abandonment of the craft doubtless owed more to his economic troubles than to any fear of punishment at the hands of the Holy Office. As has been seen, he was undergoing a financial crisis in the late 1550s and it would have been in keeping with family traditions that business interests should have weighed more heavily in his decision than any other considerations.

The Mexican Press

The fate of the Seville office was foreshadowed by the history of the Crombergers' Mexican press in the previous decade. Despite their monopolies of printing in New Spain and the export of books there, Juan's heirs appear to have taken little interest in the branch office. It may be that Brígida Maldonado was too preoccupied

[112] British Museum, MS. Add. 21447, fol. 93^{r}. Spain was not, of course, unique in this; many printers of various persuasions were executed in other countries.

[113] Francisco Fernández del Castillo, *Libros y libreros en el siglo XVI,* 2nd edn. (Mexico City, 1982), pp. 85–245.

[114] Klaus Wagner, *Martín de Montesdoca,* p. 37.

[115] Goñi Gaztambide, 'El impresor Miguel de Eguía', *passim.*

with the Seville press to nurture her late husband's project in Mexico, or she may have soon realized that it was unlikely to prove profitable, preferring to concentrate her interest in the colonies on the family's rich Mexican mines and on trade with the Americas. The number of surviving books printed by Juan Pablos indicates a derisory output, and Henry Wagner does not hesitate to call the Mexican office an utter failure.[116] As has been noted, in 1545 both the *Audiencia* of Mexico and the Viceroy complained to the Emperor that the Crombergers were not fulfilling the terms of their other monopoly by which they were to supply New Spain with books, and it was decided that, if they continued to default on their obligations, the trade would be opened up to other booksellers.[117] It would therefore appear that either the output of the Seville office was insufficient to supply New Spain, for its production was then seriously in decline, or that the colonial market was too insignificant to interest Brígida.[118] There is even some suggestion that she had decided to rid herself of the American press. The contract between Pablos and Juan Cromberger had stipulated that all books printed in Mexico were to be signed exclusively with the latter's name. Pablos faithfully observed this agreement until the first half of 1546, but from then until 1548, the year of the death of the Crombergers' associate in Mexico, Zumárraga, Pablos's colophons contained neither Cromberger's name nor his own. From January 1548 onwards, when Pablos's name first appears in the *Doctrina cristiana en lengua española y mexicana* printed for the Dominicans, colophons contain only Pablos's name. He probably obtained the monopoly of printing in New Spain in that same year.[119] These colophons indicate that the Crombergers had divested themselves of the Mexican office by at least 1548 and probably at some earlier date.[120] As Pablos had complained of dire poverty in 1547, it is unlikely that he had the wherewithal to buy out the Crombergers' interest in the press directly. However, after a period of uncertainty, a third party (perhaps the Florentine businessman Baldassare Gabiano, who in 1550 lent Pablos, who was by then styling himself 'royal printer', a substantial sum to re-equip the press) may have acquired the Crombergers' interest and eventually handed the office over to the printer.[121] If, then, Zumárraga's complaint about the inactivity of the press and lack of paper in Mexico should be dated to 1548, as was suggested in Chapter

[116] Henry Wagner, *Nueva bibliografía*, p. 12.

[117] Medina, *La imprenta en México*, i, pp. lxvii-lxviii.

[118] The small output of Antonio de Espinosa's press in Mexico (25 surviving items from the whole of the period 1559–76) suggests a very limited colonial market (see Alexandre A. M. Stols, *Antonio de Espinosa, el segundo impresor mexicano*, 2nd edn. [Mexico City, 1964], pp. 35–45).

[119] García Icazbalceta, *Bibliografía mexicana*, pp. 30–2.

[120] Medina, *La imprenta en México*, i, pp. lxix-lxx; Gestoso, pp. 115–16.

[121] García Icazbalceta, *Bibliografía mexicana*, pp. 47–8.

Three, the parlous state of Pablos's operation would have been closely connected to the Crombergers' decision to sell out.

It is not known when the Crombergers sold the Mexican press or to whom, nor is it clear whether at the same time they forfeited their monopoly on the export of books which was due to expire, in any case, in 1552.[122] Henry Wagner believed that a sale contract probably existed in the Seville archives but, despite long searches, I have been unable to discover such a document.[123] One reason for its failure to appear could be that the sale was negotiated on their behalf by an agent in Mexico, the contract being unknown because Mexican notarial documents for the relevant period are so patchy.[124] Perhaps the authorization given by Brígida to her brother, Mateo Carón, is the only trace of the transaction. Carón travelled to the New World in 1545 and was specifically instructed to negotiate with Juan Pablos. Unfortunately, Brígida was silent about what those negotiations were to involve, but they may well have been concerned with the sale of the branch office. Such an interpretation would fit with the other fragmentary evidence which survives concerning the Mexican press.

Even if the Crombergers withdrew from the venture at this time, they remained on good terms with their former employee, for in 1550 Pablos recorded that Brígida had handled a shipment of supplies for his press which had been dispatched from Seville but had been lost when the ship carrying them went down in the Caribbean.[125] Pablos continued to print until 1560, but his surviving output is not sufficient to allow us to assume that his presses were always fully occupied. When his former employee, Antonio de Espinosa, petitioned the Crown in 1558 for a revocation of the decree giving Pablos a monopoly on printing in New Spain, he claimed that Pablos's work was shoddy and his books overpriced. Indeed, Pablos was in financial difficulties at this time, for he was forced to take out a heavy mortgage on his house in Mexico City and on two black slaves he owned.[126] His situation cannot have been improved by the authorities' permission to Espinosa and others to set up as printers in Mexico. Yet Pablos was still hiring pressmen to go to the New World to work in his office just before he died in July or August 1560.[127]

[122] Others were sending books to Santo Domingo and Tierra Firme in 1549 and 1550, but the Crombergers' monopoly was for New Spain only (see documents dated 6 June 1549 [APS, Oficio 15, Libro 1 of 1549, fol. 1248^r] and 1 Dec. 1550 [APS, Oficio 10, Libro 3 of 1550, unfoliated]; and Icaza, 'Miscelánea histórica', document No. 3).

[123] Henry Wagner, *Nueva bibliografía,* p. 14.

[124] It is just possible that the contract is somewhere in the Archivo de Notarías in Mexico City. My hurried researches in that archive in the summer of 1984 failed to unearth it.

[125] Gestoso, pp. 115–16.

[126] Two documents reproduced by Millares Carlo in 'Algunos documentos referentes a tipógrafos de México', pp. 137–8.

[127] García Icazbalceta, *Bibliografía mexicana,* p. 51. On Espinosa's career in Mexico see Stols, *Antonio de Espinosa, passim.*

It is fitting that Jácome, the last owner of the Crombergers' great Seville press, and their erstwhile employee in those colonies to which the family had always looked for business should have both died in the New World at the same time. Their deaths brought to a close the history of one of the most notable printing-firms ever to be founded in Spain.

CHAPTER 5

PRINTING-SHOP, PRODUCTION, PRICES, AND PRINTING AGREEMENTS

THE previous three chapters have shown that there is a wealth of material in Spanish and Mexican archives recording the activities of the Crombergers as merchants and as prominent members of the German immigrant community in Seville. There are also documents relating to the foundation of the Mexican press and a substantial number of contracts which provide evidence for the printers' contacts with booksellers and other printers all over Spain.

There are, however, four more specialized areas of printing history on which an investigation into a major press like that of the Crombergers might be expected to cast some light: (*a*) the organization of the printing-shop; (*b*) production; (*c*) prices; and (*d*) printing agreements. Here, unfortunately, the archives are less helpful. The Crombergers' account-books and their correspondence do not, of course, survive, but there is likewise a dearth of the sort of notarial documents which might have helped to fill the gap which they have left. A close examination of the books themselves is therefore needed if we wish to study these four aspects of the Crombergers' press.

The Printing-shop

The printing-shop was attached for most of the press's existence to the Crombergers' home in the Calle de Marmolejos, and in it slaves, apprentices, and free pressmen worked side by side. There is no indication that any of the employees was a native Spaniard: the free workers of whom a record has survived were all either Italian, Swiss, or 'German'. As was usual at the time, they lived in the Crombergers' house and ate there, as is shown by the 1529 inventory of Jacobo's estate, where the following entries are found: 'Item, a table where we eat . . .; item, a table and benches where the men eat . . .; item, eight blankets for the men . . .'.[1] No evidence is available for working hours or conditions but, although pressmen throughout Europe had a reputation for drunkenness and brawling, I have found no record of such problems in the Cromberger press, nor any reliable evidence of labour disputes. The family owned four presses, and so the maximum workforce

[1] Information from two documents dated 7 June 1529 (APS, Oficio 4, Libro [2] of 1529, unfoliated).

could have been some twenty men, but it is likely to have been much smaller.[2] It is not known how many of the presses were in operation at any one time, and this cannot be calculated with any confidence, both because of the loss of editions and because we are ignorant of the size of most editions. A study of surviving printing-contracts for other offices at the time suggests that there were only about 700 copies in an average Seville edition.[3] If this is true of the bulk of the Crombergers' editions, in their busiest year the surviving products from their office would suggest that not even two of their presses were fully occupied, but this is probably yet another indication that editions have disappeared in large numbers rather than of idle presses.[4] As has been seen above, each press could have printed some 1,500 sheets (3,000 impressions) each day when working at full speed, although it is likely that such a high level of output could not have been maintained indefinitely. Juan Cromberger, at least, ran the office efficiently so that the work of compositors and pressmen was synchronized and the latter were fully employed.[5]

There were, of course, pullers, beaters, and compositors in the workshop. The Crombergers themselves may also have worked as compositors. Canons of the cathedral acted as proof-readers for liturgical editions at least, and it is not impossible that local scholars helped the printers correct not only the scholars' own editions, but some other publications as well.[6] In the later years, however, proof-reading was performed in a cursory fashion, for the books are plagued with misprints. Nevertheless, there is ample evidence that important errors were corrected as the sheets were being printed, for misprints which occur in some copies are not found in others.

Although contracts drawn up in Seville often expressly forbade it, two or more titles were frequently printed simultaneously.[7] There is evidence of sheets from different books being muddled in at least two copies of Juan's folio edition of the

[2] Febvre and Martin, *L'Apparition*, p. 197.

[3] This is considerably smaller than the average edition in the early 16th c. outside Spain (see Febvre and Martin, *L'Apparition*, p. 330). There are, of course, exceptions: the 1528 and 1540 inventories of the Crombergers' stock contain titles, especially of small, popular books, in quantities of over 1,000 copies. If the Crombergers' editions averaged only 700 copies, the number of books printed by Jacobo alone still considerably exceeded the output of a major European printer like Aldus Manutius.

[4] This calculation is based on an output of 1,500 sheets per press each day. This may be too high.

[5] Juan Pablos pledged himself to administer the Mexican press 'so that work never stops and is carried out in an orderly and synchronized manner as it is, Joan Coronverguer, in you own workshop' (see García Icazbalceta, *Bibliografía mexicana*, p. 43).

[6] Núñez Delgado may have corrected Cromberger's *Pharsalia* of 1528 (see Griffin, 'The Crombergers of Seville', p. 94). Norton mentions a humanist scholar of Zaragoza who worked as a *corrector* for Coci's press in that city (*corrector* does not, however, necessarily mean 'proof-reader') (*Printing in Spain*, pp. 73–4; see also ibid., pp. 41, 52, 83, 91, 104–5, 146–7). On proof-reading in Spain see Simón Díaz, *El libro español antiguo*, pp. 114–21, and above, p. 104, n. 18.

[7] For comparative examples of this practice in France see Martin and Chartier, *Le Livre conquérant*, pp. 300–1.

ncionero general, which he issued in 1535; in one gathering pages from a quarto book were printed on the reverse of a sheet containing pages of the *Cancionero*. That this practice was common can be seen from the dates of the editions. For example, Jacobo finished his *Pharsalia* on 22 June 1528; on the very next day he signed his edition of a Castilian translation of Boccaccio's *De claris mulieribus*, a book which would have taken several weeks to print. There is, however, no detailed evidence about the timing of the editions other than the general pattern, which appears to have been common in Spain, of printing substantial volumes and *pliegos sueltos* alternately.

The Crombergers followed the widespread practice of cutting costs by reducing the amount of paper used in later editions of titles which they had already printed. They did this in three ways: by changing the format, as with the various editions of the *Imitatio Christi*; by retaining the same format, but using smaller types, as with Chirino's *Menor daño de medicina*; or just by setting the work up in a more cramped fashion, as with Mexía's *Silva de varia lección*. Paper was expensive, but it is impossible to know how much money was saved in this way. Indeed, the lack of a complete enough series of statistics for paper prices, pressmen's wages, and of sufficient printing-contracts makes any more ambitious attempt to estimate profits futile. All that can be said is that the quality of paper, and, of course, the number of publications, declined in the late 1540s, but even Jácome did not plumb the depths reached by Seville printers in the next century. While a large proportion of their output was composed of ephemera and slight volumes, the average size of a Cromberger book was a respectable thirty-eight sheets.

There is evidence that type was cast in their workshop, but not that punches were cut or matrices struck there. In the Mexican branch office, however, punches were probably cut by Antonio de Espinosa once the workshop had been fully transferred to Juan Pablos. Although the Crombergers used a large number of wood-blocks, there is no sign that these were made on their premises. Similarly, despite their retailing books, they do not appear to have owned a bindery, although in 1536 Juan had certainly acquired binding-tools along with the stock of Guido de Labezaris's bookshops in the Calle de Génova. Much of their production must have been sold unbound. Some of it would have been purchased wholesale by the booksellers, who were particularly numerous in Seville, but there are indications that a large proportion was distributed through a wide commercial network which included the fairs of Northern Castile.

Production

Production varied widely from year to year as far as can be judged from surviving evidence, but the disappearance of editions probably lies behind many of these

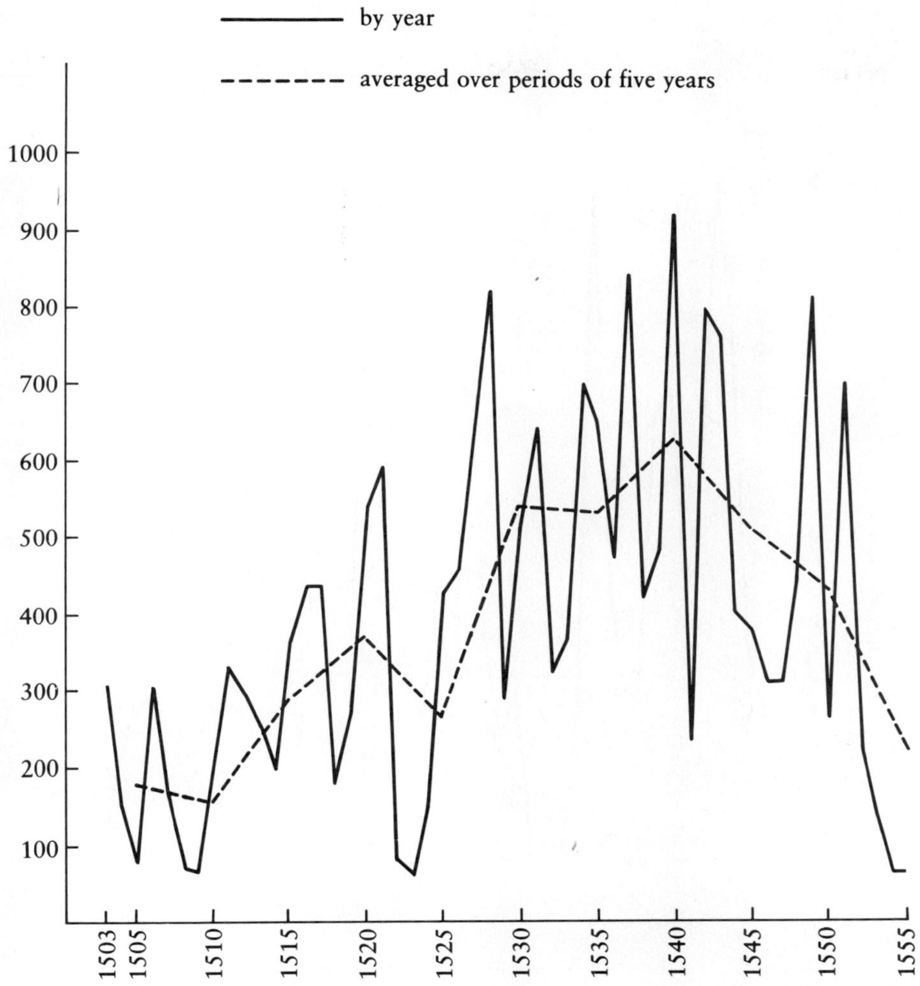

(undated editions are divided between the possible years of printing)

Fig. 3 Production by sheets contained in a single copy of all known editions

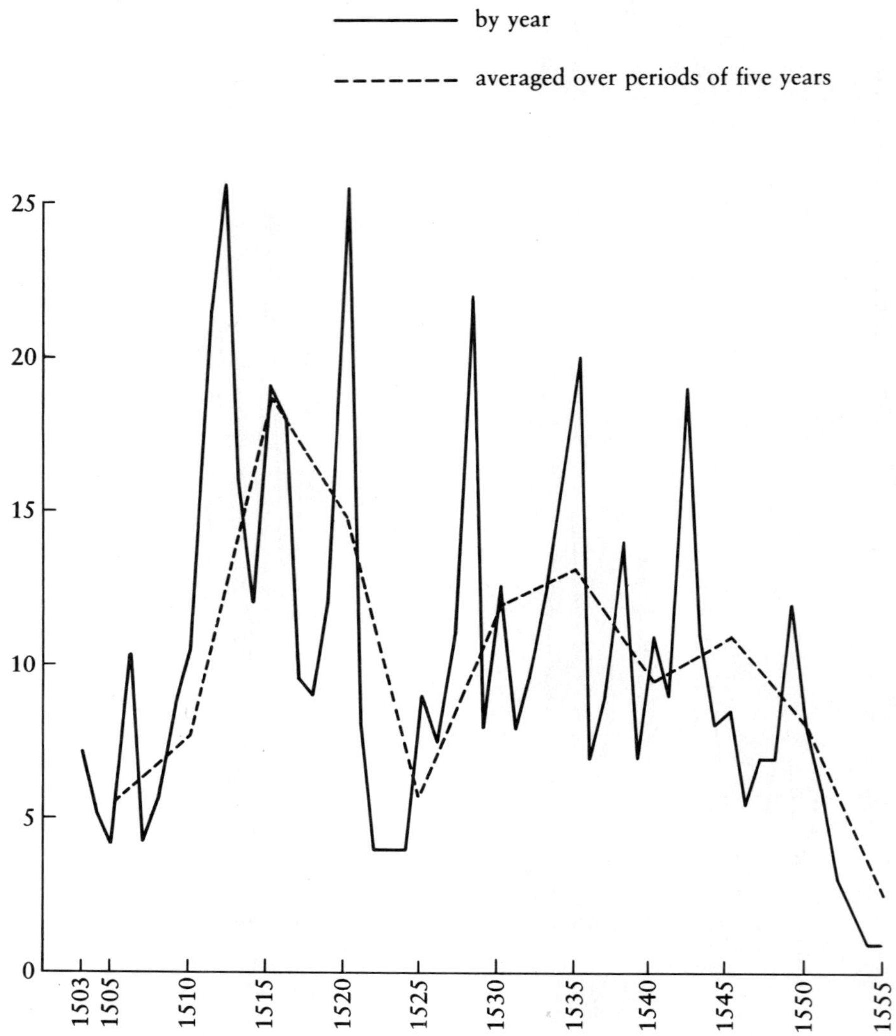

(undated editions are divided between the possible years of printing)

Fig. 4 Production by editions

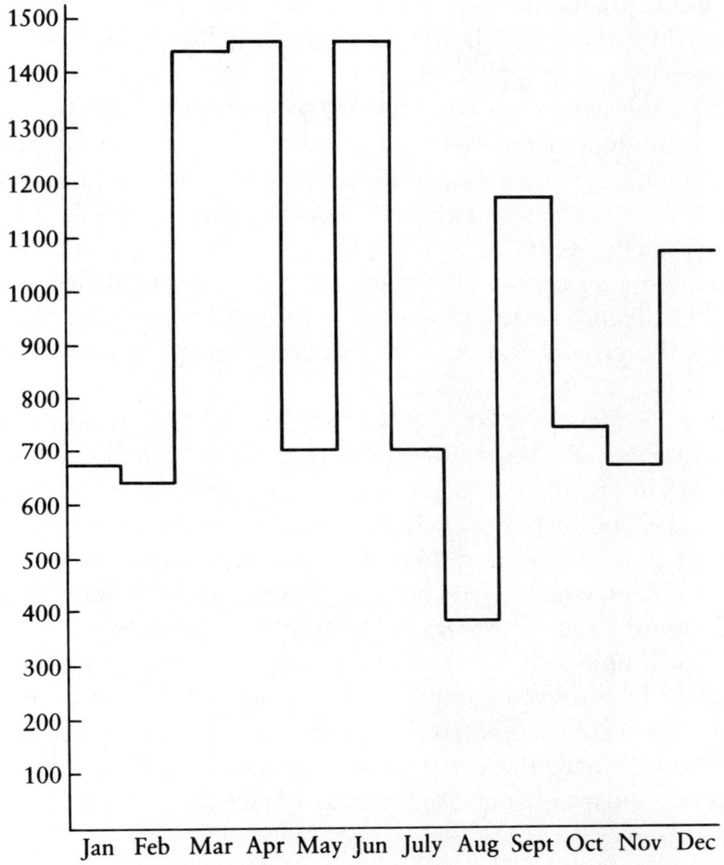

Fig. 5 Production by month: number of sheets contained in a single copy of all known editions which indicate the month in which printing was completed

fluctuations. However, Fig. 3 reveals a general trend of increasing production under Jacobo and Juan peaking at the end of their careers (1528 and 1540 respectively), while decline set in under Jácome. The number of sheets printed is a more reliable guide to levels of production, but the number of editions is, perhaps, more readily appreciated: the average number of surviving editions produced each year was over ten, while fifteen was not unusual (Fig. 4).[8] This means that the Cromberger press was prolific by Spanish standards, even if the deaths of Jacobo and Juan were reflected in temporary slumps in production (measured by sheets) in 1529 and 1541.[9] Nevertheless, it was not immune to the general vulnerability of presses to external events.[10] Fig. 3 records the drop in production during plague years at Seville (1507–8 and 1524), during the war with France in the early 1520s—a slump which is closely parallelled for similar reasons in Paris and Lyons[11]—and at the time of the implementation of damaging fiscal measures by the crown and the crisis in the Spanish economy (the early 1550s).

There is some evidence that production also fluctuated seasonally. Approximately half the surviving editions indicate the month in which printing was completed (see Fig. 5). If these editions are representative of the press's total output, a particularly interesting conclusion can be drawn from them. It has already been noted that Jacobo and Juan attended the major Spanish fairs for printers and booksellers which were held at Medina del Campo in May and October. In both months the Crombergers' output decreased, but just before the fairs, in March, April, and September, there was a burst of activity which could be explained by their hurrying to complete editions intended for sale or barter at Medina. It should also be remembered that in the spring, winds were favourable for travel to the New World; the usual season for setting sail on this route was just after the press's most concentrated period of activity.[12] A major problem

[8] Fig. 4 shows how unreliable statistics based upon numbers of editions rather than sheets are. The apparently large production of the years 1510–20 is mainly made up of *pliegos sueltos* containing a single sheet. Fig. 3 provides a corrective.

[9] The prolific output of the Cromberger press, at least by Spanish standards, can be appreciated when compared with the combined output of the Valladolid presses during the longest period of the court's residence in that city (1554–9) when it enjoyed an economic boom. Its presses produced only about seven known editions each year. Nevertheless, this is still a high figure in comparison with the output of most Spanish cities (see Bartolomé Bennassar, *Valladolid au siècle d'or: une ville de Castille et sa campagne au XVIe siècle*, Civilisations et Sociétés, 4 [Paris and The Hague, 1967], p. 516).

[10] See Febvre and Martin, *L'Apparition*, pp. 172–3.

[11] Martin and Chartier, *Le Livre conquérant*, pp. 219, 443.

[12] Shipping was to some extent erratic, however, during the first half of the 16th c. It was not until the early 1560s that merchant ships bound for the New World were organized into armed convoys. The fleet bound for Vera Cruz then sailed, in theory, in May. Although the spring voyages to the Indies could have contributed to the high production of books in March and April, it is difficult to account for the press's high level of activity in June. It must be admitted, however, that the 'spring' fleet often sailed two

faced by a sixteenth-century publisher was that of ensuring that a network of outlets would enable him to dispose as rapidly as possible of the books in which he had tied up his capital.[13] Seasonal production for fairs and the Americas may have been the Crombergers' response to that problem.

Prices

Much of the debate about who could or did read in sixteenth-century Spain hinges on two questions: how many people were literate and how many could afford a printed book? Although many inventories of nobles' libraries have now been published, the lack of any systematic study of inventories of the estates left by deceased members of other social classes in Spain means that only vague hypotheses about literacy in that country are possible.[14] Unfortunately, little useful light is shed upon the other thorny problem—the cost of books—by a study of the Crombergers, who were printing before officially fixed prices became the norm in Spain. What is clear, however, is that any confident statement made about book prices at the time is likely to be misleading. For example, Eisenberg uses Colón's *Registrum B* in his attempt to discover the average price of a romance of chivalry in the early sixteenth century.[15] Colón's entries, useful though they are for other aspects of bibliographical study, are misleading if used ingenuously to such an end. The reasons for this are several. Firstly, Colón often fails to say whether the books he bought were bound or still in quires. The cost of binding varied according to the style ordered but, as it could be more expensive than the printed sheets themselves, this is a serious omission.[16] Secondly, his books were frequently purchased in cities other than the one in which they were printed; it is impossible to calculate how much of the price he paid reflected transport costs. Finally, his books were sometimes bought long after they were printed; they could therefore be either rare and valuable items or virtually worthless remainders. For example, in November 1524 he bought two books at Valladolid which had been printed by Jacobo Cromberger in 1517: Andrés de Li's *Tesoro de la Pasión* (43 sheets) and the *Libro de medicina llamado compendio de la salud humana*

months late (see Ernesto Schäfer, 'La Universidad de los Mareantes de Sevilla', *Archivo hispalense*, 2nd ser., 5, No. 14 [1945], 271–85).

[13] Febvre and Martin, *L'Apparition*, p. 326.

[14] Cf. Carlo M. Cipolla, *Literacy and Development in the West* (Harmondsworth, 1969), pp. 60–1. Although there has recently been much research undertaken on the problem of literacy in modern and early modern Europe, there is still little information available on Spain. Conclusions about the problem in other European countries tend to be both conjectural and predictable (see Rab Houston, 'Literacy and Society in the West, 1500–1850', *Social History*, 8 [1983], 269–93).

[15] Daniel Eisenberg, *Romances of Chivalry in the Spanish Golden Age* (Newark, 1982), pp. 99–100.

[16] Irmgard Bezzel, 'Bartholomäus May (ca.1515–1576) aus Bern, ein Sammler spanischer Drucke', *Iberoromania*, 1 (1969), 235–98 (p. 242).

(33.5 sheets). While the shorter book cost him 80 *mrs*, the longer one cost a mere 8 *mrs*. In neither case do we know whether the copy was bound.[17] Again in 1524, Colón bought a copy of Cromberger's 1519 edition of Fernández de Enciso's *Suma de geografía* (38 sheets), this time at Medina. The licence printed in this edition fixed the retail price at 136 *mrs*; Colón paid only 76 *mrs*.[18] It should be clear from these examples that the *Registrum B* cannot be used as a basis for generalizations about book prices.

Another source for prices employed by Eisenberg is the 1528 inventory of the Crombergers' stock. The unbound items in this list were valued at one *mr* per sheet, irrespective of the quality of the paper or the nature of the book. But this inventory was drawn up exclusively to enable Jacobo's two heirs to divide his estate equably, and the values given to the sheets probably bear only a tenuous relation to retail prices. Only one year earlier, in February 1527, Varela had charged the local authorities 81,600 *mrs* to print 300 copies of the Seville *Ordenanzas*.[19] As this was a folio book containing 260 leaves, the cost even to the publishers was over 2 *mrs* a sheet. It is unlikely that the Crombergers would have thought of selling their books at only half this figure.

The contract with Varela is typical of those that survive for Seville printers of the period; they are agreements drawn up with booksellers, merchant-booksellers, authors, or institutions and therefore do not mention retail prices. Even where a price is printed in the preliminaries or at the end of the book, it is unreliable as a general guide because some institutions subsidized the purchaser of a particular title while, on the other hand, the holder of a privilege was sometimes allowed to charge an inflated price by the authority granting the privilege.[20]

The evidence available indicates that prices were erractic. Montesdoca's editions of Valtanás's works sold for sums ranging from 0.5 *mr* per sheet to seven times that amount at the very time in the 1550s when the Seville chapter was paying the same printer 5 *mrs* per sheet for a simple book he printed for them;[21] and at the beginning of the century two books of similar size, quality, and typographical characteristics were priced in Seville in a ratio of one to three.[22] The problem of finding out about prices and judging how accessible books were to different sorts of readers is compounded by the fact that wages and prices in general were rising throughout the period during which the Crombergers were working, but there is still no adequate study of the relationship between them, while evidence for book prices is sporadic and cannot be systematically compared with either.[23] Even the

[17] *Catalogue of the Library of Ferdinand Columbus*, Nos. 4038 and 4017.

[18] *Catalogue of the Library of Ferdinand Columbus*, No. 4092.

[19] Hazañas, i. 237–40.

[20] Norton, *Printing in Spain*, pp. 121–2.

[21] Klaus Wagner, *Martín de Montesdoca*, pp. 25, 38.

[22] Klaus Wagner, 'El negocio', pp. 71–2.

[23] Valentín Vázquez de Prada, *Los siglos XVI y XVII* (Madrid, 1978), vol. iii of id. (ed.), *Historia económica y social de España* (Madrid, 1973–), pp. 716–18.

prices we do know about can be puzzling. For example, Brocar received 2 *mrs* per sheet in a *tasa* dated 1511; his son was given only the same amount, despite inflation, twenty-seven years later.[24] Similarly, the prices of second-hand books auctioned in Seville during the first half of the sixteenth century are erractic: some were very cheap and would have been within the reach of any wage-earner; others were expensive even for new books.[25] In all these cases, the equation contains too many unknowns for us to be able to say anything more precise than that chap-books were cheap and other publications more expensive. This is to say little indeed, but serves as a warning against the headier claims that printing automatically brought the written word within the grasp of anybody who desired to purchase it.

Printing Agreements

Although contracts between Seville printers and authors or publishers can be found in some number, few survive for the Crombergers. Evidence for their printing agreements therefore has to come largely from a study of the books themselves and especially of their colophons. While in some countries at this time the name of the publisher would frequently appear with no note of the printer, if a Spanish colophon contained any information in addition to the date and place of printing, it was the identity of the printer. The publisher, if he was not the printer, would only sometimes be mentioned in addition in the colophon or elsewhere in the book. In sixteenth-century Spain various sorts of agreement were made.[26] An author could publish his work himself, merely paying the printer's labour and expenses. Sometimes the author would supply the printer with paper or give him an advance to cover its purchase. An example of publication by an author is Juan Cromberger's edition of Fernández de Oviedo's *Historia general de las Indias* (1535), which contains a letter from the author stating that the book was printed entirely at his own expense. Under such an agreement the author would normally stipulate the number of copies to be printed, the types to be employed, the format, the quality of paper, and , within certain limits, the timetable for printing. Security in the printing-shop would be demanded if the work was as yet unpublished, because there was always a risk that finished sheets would find their way into the hands of a pirate who would rapidly print the book and thus undercut the author's own edition. Under the legitimate contract the author would take delivery of all copies when printing was completed.[27] A rich man like Fernández de Oviedo

[24] Beltrán de Heredia, *Cartulario*, ii. 381, v. 445.

[25] Information from various auction documents and inventories in APS.

[26] Agustín G. de Amezúa y Mayo, *Cómo se hacía un libro en nuestro Siglo de Oro* (Madrid, 1946), pp. 17–30.

[27] Klaus Wagner, *Martín de Montesdoca*, p. 108.

could finance a handsome edition himself; a poorer author might turn to a patron for help with the cost of printing. However, despite the plethora of sycophantic dedications to be found in sixteenth-century books, patrons only very rarely financed editions.

An institution could contract for the printing of an edition, just as the *provisor* of Seville cathedral did in 1554 when Jácome Cromberger agreed to print certain liturgical books for the diocese. Here again an advance would often be given to the printer. But not all liturgical printing was arranged in this manner; the examination of the colophons of Cromberger editions reveals that, as was seen in Chapter Two, Jacobo printed the León missal in 1504 for a Genoese bookseller of Seville, Lázaro de Gazanis, and in 1528 printed two liturgical books for the Évora bookseller, António Lermet. These would have been commissioned by the booksellers, and doubtless no risks were taken by Cromberger, who was just paid for his labour and expenses.

Such arrangements were not limited to liturgical books. Colophons again show that in 1504 Jacobo printed an edition of Persius' *Satires* for the local bookseller, Juan Lorenzo, and in the previous year he had printed another edition, only part of which was commissioned by the same man. The remainder was presumably published by Cromberger and his partner at that time, Polono.[28] Likewise, authors and printers sometimes shared an edition.

Authors frequently obtained a royal privilege on the printing of their works which was valid in a particular geographical area and for a certain number of years. The author regularly sold this privilege to a publisher, often himself a bookseller. As the Crombergers combined the roles of printer, publisher, and bookseller, they acquired such privileges, as is witnessed by Juan's purchase from Guevara in 1529 of the ten-year privilege on two of the bishop's works for the large sum of 175 *ducados*. However, this isolated example should not mislead us; an author could not expect to live from his pen at this period.

A somewhat different case, that of a translator who held a privilege which he probably gave to Jacobo when he changed printers, is recorded by Norton.[29] On occasion the Crombergers sought leave from the privilege-holder to print a single edition of a work, as happened with Jacobo's *c.*1528 edition of Erasmus' *Enchiridion*. Although in this case they paid 12 *ducados* for the permission, they were not averse to infringing privileges in a most blatant manner. As we have seen, at least one author took them to court for such an abuse.[30] They, in turn, appear to have been the victims of sharp practice by another printer.[31] Most of

[28] Norton, *Printing in Spain*, pp. 133–4. Printers were sometimes paid by the publisher in copies of an edition; publishers could also pay authors in the same way.

[29] Norton, *Printing in Spain*, p. 123.

[30] See above, pp. 68, 76–7.

[31] Clive Griffin, 'More on the Spanish *Scala celi*', *The Library*, 6th ser., 5 (1983), 385–91.

the titles printed by the press were not, however, covered by a privilege, and most of its editions were not printed under contract from a publisher. The books which the Crombergers issued were mainly financed by them and published speculatively.

Far from being dependent upon contract work, Jacobo Cromberger may have commissioned other presses to work for him. Norton suggests that part of Varela's production was printed on his behalf, and the recently discovered edition of Nebrija's *Introductiones Latinae*, probably printed by Varela but definitely employing material found exclusively in the Crombergers' workshop, could be evidence for such an arrangement.[32] Moreover, there is the strange case of the Seville breviary printed in Paris at a date when Jacobo possibly had a monopoly on the production of such books; and the 1526 Spanish-language edition of St Jerome's letters printed at Valencia by Jorge Costilla but bought up by the Crombergers.[33] It has been seen that the Crombergers were always ready to diversify their commercial interests, and it would appear that their arrangements for printing their editions were no less diverse. However, there is no evidence that they regularly contracted other offices to print editions which they wished to publish. There is no parallel at this time in Seville to the situation in a city like Lyons, where rich merchant-publishers farmed work out to a host of printers who worked for them under contract.[34]

In this chapter it has been seen that, even where there is abundant archival material which can be drawn upon in the study of a press, the books themselves can never be far away from our minds. They can, indeed, complement or provide a substitute for documentary evidence, but they do, of course, have a more important intrinsic interest of their own. Part Two will therefore focus on the Crombergers' books themselves, firstly investigating what titles were printed and how they can provide the basis for a study of contemporary reading-habits, and secondly, in two technical chapters, examining the printing material which was used in them.

32 Norton, *Printing in Spain*, pp. 133–4. See the description of this edition (dated to '1513?') in App. I.

33 See above, p. 51, n. 126, and p. 67.

34 Martin and Chartier, *Le Livre conquérant*, pp. 259–60.

PART TWO

The Crombergers' Books

CHAPTER 6

THE CROMBERGERS' TITLES

THIS chapter is a survey of the 557 known editions which the Crombergers issued. Books are only published commercially when it is thought that there will be readers enough to buy them and when a publisher sees a demand, or potential demand, which is not being met. A major press's production therefore supplies the basic information for an objective study of contemporary reading-habits.[1] It may provide us with a somewhat blunt instrument, but it is one which can be instructive if only to challenge anachronistic assumptions about what readers were able or eager to buy in a particular place at a particular time.

The Cromberger press operated in a city where patronage for printers from individuals or academic institutions was almost unknown.[2] Its production was mainly unsubsidized and aimed at a potential market of readers willing and able to pay for books. The Crombergers also dominated printing in the city for the best part of the first half of the century. For these reasons, their output provides a good guide to the predilections of ordinary readers in Seville and, indeed, further afield, for it should be remembered that Seville was at that time the major centre of Castilian book-production. Some general comments have been made in the previous four chapters about the sort of books printed by the press, and about the changes of emphasis in production from one generation of the Crombergers to the next. The more detailed examination undertaken in this chapter divides the books into five categories: (*a*) devotional and spiritual works, (*b*) literature in the vernacular, (*c*) history in the vernacular, (*d*) utilitarian works, and (*e*) miscellaneous books. Not all the titles printed are mentioned individually in this survey, which discusses the whole production in broad categories, but all have

[1] Rudolph Hirsch, *Printing, Selling and Reading 1450–1550* (Wiesbaden, 1967), p. 125. The disappearance of whole editions may, of course, have distorted the picture. Uneven loss is suggested by the fact that 46 per cent of known Cromberger editions are in folio, 38 per cent in quarto, and only 13 per cent in octavo (the format of 3 per cent of editions is unknown). Folio volumes survive better than the smaller formats, which tended to be used for more popular and cheaper books. It does appear, however, that the octavo format was not used very frequently by Jacobo, although it became more popular as time went on. For a good summary of the sort of problems arising from the strict criteria I have employed for the existence of Cromberger editions, see Keith Whinnom, 'The Problem of the "Best-Seller" in Spanish Golden-Age Literature', *Bulletin of Hispanic Studies*, 52 (1980), 189–98 (pp. 191–2). It should also again be remembered that Spanish printers did not supply the whole of their domestic market; many books were imported.

[2] Eugenio Asensio in Desiderius Erasmus, *Tratado del Niño Jesús . . . con un estudio preliminar de Eugenio Asensio* (Madrid, 1969), pp. 9–10.

TABLE IV.

Statistical analysis of recorded editions printed by the Cromberger press

	editions		sheets		format				% of reprints
	no.	%	no. in one copy	%	F°	4°	8°	unknown	
Devotional/Spiritual	158	28.4	6,918	32.5	54	46	56	2	51
Romances of chivalry	39	7.0	3,473	16.3	39	—	—	—	51
Morality/philosophy	50	9.0	2,078	9.8	32	16	2	—	70
Other secular literature	71	12.7	1,457	6.8	19	51	—	1	51
Pliegos sueltos etc.	69	12.3	60	0.3	9	59	—	1	12
History	31	5.6	1,462	6.9	27	4	—	—	55
School books/classics	22	3.9	763	3.6	6	15	1	—	14
Liturgy	40	7.2	2,512	11.8	16	3	9	12	25
Law	19	3.4	756	3.6	19	—	—	—	53
Medicine	23	4.1	730	3.4	20	3	—	—	57
Almanacs, etc.	7	1.3	48	0.2	—	2	5	—	43
Theology	12	2.2	689	3.2	6	6	—	—	0
Travel/geography	14	2.5	293	1.4	8	6	—	—	29
Current debate	2	0.4	33	0.2	—	2	—	—	0
Totals	557	100	21,272	100	255	213	73	16	—

In Castilian: editions 484 (86.9%); sheets 16,977 (79.8%)
In Latin: editions 70 (12.6%); sheets 3,766 (17.7%)
In Portuguese: editions 3 (0.5%); sheets 529 (2.5%)

been taken into account when calculating the figures provided. Percentages are given first in terms of the numbers of editions printed and then (after a colon) in terms of the number of sheets used in a single copy of each edition (see Table IV). It is important to consider both figures because the number of editions, taken alone, can give a misleading impression of the press's production. A work containing five hundred folio leaves would not be distinguished from a simple quarto chap-book consisting of a single sheet. Likewise, 12.3 per cent of the surviving Cromberger editions are *pliegos sueltos*, but a calculation based on sheets gives a total of only 0.3 per cent for the output of such chap-books. Very little evidence about the number of copies in individual editions is available; the size of editions is therefore excluded from all calculations. This is a serious but inevitable shortcoming.

Devotional and Spiritual Works

'Religious' works written in, or translated into, the vernacular account for 28.4%:32.5% of the press's surviving production. Only a small number of these editions were manuals designed for the clergy: for example, guides for confessors, the vernacular *Speculum fratrum minorum*, or Fray Juan de Argomanas's *Tratado muy provechoso*. The majority were printed for secular readers. Just as Fray Luis de Granada's devotional *Libro de oración* was to become the best-seller of later sixteenth-century Spain, eclipsing in popularity the sales of all purely literary works (*libros de entretenimiento*, or 'books of entertainment', as they were known) of the period, so did devotional and mystical works capture the attention of the mass of the reading public during the first half of the century. The printing of 'religious' works has always been a staple of the presses, but this period was one of particular spiritual ferment in Spain.[3] Cardinal Cisneros's attempts to reform and raise the level of learning of the Spanish clergy in the late fifteenth and early sixteenth centuries had several consequences. He encouraged the influx of new ideas into the country and fostered both educational institutions and their handmaiden, the printing-press; the religious orders were set on the road to reform;[4] and in the 1520s the ideas of Erasmus, spread by vernacular editions of his works, caught the popular imagination in Spain to an extent unequalled anywhere else in Europe. The private reading of devotional literature (and this included reading aloud to the illiterate) was actively encouraged.

In the first half of the century Seville was the major centre of this spiritual renewal and of the production of the vernacular books which would broadcast it to the rest of the country. From its presses came the earliest works of Castilian mysticism, the first translation of Erasmus, and a host of books which would later be proscribed by the Inquisition.[5] It would be an exaggeration, however, to claim that the Crombergers had any programme of publishing mystic, reformist, or Erasmian works.[6] Rather, as one historian has neatly expressed it, 'the business

[3] The dominance of 'religious' editions in Seville printers' output is not a peculiarly 16th-c. phenomenon (see Wilson and Cruickshank, *Samuel Pepys's Spanish Plays*, pp. 29, 38).

[4] Paul J. Donnelly, 'A Study of Spanish Translations of Erasmus's *Colloquia* (1525–1536), with Special Reference to the Translations of Alonso de Virués: Together with a Critical Edition of the Versions of *Uxor Mempsigamos, Funus*, and Part of *Convivium Religiosum*' (D.Phil thesis, Oxford, 1979), 2 vols., i. 1–9.

[5] Asensio in Erasmus, *Tratado*, pp. 9–10; also see above, Ch. 4, n. 109.

[6] Such programmatic printing was not unknown at Seville (see Klaus Wagner, *Martín de Montesdoca*, pp. 37–9).

of souls was also the soul of many printers' business'.[7] Nevertheless, as the leading printers of Seville and specialists in vernacular editions, the Crombergers played an important role in the dissemination of these spiritual ideas.

A good deal of their devotional output consisted of translations of works popular everywhere in the Middle Ages: Thomas à Kempis's *Imitatio Christi*, for which evidence of at least six Cromberger editions survives; the *Meditaciones* and *Doctrina cristiana*, works attributed to St Augustine and translated into Castilian; St Jerome's letters; and translations of St Gregory's dialogues and his *Moralia*. The pietistic *Imitatio* was complemented by Spanish works like Jiménez de Prejano's immensely popular *Lucero de la vida cristiana* and Jiménez de Cisneros's *Ejercitatorio de la vida espiritual*. Frequently published despite its size, and therefore its high price, was the *Vita Christi* of Ludolph of Saxony, which had been translated into Castilian by Fray Ambrosio Montesino at the turn of the century. The Crombergers issued no fewer than seventeen editions of its four separate parts, indicating that, as elsewhere in Europe, it was a particularly widely read work. The *Vita Christi cartujano*, as it was known in Spain, had been composed in the fourteenth century, but its pietistic nature made it particularly appealing to Christians in the late fifteenth and early sixteenth centuries. Its translation into the vernacular also made the Gospels widely accessible to lay readers. Ludolph's work was the inspiration for Fray Íñigo de Mendoza's *Coplas de vita Christi*, Montesino's own original verse, and the *Retablo de la vida de Cristo* composed by the Seville Carthusian, Juan de Padilla. All of these were issued by Jacobo Cromberger, who not only printed the first edition of the *Retablo* but whose press also issued four reprints. Montesino went on to revise the Castilian *Epístolas y evangelios*, which was frequently reprinted by the family. This work made the Gospels available to Spanish readers in a less cluttered form than in the *Vita Christi cartujano*, where the Scriptures were buried among Ludolph's paraphrases and prayers. Printing therefore created a new situation for secular readers in Spain: the Scriptures, patristic works—particularly the more mystic ones—and mystic treatises themselves were now readily available in the vernacular. These books prepared the soil for the reformist ideas which blossomed in Spain, and especially in Seville, in the first half of the century.[8]

Nevertheless, not all the Crombergers' devotional publications fit conveniently into a mystic or a reformist mould. For example, they produced two editions of Clemente Sánchez de Vercial's early fifteenth-century *Sacramental* and other thoroughly traditional works including numerous editions of books of hours—often the only books to be found in the inventories of deceased individuals at Seville. The press printed at least one edition of a hagiographical *Flos sanctorum*

[7] Asensio in Erasmus, *Tratado*, p. 31.
[8] Bataillon, *Erasmo y España*, pp. 44–8.

and many thousands of indulgences. Jacobo even printed works like the little *Manual breve para informarse a devoción los cristianos menos sabidos*, which contained prayers guaranteeing those who recited them twelve thousand years' forgiveness for their transgressions. However, we should not be surprised that readers could be interested both in reformist and traditional works of devotion; trends in spirituality which are evident to those looking at the period with the benefit of hindsight were far less obvious to contemporaries.[9] Nevertheless, the bulk of the press's devotional output does reflect the spiritual ferment of contemporary Seville. Nor were the first two generations of Crombergers mere popularizers of works already printed by others. If Jacobo had prepared the soil by his early publications, Juan sowed the seed with his first editions of notable spiritual works. Important for their emphasis upon the inner life and their subsequent influence upon mystic writers were the Franciscans Bernardino de Laredo and Francisco de Osuna. In 1528 and 1530 Juan issued the first editions of Osuna's first and fourth *Abecedarios* (the third would be Santa Teresa de Ávila's favourite book), and in 1530 the first edition of his *Gracioso convite de las gracias del santo sacramento del altar* which was later banned by Fernando de Valdés. In 1535 the first edition of Laredo's *Subida del Monte Sión*, so important for Santa Teresa's spiritual progress, came from Juan's presses. Three years later Juan's friend, Fray Bernardino, entrusted him with the revised version of the work.

After Juan's death his widow, Brígida Maldonado, supervised an impressive series of reformist publications. In 1542 she issued Juan de Cazalla's *Lumbre del alma*, and in the following year she printed the first edition of the *Suma de doctrina cristiana* by Constantino Ponce de la Fuente, a canon of the cathedral and royal confessor who later fell foul of the Holy Office and ended his days in an Inquisitional gaol.[10] She appears to have gained an imperial privilege on this enormously popular work, and four months later, in 1544, issued a second edition. When the Crombergers' associate, the Erasmian Bishop Zumárraga, published a *Doctrina* on their Mexican press some two years later, he reproduced Constantino's *Suma* almost without alteration.[11] Kinder notes that the *Suma* presents a remarkable affinity with Luther's teachings, and much the same is said

[9] Asensio, 'El erasmismo', pp. 98–9.

[10] Juan de Cazalla, who was a close collaborator in Cisneros's reforms, sympathized with Erasmus' ideas and was in contact with Lefèvre d'Étaples as well as with groups of Spanish illuminists who appear to have been influenced by the *Lumbre*. The Inquisitional trial of his sister María was notorious; his nephew Agustín was burnt at the Valladolid *auto de fe* of 1558; and he himself was probably tried by the Inquisition (see Juan de Cazalla, *Lumbre del alma*, ed. with an introduction by J. Martínez de Bujanda [Madrid, 1974], pp. 30–3). In 1553 the Inquisitor-General, Valdés, ordered Cervantes de Gaete, whom we have met before, to make a private investigation into Constantino's theological opinions. For information on Constantino's *Suma* see Bataillon, *Erasmo y España*, pp. 535–41, and the notes to the description of the first edition in my App. I.

[11] Bataillon, *Erasmo y España*, pp. 825–6.

by Bataillon of the works of Fray Domingo de Valtanás who was, as we have seen, later to be arrested by the Inquisition.[12] Brígida probably printed the first edition of Valtanás's *Doctrina cristiana* at this time. Shortly before, in 1544, she reprinted his *Confesionario* which her husband had issued six years earlier.

Bataillon notes the Erasmian influence on Constantino's teachings and suggests that the success of the *Suma* may have lain behind a renewed interest among Seville printers in issuing editions of Erasmus at a time when the publication of his works in Spain was generally in decline. In particular, the rapid popularity of her two editions of the *Suma* may have persuaded Brígida to reprint a Castilian translation of the Dutchman's *Lingua* in 1543 or 1544.[13] However, the Crombergers had long been familiar with Erasmus and the Erasmianism of that group of humanist scholars who were so closely associated with Seville cathedral. Jacobo, for example, had printed many editions of works edited by the minor humanist and canon of the cathedral, Pedro Núñez Delgado, who was a successor of Nebrija in the Chair of Latin at the Estudio de San Miguel. In 1528 a neo-Latin poem composed by Delgado was appended to Cromberger's important edition of the *Pharsalia* which he printed in the style of Aldus Manutius.[14] In a posthumous collection of Delgado's epigrams which appeared in 1537, Cristóbal Núñez, a young member of the chapter, included Delgado among the humanist scholars who, he remarked significantly, formed the 'Hispalensis academia' or 'Seville Academy'.[15] Notable members of this circle were the Italian-trained founder of the Colegio de Santa María de Jesús, Rodrigo Fernández de Santa Ella—whose works were printed by Jacobo—the Erasmian archdeacon, Diego López de Cortegana, and Christopher Columbus's friend, Jerónimo Pinelo. This 'academia' was one of the foremost humanist and Erasmian circles in Spain, noted for editing Latin texts for students and making translations for a readership wider than that reached by the more scholarly Latin editions produced at Alcalá.[16] Its members appear to have been in close contact with local printers. Until 1554 the granting of licences for permission to print in Seville fell within the jurisdiction of the archbishop. Diego de Deza and his successor, the powerful patron of humanists and Erasmians, Alonso Manrique, appointed Santa Ella, Pinelo, and Constantino as his censors. These appointments resulted in a certain amount of

[12] Kinder, *Casiodoro de Reina*, p. 8; Bataillon, *Erasmo y España*, p. 544.

[13] Bataillon, *Erasmo y España*, pp. 547–8. If the date of the Cromberger edition of the *Lingua* in translation is in Nativity style, as seems probable, it was completed on 29 Dec. 1543 (22 days after the *Suma*) and not, as Bataillon assumes, on 29 Dec. 1544 (for problems of dating see the Introduction to App. I).

[14] Griffin, 'The Crombergers of Seville', pp. 69–96.

[15] Prologue by Cristóbal Núñez to his edition of Núñez Delgado's *Epigrammata* (s.l. [but recognizably not Seville], 1537). The term 'Hispalensis academia' clearly enjoyed some currency at the time. Núñez Delgado himself used it of Jerónimo Pinelo, whom he dubbed the 'hispalensis academiae praesul' (see Norton, *A Descriptive Catalogue*, No. 748).

[16] Asensio in Erasmus, *Tratado*, p. 21.

day-to-day contact between these scholars and the printers, and this contact can only have been increased by Jacobo's monopoly of liturgical printing for the diocese. Indeed, in 1520 Archbishop Deza had appointed López de Cortegana as *corrector* of the missal which Jacobo finished printing in December of that year.[17] In the following year Deza again appointed the archdeacon to supervise Cromberger's edition of the Seville breviary, which was finished in July. Jacobo had printed Cortegana's translation of Apuleius' *Metamorphoses* (better known in English as *The Golden Ass*) in about 1513, as well as his edition of the *Crónica del santo rey don Fernando tercero* (1516); in 1520 he not only printed the first edition of the Spanish translation of Barthema's *Itinerario*, a work dedicated to Cortegana by the young colleague in the chapter whom the archdeacon had asked to translate it, but also Cortegana's own translation of Erasmus' *Querela pacis*.

It was seen in Chapter Two that Jacobo's business contacts with Delgado were not restricted to printing. Cortegana likewise knew the printing families well, for he was the godfather of Varela's daughter, Isabel de Alfaro, who became Jácome Cromberger's sister-in-law. The Crombergers were not so closely identified with Erasmianism as was Miguel de Eguía at Alcalá, but they nevertheless printed an impressive list of Erasmus' works in translation: *Tratado o sermón del niño Jesu y en loor del estado de niñez* (1516 [the first edition of the first vernacular translation of Erasmus printed anywhere] and 1528), *Querella de la paz* (1520), *Enchiridion* (1528?), *Coloquios* (1529), and *Lengua* (1533 [possibly the first edition of the Castilian translation], 1535, 1542, and 1543 or 1544). In addition there were the works written by Spanish authors who were influenced by the Dutch humanist's writings, and even texts edited by Erasmus himself like Flavius Josephus' *Jewish War*, which the press issued in translation at least twice, in 1532 and 1536. The inventories of the stock of the family bookshop reveal that it contained a large number of copies of works by Erasmus, but the Crombergers also possessed a series of single copies.[18] There will, admittedly, often be a moment when a bookseller has only one copy left of a title, but these single copies may equally indicate that some of the books formed part of the printers' private library. This is especially true of the 1540 inventory, where the contents of one chest were almost exclusively single copies, among which were to be found books by Erasmus and Vives.

We know nothing certain about the Crombergers' own tastes or beliefs, and it would be imprudent to attribute to them any ideological zeal. Nevertheless, their

[17] Joaquín Hazañas y la Rúa, *Maese Rodrigo 1444–1509* (Seville, 1909), p. 275.

[18] Gestoso, pp. 36–41, 86–98. The Crombergers' stock was not, however, untypical of that of other Seville booksellers of this period. The 1532 inventory of the books held by Sebastiano de Labezaris contained multiple copies of many editions of Erasmus' works (see Klaus Wagner, 'Erasmo en el comercio librero sevillano', in Gerhard Schmidt and Manfred Tietz [eds.], *Stimmen der Romania: Festschrift für W. Theodor Elwert zum 70. Geburtstag* [Wiesbaden, 1980], pp. 451–6).

enormous output of devotional editions in the vernacular, and particularly of Erasmian and reformist works, allows us to conclude that at the very least they recognized a demand for such books. What can be stated with confidence is that they were instrumental in the diffusion of Erasmian and reformist ideas in Seville and elsewhere in Spain.

Vernacular Literature

Vernacular editions of secular literature account for a large proportion of the Crombergers' surviving products (41%:33.2%), but this category is necessarily very wide and includes (i) romances of chivalry (7%:16.3%); (ii) compendia of encyclopaedic knowledge and works which could be loosely described as 'moral' or 'philosophical' (9%:9.8%); (iii) other prose fiction, secular poetry, and drama (12.7%:6.8%); and (iv) *pliegos sueltos* (12.3%:0.3%).

(i) Romances of Chivalry

The romances of chivalry have been separated from other literature of entertainment in order to highlight their importance in the press's output. The Crombergers specialized in these works, whose extraordinary popularity in the sixteenth century not only in Spain but in Europe generally is well known. They were archaic in tone, in their ritualistic appeal to their readership, and even in their language. Works like the Arthurian *Tristán de Leonís, El caballero Cifar* (if this can legitimately be termed a romance of chivalry), and the most famous of them all, the first four books of *Amadís de Gaula*, were medieval in origin, but many others were actually written in Spain or Portugal in the sixteenth century. The Crombergers printed a wide range of both native romances of chivalry and translations into Castilian. They all appeared in the press's characteristic folio volumes printed in gothic types and announced by a monumental title-page, and most were illustrated.[19] The Crombergers issued almost the whole Amadís cycle, which finally consisted of twelve books. They also frequently reprinted many of the volumes including seven surviving editions of the original first four books. They reprinted other romances of chivalry (*Espejo de caballerías, Palmerín de Oliva, Lepolemo*, and *Renaldos de Montalbán*) several times, though only one edition each is known of *Cirongilio de Tracia, Clarián de Landanís, Lisuarte de Grecia*—the eighth book of Amadís—and *Primaleón*.

Nevertheless, although the Crombergers were the most famous printers of these popular books, they in fact risked very few first editions. Only their 1526 edition

[19] Following Eisenberg, *Romances of Chivalry*, p. 89, I have not included the short French works of the 15th c. or earlier—*Oliveros de Castilla, Partinuplés*, and *Enrique, hijo de Oliva*—as romances of chivalry. They may have been particularly popular as children's literature. For the importance of the

of the eighth book of Amadís and their 1545 *Cirongilio de Tracia* were *principes* and it appears that neither was successful, for they were never reprinted by the Crombergers or anybody else.[20] Indeed, in their edition of the ninth book of Amadís, the author, Feliciano de Silva, testily condemned the eighth book as unworthy of inclusion in the cycle.[21]

The readership of these volumes has been the subject of much debate.[22] Eisenberg is doubtless correct to say that the market for them was not composed of peasants, but then few books are printed anywhere for the illiterate. However, his conclusion, based in part on economic grounds, that their readers were almost exclusively nobles or what he calls the 'upper classes' is unproven. Contrary to his assertion, these books were not prohibitively expensive, nor were they priced, at least by the Crombergers, any differently from their other books.[23]

(ii) Morality and Philosophy

Unless uneven loss of editions has distorted the evidence, the Crombergers' surviving output indicates that the production (in terms of sheets) of works of morality, 'philosophy', and sententious compendia outweighed that of the sort of imaginative fiction and verse normally studied today (if we except the generally neglected romances of chivalry). Most of these moralistic works were either medieval in fact or medieval in tone. For example, Aesop's fables and the exemplary tales attributed to Bidpai were widely read thoughout Europe during the Middle Ages. The Crombergers issued editions of both. Further instances of traditional exemplum literature published by the press are the *Siete sabios de Roma*, which may have been read mainly by children, and the *Historia de la doncella Teodor*. Likewise, the *Proverbios* erroneously attributed to Seneca were evidently still much in demand. Seneca was, of course, considered in Spain to be a Spaniard; his works figure prominently in fifteenth-century libraries there.[24] The many

Seville presses (and particularly that of the Crombergers) in the production of romances of chivalry, see Pierre Chaunu, 'Les Romans de chevalerie et la conquête du Nouveau Monde', *Annales: économies, sociétés, civilisations*, 10 (1955), 216–28.

[20] Cervantes had, however, read *Cirongilio de Tracia*, doubtless in the Cromberger edition (see *Don Quixote*, Part I, Ch. 32).

[21] Feliciano de Silva, *El noueno libro de Amadis d[e] Gaula que es la cronica de . . . Amadis de Grecia* (Seville, 1542), fol. $+2^b$.

[22] Eisenberg, *Romances of Chivalry*, pp. 89–110.

[23] Colón's *Registrum B* shows that these romances (possibly already bound and therefore more expensive) cost about the equivalent of three days' wages of a labourer or a barber-surgeon's fee for extracting three teeth (but see above, pp. 137–9). Eisenberg is incorrect in thinking that Colón's copies cost him any more than other books of a similar size. The Crombergers' 1528 stocklist shows that they valued the romances of chivalry at exactly the same price as their other books: one *maravedí* per printed sheet. Eisenberg is also wrong to suppose that the romances were always carefully printed. Chevalier, *Lectura y lectores,* pp. 65–103, coincides with Eisenberg's views, but his arguments are highly speculative.

[24] Ladero Quesada and Quintanilla Raso. 'Bibliotecas de la alta nobleza castellana', pp. 50, 51, 54, 59; and Quintanilla Raso, 'La biblioteca del marqués de Priego (1518)', pp. 24, 80, 90.

editions of a Castilian translation of Boethius' influential *De consolatione philosophiae* and Petrarch's *De remediis utriusque fortunae*, together with *Bias contra Fortuna* and the *Proverbios*—both Castilian works written by, or at least under the name of, the fifteenth-century nobleman, Íñigo López de Mendoza, Marquis of Santillana—again testify to the enormous popularity of consolatory, moralistic, and sententious writings in Spain throughout the first half of the sixteenth century.[25] Santillana's *Proverbios* was a best-seller by any standards; the Crombergers alone issued at least eight editions between 1509 and 1548. They also printed the first, and three subsequent editions, of Hernando Díaz's abridged translation of Walter Burley's *Liber de vita et moribus philosophorum poetarumque veterum*, the original version of which had been widely read in fifteenth-century Spain.

Compendia of knowledge were equally popular. The press issued at least two editions of Alfonso de la Torre's *Visión delectable de la filosofía* compiled early in the fifteenth century from sources none of which had been written later than the twelfth century.[26] But the Crombergers' most successful publication, apart from Ludolph's *Vita Christi*, was Antonio de Guevara's *Libro áureo de Marco Aurelio* (sometimes printed with part or all of his *Relox de príncipes*). They issued at least ten editions in the space of twenty years. This didactic miscellany of information was hugely popular in the sixteenth century both in Spain and abroad, particularly in France.[27] Guevara was also the most popular Spanish author in sixteenth-century England and, indeed, the first to be translated into English prose.[28] The printing of this work provides a corrective to any suggestion that the family's association with Seville humanists and Erasmians meant that the Crombergers were uniquely wedded to the sort of work those scholars produced. In 1528, the very year in which he issued the first Spanish edition of Lucan's *Pharsalia* printed in italic type—a design which was not only closely identified with Aldus Manutius' humanist editions of the classics, but had been singled out for praise by Erasmus—Jacobo published the first edition of the *Libro áureo*. Yet Guevara's charlatanism in passing off lies of his own invention as the fruits of erudition, and his mannered Ciceronian style, were the very antithesis of the simple elegance and careful scholarship advocated, if not always practised, by Erasmus and his disciples. Nevertheless, as was noted in Chapter Two, Jacobo

[25] The Crombergers printed Alberto de Aguayo's translation of Boethius rather than the one which had been made by Pedro López de Ayala. The *De remediis* was translated by Francisco de Madrid, whose brother, Alonso Fernández de Madrid, was the translator of Erasmus' *Enchiridion*.

[26] A. D. Deyermond, *The Middle Ages*, A Literary History of Spain [1] (London, 1971), p. 146, claims that the *Visión* is 'a supreme example of Spain's cultural belatedness'.

[27] Febvre and Martin, *L'Apparition*, p. 413.

[28] Sir Henry Thomas, 'The English Translations of Guevara's Works', in Lucio Gil Fagoaga and Gerhard Moldenhauer (eds.), *Estudios eruditos in memoriam de Adolfo Bonilla y San Martín*, 2 vols. (Madrid, 1927–30), ii. 565–82 (p. 565).

could spot a potential best-seller. Although his publication of the *Libro áureo* was unauthorized and drew a sharp retort from Guevara, the book must have sold well, for in 1529 Juan thought it worthwhile investing the substantial sum of 175 *ducados* in the privilege for future editions of this work and of the *Relox*. The privilege was probably renewed when it expired ten years later.[29] In 1540 and 1549 the Crombergers even produced octavo editions of the *Libro áureo*, indicating both that this work, which they had hitherto printed in folio volumes, was now attracting a different sort of readership and that they were reacting to competition from small-format editions of the *Libro áureo* in Spanish imported from Italy and the Low Countries. At the time Juan Cromberger issued the first of his octavo editions of the *Libro áureo*, he was already selling these imported editions of the work.[30]

Another compendium of miscellaneous information which smacked of the Middle Ages, but became one of the most popular books in sixteenth-century Europe, was Pedro Mexía's *Silva de varia lección*. Mexía enjoyed an unjustifiably high reputation as a scholar in his native Seville and entrusted his first edition of the *Silva* to the presses of Domenico de Robertis, who signed it in July 1540.[31] It was an immediate success and, displaying the business acumen which has been noted previously, Brígida Maldonado moved rapidly to print the work and, probably, to secure a privilege on it. In December of the same year she issued the first edition of Mexía's enlarged and corrected version, which the press subsequently reprinted at least three times.

(iii) Other Prose Fiction, Secular Poetry, and Drama

Imaginative prose fiction and secular poetry tend to attract most attention from students of sixteenth-century letters. It may well be that such compositions are the most rewarding for literary critics, but an exclusive study of them gives a distorted picture of reading-habits at the time, even of 'literature'. At least, this is the case if the Crombergers' output is at all indicative of such habits. For example, if the romances of chivalry are discounted, such works account for a mere 6.8 per cent of the press's surviving output in terms of sheets—the same figure as for historical editions or the sum of legal and medical editions. It could be argued that fiction was so popular that the bulk of editions has disappeared because all the copies were read to pieces (for example, many sixteenth-century editions of *La Celestina* survived in unique copies), or that this sort of publication

[29] See the notes to the description of the 1540 edn. in App. I.

[30] '10 Marco Aurelios from Flanders' were listed in the 1540 inventory (see Gestoso, p. 97).

[31] Bataillon, *Erasmo y España*, pp. 637–8, characterizes Mexía as a timid Erasmian and a pedant; he dismisses any claim Mexía may have had to true learning. The *Silva* went through some 33 editions in 150 years and was translated into Italian, French, German, Flemish, and English.

survives less well than serious works which more readily found their way into contemporary libraries. All the same, available evidence from the Cromberger press should give literary critics some pause.

Among those editions which have come down to us the emphasis is upon reprints of medieval works. In 1542, the year of Boscán's death and shortly before the first edition appeared of his and Garcilaso's Italianate poetry which is often considered the dawn of the Renaissance in Spanish letters, the Crombergers issued yet another reprint of the fifteenth-century poem *Las coplas de Jorge Manrique a la muerte de su padre* of which the first edition had been printed at Seville in 1494. In 1540 they reprinted Hernando del Castillo's great collection of traditional *cancionero* poetry, the *Cancionero general*, while they also issued several reprints of original works by the fifteenth-century Castilian poet, Juan de Mena.[32] The picture is not very different for prose works. The most frequently reprinted piece of fiction to come from the Crombergers' office was Fernando de Rojas's *Celestina*, for which Jacobo soon set the pattern of printing and illustration adopted by other printers, including some in Italy. It is characteristic of them that the Crombergers should have taken over a popular fifteenth-century vernacular work like this and then become its major publishers. Indeed, Jacobo even printed a *pliego suelto* of a rhymed synopsis of the plot for a more humble readership or audience. But older medieval prose works were also frequently reprinted. There were the short chivalric stories of French origin like *Enrique, hijo de Oliva, Oliveros de Castilla*, *Partinuplés*, and *Pierres de Provenza* or the traditional medieval folk-tales like the *Historia de la linda Melosina*, issued in the quarto editions which tended to be chosen for popular books intended for a lowbrow market. Two famous amatory romances, the *Historia de duobus amantibus* by Aeneas Silvius Piccolomini (Pope Pius II) and Boccaccio's *Fiammetta*, were printed by Jacobo in translation. It has often been claimed that Diego de San Pedro's *Cárcel de amor* was modelled on these two works. This is open to question, but what is not is the popularity enjoyed by San Pedro's courtly romance.[33] The first edition had appeared at Seville in 1492 and it was frequently reprinted in Spain and abroad in the following century. Jacobo Cromberger issued four surviving editions. Such medieval courtly romances were in demand much later in the sixteenth century than some scholars would have us believe; the Crombergers' last edition of the *Cárcel* came out in 1527, but printers in Spain and the Low Countries were still producing editions at the end of the century, while the fifteenth-century *Grisel y Mirabella* by Juan de Flores was printed at least three times by

[32] Even the Castilian translation of Petrarch's famous sonnet, 'S'amor non è, che dunque è quel ch'io sento?', which is found at the end of Díaz's *La vida y excelentes dichos de los más sabios filósofos* is in 'arte mayor' rather than in Italianate hendecasyllables. Compare Chevalier, *Lectura y lectores*, p. 56.

[33] Diego de San Pedro, *Obras completas*, ed. Keith Whinnom and (vol. iii only) Dorothy S. Severin, 3 vols. (Madrid, 1971–9), i. 52–5.

the press, their last known edition appearing in 1533. In a somewhat different vein, Hernando Díaz's adaptation of Giacomo Caviceo's amatory, sacrilegious, and licentious *Libro del Peregrino* was issued by the Crombergers four times (*c.* 1515, 1527, 1544, and 1548).[34] Throughout Europe Apuleius' bawdy classical novel *The Golden Ass* was much in demand.[35] As was noted above, Jacobo printed at least one edition in translation.

A remarkable conclusion can be drawn from a survey of the Crombergers' output of the sort of literature included in this section. They had not been slow to print new sixteenth-century works of devotion or compendia of wisdom, and had even risked many first editions of such books. Yet, if their numerous romances of chivalry and, possibly, the odd tale produced in the form of a *pliego suelto* are excluded, during some six decades of printing they issued only one work of narrative fiction originally written in Castilian after 1500: the anonymous *Cuestión de amor*, and that had been composed in Italy. This provides an indication not only of the press's or the readership's conservatism; it is also a reflection of the general lack of composition of Castilian prose fiction during this period.[36]

The printing of dramatic works by the press would have been insignificant were it not for the first three Spanish editions of Bartolomé de Torres Naharro's influential *Propalladia*, a collection of plays which, although in Castilian, had been written and first published in Italy.[37] Drama was a particularly weak genre in Castile during the first half of the sixteenth century, and the native tradition is represented in the press's output by a mere handful of dramatic eclogues printed in the form of chap-books and, in the main, written by Juan del Encina. *La venganza de Agamemnón* by the humanist scholar Hernán Pérez de Oliva, but closely based upon Sophocles' *Electra*, was an anomalous publication of a 'classical' drama. It had first appeared at Burgos in 1528, but Brígida Maldonado's decision to print it thirteen years later is probably another indication of her readiness to introduce new titles into the press's traditional repertoire. There is no evidence that this innovation was a success.

(iv) Pliegos sueltos

The material cheaply distributed in the form of *pliegos sueltos*, of which the

[34] The *Libro del Peregrino* was written in the last decade of the fifteenth century and first published at Parma in 1508. It was popular in Italy, France, and Spain. On the Italian editions and the Spanish version of this work see Clive Griffin, 'Giacomo Caviceo's *Libro del Peregrino*: The Fate of an Italian Wanderer in Spain', in Anna Laura Lepschy, John Took, and Dennis E. Rhodes (eds.), *Book Production and Letters in the Western European Renaissance: Essays in Honour of Conor Fahy*, Publications of the Modern Humanities Research Association, 12 (London, 1986), pp. 132–46.

[35] Febvre and Martin, *L'Apparition*, p. 430.

[36] Ife, *Reading and Fiction*, p. 7.

[37] There are three Cromberger editions only if Norton is correct in his attribution of the 1520 edition to Jacobo.

Crombergers were leading printers, has been the most prone of all to destruction.[38] The pieces which have survived have done so by chance and may be unrepresentative of the press's production of chap-books. For example, we look in vain for surviving sheets of prayers, which were traditionally peddled by blind men, yet in 1528 Jacobo's stock contained 21,000 *pliegos* of this sort. Nevertheless, those chap-books which have come down to us offer a tantalizing glimpse of the most popular literature printed by the firm. They are varied, from works of some literary importance by authors of note to the pabulum demanded by the lowest of readers and listeners, or merely official forms and notices.[39] Most were sheets of the sort of traditional ballads or *villancicos* which were in vogue both on the streets and in the court of the Catholic Monarchs, again showing the continuing taste in the sixteenth century for medieval genres. Others contain news in verse, be it of national importance like Andrés Ortiz's rhyme about the imprisonment of Francis I after the battle of Pavia, or only of local curiosity like the sensational account of an obscure mass-murderer from Jerez de la Frontera. More literary works range from poems by Rodrigo de Reynosa to devotional verse by Fray Íñigo de Mendoza and Fray Ambrosio Montesino.[40] It has already been seen that some distinguished dramatic eclogues were printed in this form. Although prose *pliegos* appear to have been less common, some at least contained folk-tales like the surviving *Cómo un rústico labrador engañó a unos mercaderes*, letters sent by or to kings (of whom some were real and some imaginary), prognostications, or popular cures for the plague. One of the few surviving prose works of note printed by the Crombergers in this form is Diego de San Pedro's *Sermón*.

History

History, or pseudo-history, written in or translated into Castilian, accounts for 5.6%:6.9% of the Crombergers' surviving output. With the exception of the popular *Crónica del Cid* and the *Crónica del conde Fernán González*, historical works were printed in folio editions and were frequently substantial volumes. There

[38] Antonio Rodríguez-Moñino, *Diccionario bibliográfico de pliegos sueltos poéticos (siglo XVI)* (Madrid, 1970), pp. 18–22. I have listed all *pliegos* under 'vernacular literature'; some could, however, have been included under other headings.

[39] F. J. Norton, 'Some Spanish Printed Bifolia of the Early Sixteenth Century: Francisco de Peñalver's *Carta de la gota* and its Antecedents', in Dennis E. Rhodes (ed.), *Essays in Honour of Victor Scholderer* (Mainz, 1970), pp. 271–5.

[40] Antonio Rodríguez-Moñino, *Los pliegos poéticos de la colección del Marqués de Morbecq (siglo XVI)* (Madrid, 1962), pp. 46–8, attributes to the Cromberger press a *pliego* containing two poems (reproduced on his pp. [157]–164). In the first a black slave is solicited by a white noblewoman. A close examination of the types used in this *pliego* reveals that this is not a Cromberger product and may not have even been printed at Seville. This affects somewhat the argument in P. E. Russell, 'Towards an Interpretation of Rodrigo de Reinosa's "poesía negra" ', in R. O. Jones (ed.), *Studies in Spanish Literature of the Golden Age Presented to Edward M. Wilson* (London, 1973), pp. 225–45 (p. 231).

were only three classical histories, all published in translation: Flavius Josephus' account of the Jewish War, Quintus Curtius Rufus' life of Alexander the Great, and Herodian of Syria's history of Rome from the time of Marcus Aurelius to Gordian III. The remainder were mainly fourteenth- or fifteenth-century Castilian chronicles—an indication of Castilians' obsession with their own medieval history. Several had already been printed at Seville in incunable editions, and the Crombergers were clearly catering for an established market for such works. The most popular (five editions) appears to have been Diego de Valera's *Crónica de España abreviada*, the first edition of which had appeared at Seville in 1482, making it one of the earliest books printed in the city. As the title suggests, it is a potted history of Spain, and it is also prefaced by a geography of the world as it was known to Valera when he wrote the work in 1481. By the time it was first published by Jacobo in 1517, new discoveries had made its geographical information look decidedly old-fashioned; when Jácome issued the last Cromberger reprint in the 1550s, it was ludicrous. The other 'historical' work which was frequently reprinted by the press was Guido delle Colonne's famous thirteenth-century *Historia destructionis Troiae*, which, together with a host of histories of Troy, had been popular in Europe throughout the Middle Ages. A Castilian version had long been available, but the text printed by the Crombergers had been retranslated by the local scholar Pedro Núñez Delgado.

It is some measure of the lack of interest among readers, even in Seville, both in contemporary history and in the Indies that the only work of this nature concerned with the Americas to be printed by the major office in the city was the first edition of Gonzalo Fernández de Oviedo y Valdés's *Historia general de las Indias* (1535). Even then, Oviedo complained that he had been obliged to bear all the printing costs himself, for Juan Cromberger was evidently unwilling to invest in a work which he considered unlikely to sell. The large number of surviving copies of this edition which contain a signed dedication by the author—suggesting that he gave them away free of charge—and the fact that the work was never reprinted at Seville both indicate that Juan had judged the market correctly.

Utilitarian Works

From the earliest days printers recognized that they would make a profit if they produced the sort of works which were needed by professionals rather than mere leisure reading. The earliest editions were of liturgical books or students' texts, the latter always providing a most profitable branch of publishing.

Apart from liturgical books, almost all the Crombergers' Latin editions were intended for students and were the standard teaching texts of the day (3.9%:3.6% but probably exceptionally prone to loss): for example, the *Aurea expositio hymnorum*, Verino's *Liber distichorum*, Persius' *Satires* with Nebrija's commentary, and Sedulius' *Paschale*, as well as the curriculum book and favourite in Spain, Lucan's *Pharsalia*. Local pedagogues like Santa Ella and Delgado entrusted their editions to Jacobo. More important, however, was Antonio de Nebrija himself, whose books became the standard manuals in Spain for the teaching of Latin and, as was shown by Brocar and Eguía, were a lucrative source of income for printers. His *Introductiones Latinae* were particularly widely used and were evidently the apple of their author's eye, for he insisted on correcting proofs of this work himself.[41] Nebrija, who was an Andalusian, had long been associated with Seville, and in 1513 was in the city just before taking up a chair at the new University of Acalá. It was then that Jacobo Cromberger (or possibly Varela using Cromberger's material) printed an edition of the *Introductiones*, doubtless in co-operation with the author. Jacobo had, it should be remembered, already printed two editions of Nebrija's Spanish–Latin dictionary. But the press also issued more humble school-texts like Sulpitius' *Doctrina mensae* and, most probably, the widely used *Catón*, which figures so prominently in the 1540 inventory of Juan's stock.[42] Even humbler were the vernacular *pliegos* which were used to teach children to read. Jacobo printed large numbers of ABCs, but it is often forgotten that little books, such as the story of the Cid or the *Siete sabios de Roma*, and the simple sheets of ballads which poured from his presses, would also have been used, as such rhymes were in Spain right up to the nineteenth century, by children learning to read.[43]

By far the largest category of utilitarian Latin editions printed by the Crombergers (7.2%:11.8%) was composed of liturgical works, for which the family enjoyed a justifiably high reputation. Local variations in ritual meant that different editions were required in each diocese, thus providing the firm with a steady stream of work. The Crombergers printed breviaries, missals, diurnals, manuals, processionals, and orders of baptism for Seville, Córdoba, Jaén, Badajoz, Évora, and León, as well as editions for religious orders. They also produced collections of constitutions and synodals in the vernacular for various dioceses including Málaga, and several editions of the Roman psalter. Likewise, the press issued books of hours in Latin, Castilian, and Portuguese (these have been calculated under the heading of 'Devotional and Spiritual Works' when in the vernacular). As has already been noted, contracts for service books were particularly welcome to printers, as they did not run the risk of finding their capital tied up in unsold editions, and they were frequently given an advance on their expenses. The Crombergers must have produced many such editions which have disappeared, for the Council of Trent and the prohibitions contained in

[41] Information kindly given to me by F. J. Norton.

[42] Gestoso, pp. 90, 93.

[43] Rodríguez-Moñino, *Diccionario,* pp. 18–20; also see above, p. 124.

various *Indexes* resulted in old editions' becoming obsolete. They were discarded in large numbers and frequently only survive as the scrap paper used by binders or as the sheets of second-hand vellum bought up by lawyers in the late sixteenth century as folders for their documents. Even fewer copies of indulgences printed by the family still survive, but it is known that they were produced in large numbers under contract from the ecclesiastical authorities.

The other important utilitarian editions were legal and medical works (3.4%:3.6% and 4.1%:3.4% respectively). With the exception of the *Ordenanzas reales de Castilla* and the two prestige editions of the Portuguese *Ordenações manuelinas*, these were slight folio volumes containing fewer than fifteen sheets. None was the sort of textbook written in Latin which is frequently found in the inventories of Spanish lawyers of the period and which were imported from foreign presses. The Cromberger editions were all in the vernacular and were simple manuals or collections of national or local laws for everyday use by semi-professionals or laymen. The same is true of the medical works, which, with only a few exceptions, were printed in Castilian and were practical guides for surgeons, apothecaries, or even the patients themselves. Reprints of a mere three titles account for approximately 60 per cent of the Crombergers' output of medical editions. They were the *Tesoro de los pobres*, a thirteenth-century do-it-yourself guide for the sick apparently commissioned by Petrus Hispanus (Pope John XXI) from one 'Master Julián' or 'Juliano'; numerous reprints of Alfonso Chirino de Cuenca's *Menor daño de medicina* composed in the early years of the fifteenth century; and the more modern *Modus faciendi*. Despite its title, the last of these was largely in the vernacular and was based on early Arab treatises. It was the work of Fray Bernardino de Laredo, a famous Seville doctor who was in charge of the pharmacy at the Friary of San Francisco del Monte near Seville, a friend of the Crombergers, and, as has been seen, a spiritual writer of considerable importance. Jacobo's *Modus faciendi* printed in 1527 was probably the first edition of this work.

The remaining practical books issued by the press (1.3%:0.2%) were little almanacs, which must often have been found among peddlars' wares, like Andrés de Li's *Repertorio de los tiempos*, of which three surviving Cromberger editions are known but doubtless many more printed; Roberto de Nola's famous collection of recipes, the *Libro de cocina*, printed by Juan in 1538; or Fray Juan de Ortega's entertaining introduction to mathematics, the *Tratado sutilísimo de aritmética y de geometría* printed in 1534 and 1542. Ortega's work may have been a school-book or may have been intended for a more general readership.

Miscellaneous

The remaining 5.1%:4.8% of the press's surviving output does not fall neatly into a single category. Calculations based upon the number of sheets contained in a single copy of each edition make theology written in Latin appear relatively important. However, a single theological work, Archbishop Diego de Deza's *Novarum defensionum doctrine beati Thome de Aquino super primo [to fourth] libro sententiarum questiones*, accounts for the bulk of those sheets. It is the longest book ever printed by the press. It contains almost one thousand folios and was a major undertaking for Jacobo, who printed it over a period of three years. Norton suggests plausibly that such an uncommercial work would have been printed under contract from the archbishop.[44] If this is true, Jacobo would have acted as no more than the craftsman and would have been paid accordingly by Deza or a third party. It was doubtless in Cromberger's interest to work directly for the archbishop; not only did it mean a guaranteed income for his labour, but the monopoly of liturgical printing for the diocese, which Jacobo had enjoyed for several years, was also in Deza's gift. In about 1517 Jacobo was not averse to printing another Latin work by the same author.

As might be expected, given the importance of Seville as a centre for maritime trade and exploration, several geographical works, navigational treatises, and travel-books, all in Castilian, were printed by the press. In 1535 Juan Cromberger printed the first known edition of Francisco Falero's *Tratado de la esfera y del arte de marear*. Some books were more directly concerned with the Indies. In January 1511 the expatriate Italian humanist, Petrus Martyr Anglerius, whom Elliott describes as 'the first popularizer of America', was granted a royal privilege on his writings.[45] Three months later an edition of his *Opera* was signed by Jacobo, who must have been approached by the author to print them for him. Part of this volume is a description of the Caribbean accompanied by a crude woodcut map. Eight years later Jacobo printed the first of the two Cromberger editions of Martín Fernández de Enciso's *Suma de geografía*. Enciso was an early explorer of Central America, and his *Suma* provided navigational information about both the Old and the New Worlds as well as a description of the Indies larded with fanciful tales. And in 1522 and 1523, as was noted in Chapter Two, Jacobo issued the first editions of Cortés's second and third letters sent from Mexico to Charles V.

The Crombergers were the most important printers in a city which was radically affected by its monopoly of trade with the Indies. Nevertheless, the number of works printed by them which dealt with the Americas is unimpressive.[46] Indeed, most of their travelogues were written about the Old World.[47] Jacobo printed the first Spanish edition of the romanticized thirteenth-century travels of Marco Polo;

[44] Norton, *Printing in Spain*, p. 12.

[45] For Martyr's view of the New World see J. H. Elliott, *The Old World and the New 1492–1650* (Cambridge, 1970), pp. 25–6.

[46] Fernández de Oviedo y Valdés's *Historia general de las Indias* is calculated above under 'history'.

[47] Compare Whinnom, 'The Problem', p. 196.

the first Spanish edition, and a subsequent reprint, of Lodovico Barthema's account of his journeys to the Middle and Far East; and at least two descriptions of the Holy Land: the popular *Misterios de Jerusalén*, which the press printed three times, and the pietistic *Verdadera información de la Tierra Santa* written by the Franciscan, Antonio de Aranda. Similarly, much of the appeal of Caviceo's *Libro del Peregrino* probably lay in its accounts of travel to the Middle East. However attractive the Indies proved to soldiers and missionaries, merchants and bureaucrats, they appear to have met with general indifference among the purchasers of printed books. It was, perhaps, only with the furore caused by the 'Affair of the Indies' in the mid-century that Jácome Cromberger attempted to capitalize upon a sudden explosion of popular interest in the New World. Even then, he was cautious (see above, p. 105).

A survey of the Crombergers' books allows several conclusions to be drawn. First, it was noted in Chapter One that incunable printing at Seville was remarkable even for Spain because of the large proportion of editions printed there in the vernacular. The percentage of sheets printed in the vernacular by the Crombergers was even higher: 82.3 per cent. This rises to an extraordinary 93.4 per cent when liturgical editions in Latin are discounted. Together with the large number of devotional works issued, these figures provide a graphic illustration of two major intellectual and spiritual currents in early sixteenth-century Spain: the interest even among humanist scholars in the vernacular, and the reform.[48] Second, if devotional works, liturgy, and theology are added together, they form about half of the press's output. Third, the readership for which the press catered was generally conservative: prose fiction, verse, history, medicine, moralistic works, and compendia of wisdom were all dominated by works which had been written in the Middle Ages or, as was the case with most of the romances of chivalry, were sixteenth-century compositions cast in a medieval mould.[49] Many of the Crombergers' books were reprints of incunable editions, while an even larger number were constantly reprinted to satisfy an unchanging demand. Over 36 per cent of their editions were reprints of books which they had already issued themselves. Far from being exclusively, or even primarily, the 'agent of change' which some historians claim, printing reinforced conservative tastes.[50] Fourth, the

[48] Donnelly, 'A Study', i. 1-14.

[49] The same appears to be true of the production of printers throughout Spain in the Golden Age (see Whinnom, 'The Problem', p. 193).

[50] Dennis Hay remarks on this conservatism everywhere in Europe at this period and rightly rejects the idea that printing immediately brought about a revolution in taste and ideas; see his introduction to John Carter and Percy H. Muir (eds.), *Printing and the Mind of Man: A Descriptive Catalogue Illustrating the Impact of Print on the Evolution of Western Civilization During Five Centuries*, 2nd, rev. and enlarged edn. (Munich, 1983), pp. xv–xxxiv (pp. xxii–xxiii). There is a notable overlap between the titles issued

traditionally popular books printed by the Crombergers were similar to, and often even identical with those in demand elsewhere in Europe at the time.[51] As is suggested by the large number of translations printed there, reading-habits in Spain—at least so far as production in Seville is any indication of them—were not unique.[52]

However, this conservatism was not unremitting. It has been seen that Spanish readers were quick to demand editions of new reformist works in the vernacular. Even secular literature was not uniformly traditional. It was characteristic of Brígida Maldonado, for example, that she should have capitalized on the popularity of another new work, Boscán's translation of Castiglione's *Il cortegiano* (first printed at Barcelona in 1534), and issued an edition of it. Jácome subsequently issued two more Cromberger editions. In his prologue to his friend's translation of this fundamental Renaissance work, Garcilaso de la Vega dismissed in Olympian fashion the very literature which formed the backbone of the Crombergers' production. Clearly the market was perfectly capable of absorbing both the refined Castiglione and the medieval literature which his admirer, Garcilaso, so deplored in his own country.

With the exception of Jácome, the Crombergers were cautiously innovative, but most of their works were traditional, for they were primarily businessmen and knew their markets well. They were opportunists, reprinting new books which had already proved popular in editions printed by other presses like Mexía's *Silva*, Erasmus' *Enchiridion*, or Castiglione's *Il cortegiano*; they were occasionally the first to recognize a potential best-seller like Guevara's *Libro áureo* or Constantino's *Suma*; and they were eager to print service books or indulgences for which payment was guaranteed. But the greater part of their repertoire consisted of a virtually static list of tried and tested favourites. They covered almost all areas from jobbing printing to theology, from beautiful liturgical books printed on vellum to cheap *pliegos sueltos*. It is indicative of the state of Spanish printing at the time that a press which did not have access to patronage could not afford to specialize too much in a single kind of printing. Risks had to be spread.

by the Crombergers and those found in 15th-c. noble libraries in Castile (see the articles referred to above on p. 53, n. 134). What tend to be missing from the Crombergers' output are the classics, while the romances of chivalry which loom so large in their production are generally absent from the inventories of 15th-c. libraries despite incontrovertible evidence that the noble owners of these libraries were familiar with the romances (see Lawrance, 'The Spread of Lay Literacy', p. 90).

[51] Febvre and Martin, *L'Apparition*, pp. 381–3, 389–92, 413–14, 429–33. There are considerable similarities between the Crombergers' output and that of their contemporaries in England (see Bennett, *English Books and Readers 1475–1557*, pp. 185–97).

[52] Spanish literary production did vary widely from region to region and even city to city (see Antonio Rodríguez-Moñino, *Construcción crítica y realidad histórica en la poesía española de los siglos XVI y XVII*, 2nd edn. [Madrid, 1968], p. 56).

CHAPTER 7
TYPES

Punch-Cutting and Type-Founding

PRINTING in Europe became possible only when a series of technological advances had been made. The manufacture of paper from rags had by the fifteenth century provided an economical substitute for expensive vellum, and paper was regularly used for manuscripts; it was upon paper that most of the printing of books even in the earliest years was carried out. The press itself had by the time of Gutenberg already been developed for printing multiple copies of religious pictures and playing cards from blocks. But it was the invention of moveable metal types and the oil-based inks which could be used upon them which signalled the beginning of printing as we think of it. This in turn depended upon the invention of the adjustable mould, in which large numbers of pieces of type could be cast on bodies of the same height from face to foot, the same point size, but of varying breadth to allow for the regular spacing of letters of different widths.[1] The skills of punch-cutting, die-making, and casting had, however, long been practised by goldsmiths and artisans engaged in the production of stamps for medallions, coins, and seals. The processes involved in the manufacture of printing types were merely an extension and adaptation of these skills. A short introduction to some of the technical aspects of type-production is provided here for the non-specialist reader who is nevertheless interested in the evidence which a close study of types used by an individual press can give.

Letters of uniform design (a fount) were each engraved in reverse on the end of an iron or steel punch, which was then driven into a warm metal block (normally copper), thus making a primitive mould (a rough strike). Once this rough strike had been justified by a skilled craftsman it became a matrix. The matrix was placed at the bottom of an adjustable mould and molten type-metal was poured in; the type-founder would shake the mould to ensure that it had been completely filled and then the new piece of type was ejected. The individual pieces were later finished by hand.[2]

The process was a long and complex one. A type-maker who supplied

[1] For an account of the skill, expense, and secrecy involved in the invention of the first adjustable mould, see Seán Jennett, *Pioneers in Printing* (London, 1958), pp. 7–11.

[2] For a detailed explanation of the process of type-manufacture from punch-cutting to casting see Joseph Moxon, *Mechanick Exercises on the Whole Art of Printing [1683–84]*, ed. Herbert Davis and Harry Carter, 2nd edn. (New York, 1962), pp. 87–190.

Christopher Plantin estimated in 1576 that each punch would take him about a day to cut; the justification of the matrices was a time-consuming task; and the production of a fount containing sufficient numbers of all the individual letters in both upper and lower case, as well as punctuation marks, abbreviations, tied letters, and paragraph marks, could well take up to six months.[3] The cost of the skilled labour, in addition to the type-metal and the capital investment in equipment, meant that types were expensive and constituted the principal item in the outlay of setting up a printing-office.[4] Once a printer had acquired, cast, or commissioned a type it would also need to be replaced from time to time or, if the matrices were still available, recast. Recasting was a lengthy but indubitably less costly process than cutting a new set of punches. At Valencia in 1492 a type-founder, using matrices which were already available, took three months to produce enough type to set up five formes simultaneously, and in Seville a similar operation took some two months in 1545.[5] The frequency with which a fifteenth- or sixteenth-century type would have to be recast is a question to which there is no simple answer. As is so often the case with research into the history of printing during the early period, too little statistical material is available to allow anything more than informed speculation. The wear of type depended on the frequency and care of its use, the size of the editions printed, and the quality of the metal employed in the casting, but little is known about these factors. One investigation into the durability of fifteenth-century type concludes somewhat lamely that 'the incunable printer would normally expect his type would last for several tens of thousands of impressions . . .; incunable type was fairly durable'.[6] However, if a piece of type were set some thirty or fifty times in a substantial book which was printed in a run of one thousand copies, those 'several tens of thousands of impressions' could easily be made in the production of a single edition.[7]

Against this must be set the improvement in the durability of type made in the sixteenth century with the introduction of antimony, or alloys of antimony, into the mixture of lead and tin of which type-metal was composed. The mining of antimony appears to have begun at the beginning of the century.[8] Indeed, the sale

[3] Michael Pollak, 'The Durability of Fifteenth-century Type', *Library Quarterly*, 40 (1970), 371–90 (p. 377).

[4] Throughout the 16th c. the acquisition of types was by far the most expensive item in setting up a printing-shop (see Gilmont, *Jean Crespin*, p. 50, and below, pp. 172–3, n. 37).

[5] Konrad Haebler, 'Typefounding and Commerce in Type during the Early Years of Printing', *Ars Typographica*, 3 (1926), 3–35 (p. 26); Klaus Wagner, 'Los impresores', p. 141.

[6] Pollak, 'The Durability', p. 389.

[7] Pollak, 'The Durability', p. 386. A sample book examined by Pollak contained some 3,000,000 letters, which were printed from a fount estimated to contain from 60,000 to 100,000 pieces of type. Wear would, of course, depend on the number of pieces and therefore on how frequently a single piece was used. One fount in 16th-c. Seville contained 44,000 pieces (see Klaus Wagner, 'Los impresores', p. 141). Febvre and Martin, *L'Apparition*, p. 165, give average figures of 60,000–80,000.

[8] Febvre and Martin, *L'Apparition*, p. 75.

of old types by prosperous city printers to their country cousins who continued to use them in a worn state for many years, as happened in sixteenth-century England, suggests that types of the period were not worn out in the printing of a single edition. Nevertheless, in 1570 Paulus Manutius claimed that he needed to cast new characters for every book he produced and that if he did not do so, they wore out just as he was in the middle of printing the edition.[9] No doubt this apparently contradictory evidence for the durability of type can be explained not only by the various alloys used but by differing standards among printers. Types which Manutius considered beyond use would be perfectly acceptable to a jobbing printer. Thus when the contract for the foundation of a press in Mexico was drawn up, Juan Cromberger obliged his employee, Juan Pablos, to swear that, 'when the types which I am transporting with me are so old that they can no longer be used, I must melt them down and sell the metal. I am forbidden to give any of them away or sell them'.[10] Cromberger was here clearly trying to prevent a potential rival challenging his monopoly of printing in the New World. He realized that material which, by his own high standards, was worn out could still have been put to good use by a less particular printer. What is certain from all this is that types which were in constant use had to be recast frequently if a sharp impression was to be obtained from them. This makes the dating of unsigned books on the evidence of the wear of types a hazardous enterprise. A more reliable method of dating is based upon modifications made to founts. Some of these were evidently due to recasting, such as the reduction of the size of the body on which a certain face was cast at a particular date, while others involve the introduction of a new form of letter or other sorts which probably coincided with the recasting of a type (see the notes on Crombergers' Types 2 and 10 in Appendix Two).

It was formerly assumed that until the sixteenth century every printer made his own types from punch-cutting to finishing off the cast letters, and that every type was therefore unique, even though it was admitted that founts changed hands as they were sold, hired, inherited, or pledged. Haebler affirmed that this was the common practice of the early printers, maintaining that all the stages of type manufacture were part of the skills which every master-printer would have possessed.[11] However, this is not entirely true even for the fifteenth century. Sub-contracting out to specialists was already taking place in the 1470s.[12] Types derived from a single set of punches were used simultaneously by a large number of printers working in different German cities from about 1480 onwards. By at

[9] Febvre and Martin, *L'Apparition*, p. 74. Some printing contracts in 16th-c. Spain stipulated that new types be cast for a particular edition (see Abizanda y Broto, *Documentos*, i. 342–56).

[10] *Documentos para la historia de la tipografía americana*, pl. 3.

[11] Haebler, 'Typefounding and Commerce in Type', p. 35.

[12] Lowry, *The World of Aldus Manutius*, p. 11. For information on the 15th-c. trade in matrices in Northern Europe see Martin and Chartier, *Le Livre conquérant*, p. 205.

least the first decades of the sixteenth century, types from the same punches travelled large distances: the Greek alphabet which Froben used at Basle, for instance, can be traced in books printed in centres as far from Basle as Paris and Cambridge.[13] When Thomas Platter set up his own independent press in sixteenth-century Basle he was supplied with matrices by punch-cutters who also cast the type for him; he seems not to have possessed the skills to do it himself. One of his suppliers was equipped with punches for a variety of founts and would have specialized in producing matrices for printers.[14] Later in the century Plantin certainly engaged in the selling of matrices, but he neither made his own punches nor cast his own type. Rather, he had matrices and punches produced for him and then put the work of casting out to specialized type-founders while retaining ownership of the punches or matrices.[15] There is some evidence that long before Plantin's day there was a trade in matrices at the Frankfurt book fair and that printers were supplied with the means of making type from this source.[16]

As far as Spain is concerned, there is less evidence about type-production than has been discovered for France, Germany, and the Low Countries. However, some punches were certainly cut there in the fifteenth century. One of the first Valencian printers, Alfonso Fernández de Córdoba, was a silversmith by training and appears to have been skilled in punch-cutting and type-founding.[17] Norton also maintains that, although the first immigrant printers to Spain came equipped with types or the means of casting them, new designs were being cut in that country soon after their arrival.[18] It is possible that a type used by Antonio Martínez in 1486 was cast in Spain or even Seville.[19] In the late fifteenth century the immigrant printers Meinardo Ungut and Stanislao Polono had at least one new type cut for them in Spain, probably in Seville. This is the fount which would later pass to Jacobo Cromberger as his Type 11 and which did not make its first appearance in books printed by his predecessors until Polono used it in 1500. Indeed, it may well be this type which is referred to in the inventory of Ungut's estate as 'a vernacular type which is still being made'.[20]

In the period which concerns us more closely, there is no direct evidence for punch-cutting in Seville and I suspect that, as the century wore on, printers there, as elsewhere in Spain, mainly relied upon imports or upon the services of specialized punch-cutters who were probably itinerant. This lack of evidence for punch-cutting and even for the existence of the punches themselves is mirrored in contemporary

13 Carter, *A View*, p. 109.
14 Platter, *Autobiographie*, pp. 109–10.
15 Carter, *A View*, pp. 9-10, 95, 97.
16 Carter, *A View*, p. 66.
17 Haebler, 'Typefounding and Commerce in Type', p. 19.
18 F. J. Norton, 'Typographical Evidence', p. 99.
19 Sheppard and Painter, 'Introduction to the Presses', p. liv.
20 This document is reproduced in Tenorio, 'Algunas noticias'.

documents from Barcelona. While there are frequent references to punches in the contracts and inventories of Barcelona printers in the fifteenth century, with one exception only matrices are mentioned in the sixteenth.[21] By the middle of the sixteenth there were specialized type-founders in Barcelona, and this suggests that by then printers there generally no longer even cast types themselves.[22]

Unfortunately, no contracts for the purchase from a supplier of punches, types, or even matrices in Seville have been found for the first half of the sixteenth century. The only reference to a type-founder working for a single press is to Antonio de Espinosa, who, in 1550, signed a contract in Seville in which he agreed to go to Mexico to work in Juan Pablos's printing-house as a 'fundidor [i.e. type-founder] e cortador', the latter term implying that he was to cut punches as well as to cast type.[23] It is likely that he did make punches in Mexico as Juan Pablos's stock of types was radically changed after his arrival, but, since this is a case of a colonial press set up thousands of miles away from any potential specialized supplier of punches or matrices, it is not necessarily typical of practice in Spain.[24] In 1532 a contract was drawn up in Seville between Juan de Virida and Juan de León, the latter being one of the many men of that name who appear in documents connected with printing in the city. He was a Frenchman and normally called himself a printer. In that same year, however, he was referred to as 'Juan de León, Frenchman, fashioner of letters for printing' suggesting that he was a specialized punch-cutter.[25] At no time, however, is he found practising this skill. Two documents show that, even if there were no punch-cutters in the city, type-founders could be found there around the middle of the century. In 1545

[21] Madurell and Rubió, *Documentos*, documents Nos. 13, 65, 73, 84, 188. The one exception at Barcelona was Claudi Bornat, whose punches are recorded in an inventory of 1572 (see José María Madurell Marimón, *Claudi Bornat* [Barcelona, 1973], p. 202). At Zaragoza Coci possessed his own punches in 1537, but his material was based on that of his 15th-c. predecessors, the brothers Hurus, and he may well have inherited their punches (see Abizanda y Broto, *Documentos*, i. 308). See also Cruickshank, 'Some Aspects'.

[22] José María Madurell Marimón, 'Antiguos fundidores de letras en Barcelona', *Gutenberg-Jahrbuch* (1970), 289–97 (p. 289).

[23] Gestoso, pp. 115–17. When Juan Pablos's office depended upon the Cromberger press at Seville, he was supplied with ready-made types (see Gestoso, p. 62).

[24] Douglas C. McMurtrie, 'The First Typefounding in Mexico', *The Library*, 4th ser., 8 (1927), 119–22 (p. 121). Espinosa, a native of Jaén, was the first important Spaniard to print in Mexico. He had previously worked at Alcalá, Granada, and Seville, where, in 1551, he was making preparations to leave for Mexico (see Klaus Wagner, *Martín de Montesdoca*, p. 106). After serving Pablos in Mexico, he returned to Spain to challenge the monopoly of printing in the New World which Pablos had doubtless obtained in 1548. His petition was successful, probably because of powerful friends at court, and he was back in Mexico by 1559, where he soon began printing in his own independent office (see Eduardo F. Araujo, *Primeros impresores e impresos en Nueva España* [Mexico City, 1979], p. 34, and Stols, *Antonio de Espinosa*, pp. 7–11). Henry Wagner, *Nueva bibliografía*, pp. 18–19, suggests that the word 'cortador' refers to a cutter of wood-blocks. I have not seen it used in this way in contemporary documents, but Stols (p. 13) believes that Espinosa did, indeed, cut his own blocks in Mexico.

[25] Document dated 26 June 1532 (APS, Oficio 4, Libro 1 of 1532, unfoliated).

the Seville printer Simón Carpintero provided the metal for a type-founder to make him a fount, while in 1560 a contract states that certain matrices were to be examined by one 'Guillermo de Tello, type-founder' or by 'any other craftsman trained in this skill'.[26]

The lack of sale contracts suggests that, even if there was a trade in punches, matrices, and types, it was not particularly active. Indeed, such silence implies either that Seville printers travelled to centres where these items were manufactured and there signed sale contracts, or that they did not regularly purchase such material, being content merely to take over the equipment of firms which had ceased to print and base their stock of types upon that of their predecessors. This was certainly the case with the two major printers in the city during the early part of the sixteenth century: Jacobo Cromberger and Juan Varela de Salamanca. They acquired the stocks of the leading companies of the fifteenth century, the partnership of Ungut and Polono and that of the 'Compañeros alemanes', and then added little to them. When they did renew their material, the new types tended to be shared with other Spanish printers: Cromberger's Types 1, 10, 13, 14, and 17, which had not been inherited from earlier Seville presses, are remarkably similar to types used elsewhere in the country. It appears that these types were shared and that rough strikes or justified matrices were acquired by several printers from a small number of suppliers.

It is not known whether the suppliers were punch-cutters who made matrices for many presses or whether a major printer acquired the punches and then supplied his colleagues with matrices or even types. The acquisition of matrices would, of course, have been much more economical for a printer than the purchase of expensive punches; and it would have made commercial sense for Spanish printers, who mainly used unremarkable gothic types at this period and would scarcely have required the sort of monopoly of them offered by ownership of the punches. Skilled punch-cutters demanded high wages, as can be seen from the extraordinary sum promised to Espinosa in 1550: he was contracted for three years at 150 *ducados per annum* even though he was still under 25 years old.[27] By the end of the century Seville printers would not even be acquiring matrices from elsewhere, but buying ready-cast types from Madrid.[28] It is thus unlikely that printers working in the city during the first half of the century cut their own punches or employed a skilled resident of Seville to do so for them. However, there is at least some positive evidence that they recast their own types in inherited fifteenth-century matrices and from newer ones which they had acquired themselves.[29]

[26] Klaus Wagner 'Los impresores', p. 141, and his *Martín de Montesdoca*, p. 126.

[27] Gestoso, pp. 116–17.

[28] Wilson and Cruickshank, *Samuel Pepys's Spanish Plays*, p. 8.

[29] Norton, *Printing in Spain*, p. 35, suggests that a founder was supplying material to a number of

Ungut possessed equipment which enabled him to cast his own type, and this material passed to Jacobo Cromberger when the latter married Ungut's widow.[30] It appears that Jacobo himself was skilled in type-casting, at least if the epithet 'moldero' which was applied to him in a document of 1515 refers to his expertise in the use of moulds for casting molten metal.[31] These 'moldes' are again mentioned in the important document of 8 September 1525 which records Jacobo's gift of his printing-office to Juan: among the material which passed from father to son was 'all his equipment, types, "moldes", and everything required to engage in the "oficio de molde"'.[32] More explicit are entries in the division of Jacobo's estate between his two heirs Catalina and Juan in 1529: 'item . . . a large Portuguese copper cauldron which is full of type . . .; item, this agreement does not include the printing-shop, presses, equipment, type-metal, types, matrices, figures [i.e. probably the woodcuts], and everything else pertaining to the said art of printing'.[33] The matrices which belonged to Jacobo do not form part of the estate to be divided because they had already passed to Juan in 1525; it is significant that they are singled out here for mention because this implies that they were a valuable possession and therefore that they were still being used. Equally suggestive is the silence concerning any punches which the press might have had. This implies that Jacobo, and after him Juan, owned only matrices, and would go some way to explaining why other presses shared some of the types they employed.

As late as 1555 there is evidence that Jácome Cromberger had the wherewithal to cast type or to have it cast for him. In that year Brígida Maldonado sold to her son Jácome and his wife the presses on which he had been printing for many years. The equipment is listed as follows: 'three printing-presses with their equipment, types, other tools, matrices, and original books [i.e. printers' copy], all valued at . . . 400 *ducados*'.[34] Proof that matrices were not merely part of the out-dated clutter accumulated by a press such as that of the Crombergers which

Spanish printing-offices in the late 15th and early 16th cc. He may really mean a cutter, for the fact that printers at this time often possessed matrices suggests that they did the casting in their own workshops.

[30] Hazañas, i. 29, 87–8; Tenorio, 'Algunas noticias', p. 637.

[31] Hazañas, i. 103. Jacobo Cromberger had frequent personal and business contacts with silversmiths. He would have shared with them a skill in casting metals, and they may even have helped in the process of type-founding, but I have found no documentary evidence of such co-operation. The early printers in Germany and elsewhere were often associated with goldsmiths.

[32] Hazañas, i. 130 (his wording, not transcribed directly from the document). The word 'molde' may, however, have a different meaning; in 1507 a printer of woodcuts in Seville also talked of 'all the picture "moldes" as well as those on which is mapped the entire kingdom of Granada . . . and, in addition, a press for printing picture "moldes"' (see Gestoso, p. 13). Here 'molde' seems to refer to woodcuts. The word has this meaning in a contract signed at Seville in 1551 between a manufacturer of playing-cards and a craftsman who describes himself as a 'cutter of "moldes" for cards' (see Gestoso, p. 120). The 'oficio de molde' mentioned in the transfer of the Cromberger press from Jacobo to Juan may simply mean printing.

[33] Gestoso, pp. 43, 55.

[34] Document dated 20 Apr. 1555 (APS, Oficio 15, Libro 1 of 1555, fols. 781^r–784^v).

had been operating for a period of over fifty years is provided by a sale which we may reasonably suppose was more business-like than that between Jácome and his mother. Again in 1555, Varela's heirs, one of whom by marriage was Jácome Cromberger, sold equipment which had come from Varela's estate to the Seville printer Gregorio de la Torre. Although the sale contract has disappeared, a document survives in which Torre promises to pay the heirs '325 gold *ducados* . . . which I owe you for two printing-presses with their equipment, eight chases, nineteen type-cases, eighteen wooden imposing surfaces, a certain quantity of wooden ornamental initials, 1,500 *libras* [690 kg.] of type-metal, twelve sets of matrices for the said types, and other printing equipment which I am purchasing from you'.[35] A printer would certainly not have bought expensive equipment which was of no use to him; Gregorio de la Torre must have intended to use these matrices well into the second half of the sixteenth century to recast the type which he now owned.

A more economical method of acquiring new founts was to hire matrices and then cast type in them. In 1556 a Cordoban printer, known as Juan Bautista, paid one *ducado* for every one hundred letters he cast in '361 Greek matrices for making certain Greek types' which he had borrowed from the Seville printer, Pedro de Luján.[36] This contract also contains a clause by which Juan Bautista was obliged to pay six *ducados* for every matrix he failed to return to its owner. Although this figure was probably inflated to dissuade Bautista from losing or selling the equipment, because Greek matrices were something of a rarity in Spain, it nevertheless indicates the high cost of matrices in Seville at this time. Similar evidence is available in 1560, when Simón Carpintero borrowed matrices in which to cast type for a missal of Cordoban use which he was to print.[37]

[35] Document dated 29 Nov. 1555 (APS, Oficio 5, Libro 4 of 1555, unfoliated). He was given two years to pay off the debt and seems to have done so, as only half the sum was still owing on 7 Aug. 1556 (APS, Oficio 5, Libro 2 of 1556, unfoliated). As Varela probably stopped printing in 1539, it is surprising that he had not sold his presses then. Hazañas, ii. 132–3, believes that he must have done so. If this were the case, however, the origin of the two presses sold to Gregorio de la Torre would be mysterious unless they were Jácome's, but it is unlikely that Jácome's presses would have been the property of Varela's heirs. The presses sold to Torre are probably the ones which were installed in the famous Monastery of San Francisco in Seville, where he had been obliged to live as a friar after a chequered career in Spain and France (see Gestoso, p. 134).

[36] Gestoso, p. 119.

[37] Martín de Montesdoca hired these matrices from Juan Gutiérrez, a book-merchant who later became a printer, and had to pay from 2 to 10 *ducados* for every one lost or damaged; he then lent them to Carpintero. He also sold matrices for music to Gutiérrez (see Klaus Wagner, *Martín de Montesdoca*, pp. 126–7). This sale is evidence for the high cost of types compared with presses: Torre bought two presses with a large quantity of type for 325 *ducados*; Gutiérrez bought one press with one fount for 32 *ducados*. In 1540 a printing-press was sold in Seville for a mere 12 *ducados* (see Gestoso, p. 100). In 1545 a printer paid 85 *mrs* per 1,000 letters for the casting of a type for which he provided the metal (see Klaus Wagner, 'Los impresores', p. 141). In 1539 the University of Salamanca offered to lend 700 *ducados* to a printer willing to settle in the city to print Latin and Greek works. The sum was so large, it is implied, because

It seems, then, that sixteenth-century printers in Seville did not cut punches. They were owners only of the matrices in which they cast and recast their types, and they continued to do so well after the middle of the century. While the examination of local archives provides a certain amount of information on the subject of types, a close examination of the books printed by the Crombergers allows us to record the number, design, and evolution of the types used in their office.

Type Design

Types can be divided into two main families: gothic and roman. The first printers not unnaturally attempted to make their books look as similar as possible to the manuscripts which they sought to replace. Manuscript hands differed, however, from one country to another and also depended to some extent upon the nature of the work which was being written or copied. So firmly had gothic hands been associated with liturgical manuscripts that, even in the late sixteenth century, such books were still being set up in the gothic types which had largely been replaced by roman and italic founts in other sorts of books; in some countries gothic resisted even beyond the end of the sixteenth century, while printers in others rapidly switched from gothic to roman early in the century for almost all kinds of edition.

The first books to be printed in Germany were in German gothic types. The loose term 'gothic' should, however, be more closely defined. The first books were in the angular 'textura' gothic, but this is only one of three members of this family. The 'rotunda' had a more successful history and spread throughout Italy, France, and Spain and generally continued well into the sixteenth century. The third of these designs, the 'bastarda', which was based on more cursive book-hands, prospered in Northern Europe in the first half of the sixteenth century, while in Germany it was still popular in the middle of the twentieth century.[38]

Alongside the gothic types were the romans based upon Italian humanist hands and first appearing in Germany and Italy in printed books in the 1460s. The hand from which such types were derived was not familiar to the majority of readers, and works printed in roman in the fifteenth century tended to be the learned classical texts which were associated with that hand. A descendant of roman, the italic type based upon informal Italian chancellery hands, did not make its first

of the high cost of providing the office with suitable types (see Beltrán de Heredia, *Cartulario*, ii. 589–91). Evidence from France gives a similar picture: in a Parisian inventory of 1523, a printing-press was valued at about 9 *livres* while a fount was worth 36 *livres* (see Febvre and Martin, *L'Apparition*, p. 164, and Martin and Chartier, *Le Livre conquérant*, p. 279).

[38] Gaskell, *A New Introduction*, p. 16; A. F. Johnson, *Type Designs: Their History and Development* (London, 1934), p. 14, has an additional category, the 'fere-humanistica'.

appearance until the beginning of the sixteenth century in a type commissioned by Aldus Manutius at Venice.

When we turn to fifteenth-century printing in Spain, we find that the first books were printed in roman, but that printers soon transferred most of their production to gothic and this type went on to drive out roman almost everywhere in Spain for the whole of the first half of the sixteenth century. It is the gothic which must be central to any study of the typography of Spanish printers at this time. It accounts for all but three of the twenty types used by the Crombergers. Updike attributes the distinctive quality of Spanish printing of the fifteenth and sixteenth centuries particularly to its types and claims that this distinctiveness is a natural product of 'subtle influences emanating from the soil and skies of Spain itself'.[39] We can do better than this. The reasons for the predominance of gothic over roman, and of the rotunda over the other sorts of gothic, will be treated separately.

Naturally enough, the 'general reader' of fifteenth-century printed books probably preferred a type which more closely resembled the book-hands to which he was accustomed, and he may well have found roman letters difficult to read.[40] However, gothic would have been equally familiar to this general reader in countries other than Spain where roman fared better, and some further explanation is required. Bohigas maintains that the answer is simple: the Renaissance did not make the impact in Spain that it had in Italy, where roman type—the face which reflected the new classical scholarship, based as it was on the inscriptions of antiquity—had originated.[41] While a printer like Aldus Manutius, whose editions typified the New Learning, possessed no gothic founts at all, printers in benighted Spain had to adapt to local conservative tastes for gothic, or go under. But this explanation does not account for the anonymous fifteenth-century printer at Salamanca who continued to use roman types when they had been all but banished in the rest of the country, for the fact that most Spanish printers of incunables possessed at least one roman type and used it intermittently, nor for the introduction of such a type by Brocar in Alcalá at a time in the sixteenth century when other Spanish printers, particularly at Seville, were abandoning it.[42]

Although conservatism of taste and the marked nationalism of late fifteenth-century Spain may have had something to do with the rejection of roman type, an examination of the sort of titles produced on Spanish presses provides a more convincing explanation of this phenomenon.[43] As was noted in Chapter One, the

[39] Daniel Berkeley Updike, *Printing Types: Their History, Forms, and Use*, 2nd edn., 2 vols. (Cambridge, Mass., 1937), i. 103.

[40] Norton, 'Typographical Evidence', p. 99.

[41] Pedro Bohigas, *El libro español: ensayo histórico* (Barcelona, 1962), p. 156.

[42] Norton, *Printing in Spain*, p. 42.

[43] At least one historian of the printed book believes that regional or national prejudices did not govern the choice of type design as long as unfamiliar founts did not create difficulties of comprehension (see Hirsch, *Printing*, p. 7).

percentage of vernacular books printed in Spain was extraordinarily large. Editions of classical works in Latin were traditionally printed in roman, and the unusual press at Salamanca which insisted on employing roman founts not only specialized in such editions, but was also closely associated with the leading Spanish humanist of his day, Antonio de Nebrija. Similarly, in the first years of the sixteenth century Jacobo Cromberger used roman types, but only to any great extent in two books. One was the *Ode in divae Dei genitricis laudes* of Fernández de Santa Ella, the major humanist scholar in Seville, and the other Nebrija's edition of Persius' *Satires*. Likewise, Brocar used them for his more academic Latin texts.[44] Printers were conscious that a certain sort of public, a certain sort of work, and even a certain language required a particular kind of type.[45] But this consciousness is seen not only in the pages of the printed book. In some contemporary documents gothic notarial hand is used for the vernacular, but the occasional words in Latin are written in a neat italic.[46] When Charles V undertook his Tunis campaign in 1535 he commissioned an artist to accompany his troops and depict important battle-scenes. These graphic records were later transformed into the set of magnificent tapestries, the *Conquest of Tunis and La Goletta*, which now hang in the Alcázar of Seville. Each tapestry is accompanied by narrative passages in Latin and Spanish. Predictably enough, the Latin was woven in roman letters, the Spanish in gothic.

There was, then, a distinction made in the first half of the sixteenth century between those books which were expected to appear in roman and those which were to be printed in gothic. This distinction is illustrated by the description of that type which was being manufactured for Ungut at the time of his death: 'a vernacular type [i.e. specifically for printing in Castilian] which is still being made'. As Spanish printers concentrated upon vernacular and liturgical printing, their books naturally used gothic types. The predominance of gothic in Spain has more to do with the sort of books produced there than with any 'innate national predilection' for that design.

Once the pattern of gothic printing had been established, not only would the vernacular reading-public's taste have been reinforced, but economic considerations would have militated against any typographical experiments. At a time when most Spanish printers operated under the threat of bankruptcy, they

[44] Norton, *Printing in Spain*, p. 42.

[45] Lowry, *The World of Aldus Manutius*, p. 130, shows that this consciousness also existed in Venice, at least at the beginning of the 16th c. Robert Brun, *Le Livre français* (Paris, 1948), p. 41, similarly notes that, 'To the reading public [of 16th-c. France] gothic was identified with certain books like devotional works, novels, or chivalric literature . . . The same applies to legal works'. See also Febvre and Martin, *L'Apparition*, pp. 110-11.

[46] Printers thought of themselves as 'writing' a page mechanically. At Seville in 1529 a printer or pressman, Cristóbal Bernal, was referred to as a 'scribe who uses types' (see document dated 7 Jan. 1529 [APS, Oficio 4, Libro 1 of 1529, fol. 136^{r-v}]).

would have been reluctant to incur the expense of having designs of doubtful appeal cut. Indeed, most of Jacobo Cromberger's typographical innovations can be explained by commercial considerations. Brocar was able to invest large sums in the cutting of new faces because he had, in Cardinal Cisneros, a rich patron. Other sixteenth-century printers in Spain were not so fortunate and largely contented themselves with reusing the old matrices which they had acquired from their fifteenth-century predecessors. Eventually Jacobo Cromberger would use gothic types even in books of the sort which formerly would have been printed in roman, and would discard his roman type for everything but the humble task of printing folio-numbers in some of his liturgical books.

It was not until the middle third of the sixteenth century that Spanish printers generally introduced into their repertoire italic types, and above all the new romans, which had been made familiar by their appearance in popular works such as those of Erasmus. In the 1530s Juan Cromberger responded to this fashion and acquired a small roman fount (Type 19) but even then this type was not a success. A similar failure was his father's introduction of another foreign type, the italic, which will be discussed later in this chapter.

So strong was gothic in Seville that roman types may not have been available to most printers even many years later. In June 1550 Pedro de Luján agreed to print an edition of Gutierre de Trejo's *In sacrosancta Jesu Christi quattuor Evangelia*. In February of the following year he reiterated his promise, but still the book did not appear. One reason for the delay was doubtless the difficulty of obeying the stipulation that the edition be printed in 'two ancient Latin types [i.e. roman]'.[47] Eventually, in 1554, Luján produced the book, but in a shabby gothic type, not in roman. There are, then, reasons other than the 'influence of the Spanish soil and skies' which help to explain the resistance of gothic types in Spain.

It was not, however, just any sort of gothic type which was used. The rotunda soon became almost the only form seen in Spanish printing and, indeed, so popular was it in Spain during this period that it became identified with that country in the minds of foreigners. Plantin called one of his rotunda types the 'Castillane' and another the 'Canon d'Espagne'.[48] Again, the reason for the success of this design is not merely some inexplicable attraction felt for it by contemporary Spaniards nor necessarily its resemblance to local hands, although it was, indeed, a popular book-hand in Castile, where it was normally termed 'letra de juros'.[49]

[47] Gestoso, p. 118. On the even more drastic lack of Greek types in Spain, see Gil Fernández, *Panorama social*, pp. 586–94.

[48] Carter, *A View*, p. 16, n. 2; Johnson, *Type Designs*, p. 23; H. D. L. Vervliet, *Sixteenth-century Printing Types of the Low Countries* (Amsterdam, 1968), pp. 178–9.

[49] Millares Carlo, *Introducción*, pp. 49–50. Juan de Yciar, in his *Recopilacion subtilissima intitulada orthographia practica* (Zaragoza, 1548), calls this book-hand 'letra redonda'.

Printing came to Spain from Italy, and from Italy came the early roman types in which the first Spanish incunables were produced. But the rotunda gothic which soon replaced them was also Italian in design. Indeed, so closely identified with Italy was the type that German printers of the early period referred to it as 'litterae venetae', for it was in Venice that most of the innovations in type-design during the late incunable period took place.[50] The rotunda soon became the accepted gothic type in Mediterranean countries. It spread to Lyons in 1482, travelled from there to the south of France, and thence to the Iberian Peninsula.[51]

The use of the rotunda in late fifteenth-century Seville was only natural. The major printers in the city from the 1490s onwards were the 'Compañeros alemanes', and Ungut and Polono. The 'Compañeros' are thought to have come from Venice. Polono and Ungut had previously worked in Naples and brought with them at least four types which had been used in Moravus' Neapolitan office. As the types used by these firms formed the core of the material with which Varela and Jacobo Cromberger began their independent careers, it comes as no surprise to find that it is the Italian style of gothic which holds sway in the city during the sixteenth century.

It is fortunate for us that the bulk of the Cromberger press's material was gothic, because peculiarities and innovations in this type are more readily perceptible than in roman or italic. Although much of their production is unsigned and undated, a close examination of their types enables works printed by this firm to be indentified and also dated with a certain degree of precision.[52]

The Crombergers' Types

The full range of types used by the Crombergers is illustrated in Appendix Two with detailed notes on modifications and dating. An examination of these types reveals significant characteristics of the press. Of the twenty types employed seventeen are rotunda gothics, two roman, and one italic. As the roman types appear only very rarely and the italic but once, Bohigas is correct to label the press as 'remarkably gothicizing'.[53] In this the Cromberger press is representative of most Spanish printing during the first half of the sixteenth century, but it may have been even more insistent on the preservation of the gothic than contemporary printers in some other Spanish cities. The explanation for this is the popular nature of Seville printing in general and of the Cromberger press in particular.

[50] Johnson, *Type Designs*, p. 20.

[51] Carter, *A View*, p. 52.

[52] Norton, 'Typographical Evidence', *passim*. Norton's method was adapted from that developed by Bradshaw and Proctor for the study of incunables. It is still useful for books printed long after the 15th c. and gives reliable information about the products of a press like that of the Crombergers.

[53] Bohigas, *El libro español*, p. 171.

In Chapter Six the continuity of taste in secular literature between the fifteenth and sixteenth centuries was noted. The physical presentation of these works was equally traditional, and the Crombergers' large output served to fulfil the expectations of the reading public and reinforce its conservatism. Their folio editions of romances of chivalry, for example, which were illustrated with massive title-page woodcuts and printed entirely in gothic types, set a pattern followed everywhere in Spain during at least the first half of the sixteenth century. If the readers of secular literature were conservative, the Crombergers provided them with the books they loved, and ensured that these books would be presented in a familiar manner.

Of the twenty founts employed over a period of some sixty years, eight derived from the stock of Polono and Ungut (Types 3, 6, 7, 8, 9, 11, 15, and 16) and at least three more came from the 'Compañeros alemanes' (Types 2, 4, and 5). At least four of the types originally owned by Ungut and Polono date back to the 1480s, when they were used in Italy (Types 7, 9, 15, and 16). The long period during which Types 7 and 15 lay neglected in Jacobo's office does, incidentally, suggest that he was not short of type or, at least, of unused matrices. Of the remaining types, at least three derive from fifteenth-century founts used by other printers in Spain or are based on the designs of such founts (Types 1, 10 and 13), while two more derive from types recorded in the first decade of the sixteenth century (Types 14 and 17). This leaves only four new ones, of which two are traditional in design (Types 12 and 20) and only two could be said to belong firmly to the sixteenth century (Types 18 and 19, the first an italic and the second a small roman). However, if the Crombergers' stock of types was never revolutionized, neither was it stagnant under Jacobo's management or his son's.

Jacobo was not content merely to reuse inherited types. Rather, he was sensitive to contemporary taste and adapted to it even though his modifications were largely confined to the gothic types which were so popular in Spain. From 1504 until his death in 1528 he modernized his material by abandoning six outdated founts (Types 3, 5, 6, 9, 11, and 16) and adding eight new ones (Types 1, 10, 12, 13, 14, 17, 18, and 20). In at least three of these cases he adopted the fashionable founts of other Spanish cities (Types 1, 13, and 17). He might not have been as innovative in this modernization as Brocar was but, as we have seen, neither did he have access to the sort of patronage which Brocar enjoyed. Rather, he acquired types which had already been tried and found successful by Brocar and other Spanish printers. This is shown by the large number of printers who acquired some of these new types. As well as adding new founts to his equipment and abandoning old ones, Jacobo also modified five types, some of them radically (Types 2, 4, 8, 13, and 16). He was not, therefore, unremittingly conservative in his material.

Indeed, he made one very important innovation to his stock of types which deserves further comment. It had long been thought that in 1528 he printed a text which was most unusual for a Seville press of this period: the first edition printed in Spain of Lucan's *Pharsalia* and the only one to appear there in Latin during the sixteenth century. What was not realized until two unknown copies of this edition came to light in the Biblioteca Nacional at Lisbon was that this book had been set up in italic type. The type employed was an imitation of the Aldine italic, which was certainly known in Seville by this date from imported books. The colophon of this edition clearly states that it was printed by Jacobo in 1528, a claim confirmed by both internal and external evidence. This date makes his *Pharsalia* the first book in which italic types were employed in Spain. The use of this type, together with the octavo format, in an edition of a classical work which is an indirect copy of the Aldine *Pharsalia* printed at Venice in 1502 shows that this edition was a daring venture for Jacobo. Vervliet has remarked that 'more than any other kind of letter [italic type] was the emblem and blazon of new ideas, the garment in which the Renaissance showed herself to an enthusiastic following'.[54] By 1500 Aldus, who was the foremost typographical exponent of the New Learning, had commissioned the first italics to be made anywhere, and they had been specifically designed for his projected series of octavo editions of the classics. This series had travelled rapidly throughout Europe and was equally rapidly pirated in many cities. However, Seville was hardly a centre for the publication of classical texts in Latin, let alone for editions employing innovative types and formats.

A clue to the reasons behind Cromberger's decision to issue an edition inspired by Aldus is given by a short neo-Latin poem printed on some blank pages which followed the colophon. The poet was the same Pedro Núñez Delgado who has been encountered before in this study and who, as has been noted, was associated with Seville humanists and pedagogues, many of whom were imbued with Erasmian ideas. Jacobo's edition of the classical text appears to have been addressed to this circle or their students. In can therefore be appreciated why he used it to experiment with a style of printing copied from Aldus, who was so closely associated with humanist ideas and had been Erasmus' host and publisher on several occasions. Indeed, Erasmus had singled out Aldus' italic types for special praise in a letter of 1507.[55] What is, however, striking is the contrast between this edition of Lucan and the bulk of the Cromberger press's output. Nevertheless, the inclusion of the poem suggests that Núñez Delgado may have had a hand in the edition, if only to encourage Cromberger to issue it and to do

[54] Vervliet, *Sixteenth-century Printing Types*, p. 68.

[55] *Opus Epistolarum Des. Erasmi Roterodami*, ed. P. S. Allen *et al.*, 12 vols. (Oxford, 1906–58), i. ep. 207, p. 439.

so in a format and a type which looked modern in a way that only imported books had done until then, and also offered a virtual guarantee of a text edited to the highest standards of contemporary scholarship.[56] Cromberger may also have considered that the time was ripe, only one year after the Sack of Rome and at a moment when imports from foreign presses were being interrupted by the international crisis, for a venture into the market which they had dominated. He may have believed that, if he was to succeed with the experiment of printing octavo editions of the classics in the new Italian type, he would do well to choose the *Pharsalia* to inaugurate the series. It had not only been published by Aldus, but was a curriculum-text long popular in Spain, and its author was considered by Spaniards to be a compatriot. The *Pharsalia* would have been an appropriate text to publish at a time of intense Spanish nationalism, especially as it told a story of internecine strife taking place in Rome while her true enemies looked on—a situation which could scarcely fail to strike a chord among the Spanish subjects of that new Christian Roman Emperor, Charles V, whose conflict with Pope Clement VII had to be seen against the back-cloth of the real threat to Christendom, the Turk.[57] Nevertheless, the introduction of this new italic type into Jacobo's material—the only change which shows him leading rather than following typographical fashions—was a failure. Only the one title is known to have been printed by him in these types and months later many copies of his edition were still unsold.[58] Jacobo died soon after his *Pharsalia* was published, and it is significant that Juan did not repeat the experiment. The dominance of gothic type was not to be challenged.

Juan Cromberger introduced far fewer changes to the typographical material which he inherited than his father had done. Apart from a minor modification to Type 7 and some changes to Type 2, which may anyway have taken place before Jacobo's death, he preserved all the types which had come down to him and introduced only one other: the roman Type 19. The acquisition of a small roman of the sort used in a school-text like his 1535 edition of Verino's *Liber distichorum* is a sign of the belated revival of roman in Spain, but this type also seems to have been unsuccessful, either because Juan could not break into the market for roman texts, which were being regularly printed in cities such as Alcalá, or because the sort of markets for which he catered were not ready for such innovations.

As might be expected, Jácome's years as the head of the family press were ones of typographical stagnation. The period 1540 to 1560 therefore shows no innovation in the press's types. Jácome maintained the stock inherited from his

[56] Although Lowry, *The World of Aldus Manutius*, pp. 217–49, indicates that the editing of Aldus' octavo editions of the classics may not have been as scholarly as was once thought, by the lights of his age the standard of editing was high, and among contemporaries Aldus' reputation was unparalleled.

[57] For a detailed account of this edition and its italic types, see Griffin, 'The Crombergers of Seville'.

[58] Document dated 7 June 1529 (APS, Oficio 4, Libro [2] of 1529, unfoliated).

father, but used it in a more worn state than had ever been seen before, possibly because he did not know how to recast it and there was no skilled artisan to do it for him, or because finances were exhausted. Only one fount was altered during this period; in 1543 Type 10 was cast on a smaller body, but even this recasting was made when Brígida Maldonado, rather than Jácome, was in charge of the press.

The development of the Crombergers' typographical material therefore parallels the rise and decline of the press's fortunes over three generations. The range of its types was sufficient for any sort of work a leading printer had to undertake. For example, liturgical printing required material with which to print music; the Crombergers owned two sizes of staves and notes (see Appendix Two). They would have possessed sufficient numbers of pieces of type in each fount to keep the presses employed when production was in full swing. The minimum quantity of type required for each press in a sixteenth-century printing-shop, if there were to be no pauses in production, was the amount sufficient to fill three or four formes. Two formes would normally be printed each day, one in the morning and one in the afternoon and evening. The remaining one or two formes of type would not be available for setting, either because the types were being washed and distributed back into the cases or because they were standing locked in the formes awaiting the correction of first or second proofs. Obviously, the amount of type required and the speed with which a forme was printed would depend on the size of the edition being run off.[59] Unfortunately, the precise amount of type possessed by the Crombergers is impossible to ascertain. Nevertheless, the occasional references to the weight of type owned by other Seville printers suggest that the quantity increased over the prosperous first half of the century. In 1499 Ungut left three presses and six *quintales* [276 kg.] of type, or 200 *libras* [92 kg.] for each press.[60] In 1555, when Gregorio de la Torre bought Varela's presses from Jácome Cromberger and Varela's heirs, 750 *libras* [345 kg.] of type-metal per press, as well as a large quantity of cast type, was sold with them.[61]

The range of types employed by the Crombergers in their branch offices was naturally more restricted. If they really did print in Portugal, the itinerant press there carried only two founts: Types 2(*c*) and 7(*a*). Before the Mexican branch became independent of the Seville house its material included Types 1, 2, H, and 8(*a*).

If any pattern can be discerned in the typographical modifications made, especially by Jacobo, it is not only one of updating the press's material within

[59] Gilmont, *Jean Crespin*, pp. 50–3.

[60] Tenorio, 'Algunas noticias', p. 636.

[61] Document dated 29 Nov. 1555 (APS, Oficio 5, Libro 4 of 1555, unfoliated). Gaskell, *A New Introduction*, p. 38, gives the average weight of type per printing-press in the 16th c. as between 225 and 900 kg. At Zaragoza in 1579, six cases of type weighed 340 *libras* [156.4 kg.] (see Abizanda y Broto, *Documentos*, i. 357). See also Cruickshank, 'Some Aspects'.

the limits of the conservative use of gothic in Spain, but also more straightforwardly one of development towards the use of smaller text-types. Types 10 and 13 are smaller alternatives to Type 8 for folio and quarto books respectively. This is paralleled by increasingly compressed setting. It is reasonable to suppose that this development was the result of the need to economize on the cost of paper, which was expensive and was not always readily available in Seville.[62]

However, a final note of warning should be sounded. While frequent occurrences of some types enable inferences to be drawn about typographical developments and the dating of undated editions, some other founts are recorded in only one edition (Types 5, 9, 18, and 19). Given the large investment involved in acquiring a new type—if, indeed, it was bought rather than just hired or borrowed—it is unlikely that such founts were not used in other books all trace of which has disappeared.

[62] Although the cost of the paper would eventually be passed on to the purchaser, it was in the interests of the printer to economize not only to make his product more competitive, but also to reduce the amount of his own capital invested in an edition. When Crespin financed editions himself at his Genevan press, he used small types; when others were paying, he tended to use larger types (see Gilmont, *Jean Crespin*, p. 79).

CHAPTER 8

WOODCUTS, ORNAMENTAL MATERIAL, AND ORNAMENTAL INITIALS

Woodcuts

THE early printers in the Iberian Peninsula brought with them from Germany, Italy and France all the tools of their trade; these included the wood (or occasionally metal) blocks from which illustrations were printed. When they wanted to increase their stock of woodcuts, they tended either to copy foreign designs or to import blocks from abroad.[1] This was natural enough; printing was a new art in Spain and many of the first printers still had close commercial contacts through the great trading companies with their homelands, especially Germany, where the craft was considerably more developed. Not only was printing with movable types a late arrival in Spain, but the art of woodblock-cutting had developed much later in Spain and England than it had done in Northern Europe and Italy; little of the illustrative material used in Spanish incunabula until about the 1490s is of native design compared with that encountered in the early books printed in Germany, France, Italy, and the Low Countries.[2] The Zaragozan printers Hurus even took a woodcutter from Constance to work for them in Spain.[3]

Nevertheless, in the final years of the fifteenth century a particularly 'Spanish' style of woodcut illustration emerged, and this remained virtually unchanged until the mid sixteenth century. The conservatism of Spanish printing, which has been remarked upon before in connection with the sort of books issued and the continued use of gothic types long after most other countries had relegated them to the most popular products of the presses or to a few special sorts of publication, can also be seen in the style of Spanish book-illustration, which developed little from the incunable age.[4] Whatever the reasons for this conservatism, which is

[1] Norton, *Printing in Spain*, pp. 69–70; and Lyell, *Early Book Illustration*, pp. 38–40.

[2] Arthur M. Hind, *An Introduction to a History of Woodcut with a Detailed Survey of Work Done in the Fifteenth Century*, repr. of 1935 edn., 2 vols. (New York, 1963), ii. 741; Marlene Chambers, 'Early Printing and Book Illustration in Spain', *The Library Chronicle*, 39 (1973), 3–17 (p. 4).

[3] Millares Carlo, *Introducción*, p. 127.

[4] In this, Spanish printing was somewhat similar to English printing at this time. Woodcuts used in England were often either imported from the Continent or cut in England by foreign craftsmen. The relative backwardness in the pictorial arts in both Spain and England meant that 15th-c. blocks did not look incongruous in the 16th c., while in France and Germany such old-fashioned woodcuts were generally no longer in use. In Italy old blocks continued to appear, but only in books printed for the popular

probably better explained by the shaky finances of the printers than by any reference to an undefined 'Spanish taste' in book-illustration (especially as most of the printers were foreigners anyway), it cannot be denied that a Spanish incunable or early sixteenth-century book is easily recognized. This is particularly true of the title-pages of folio or quarto volumes. They frequently contained large blocks cut with bold lines representing a coat of arms or the hero of the work in combination with a minimum of letterpress in large gothic types or in the form of a simple xylographic title. The heraldic title-page was far more popular in Spain than elsewhere in Europe.[5] The large woodcut and title are normally enclosed within a border made up of several pieces printed from parts of blocks which previously formed a four-sided compartment, but were subsequently cut up and used in a haphazard manner. The page is thus given a robust and monumental appearance in comparison with the more delicate Italian title-pages. Although the woodcuts used in Spain are somewhat crude, they combine well with the heavy gothic types employed.

The percentage of early Spanish books which are illustrated is a point of some dispute. Lyell claims that it is high, saying that some two hundred of approximately nine hundred incunable editions which came from the Spanish presses were illustrated, while Bliss maintains that printers in Spain were more sparing with their blocks than were their colleagues in other countries.[6] What is certain is that we never see in Spain a book as richly illustrated as Anton Koberger's *Nuremberg Chronicle* printed in that city in 1493, which used over six hundred cuts. Nevertheless, a high proportion of Spanish books contain a woodcut on their title-page even if the printer was subsequently sparing with illustrations within the text. Statistics, which are often unreliable in the study of the early printed book because so many editions have disappeared, are even more misleading in the case of illustrated books, for these have suffered from even greater loss through being cut up for their woodcut pictures.[7]

Despite the absence of reliable statistics, it does seem that Seville was an important centre for book-illustration. We have seen that the first surviving illustrated book to be printed in Spain was the 1480 *Fasciculus temporum* which came from the Seville press of Bartolomé Segura and Alfonso del Puerto; in this

market (see A. W. Pollard, 'The Transference of Woodcuts in the Fifteenth and Sixteenth Centuries', *Bibliographica: Papers on Books, Their History and Art*, 2 [1896], 343–68).

[5] A. W. Pollard, *Early Illustrated Books: A History of the Decoration and Illustration of Books in the 15th and 16th Centuries* (London, 1917), p. 211. The first incunables, like the manuscripts they imitated, had not used decorated title-pages, nor indeed title-pages at all, but the early printers soon realized the potential of the title-page and, by the end of the 15th c., it was common in printed books.

[6] Lyell, *Early Book Illustration*, p. 3; P. D. Bliss, *A History of Wood-engraving* (London, 1964), p. 89.

[7] This is true of a great Spanish library, the Biblioteca Colombina of Seville, which has suffered serious depredations over the centuries and, in particular, has been robbed of illustrated books or the illustrations from such books.

edition those early printers employed a dozen stock metal cuts. Although Spanish incunables are not noted for the originality of the designs of their illustrations, if there is originality anywhere in Spain, it is to be found in Seville; some blocks cut in that city in the sixteenth century were imitated not only in other Spanish cities, but also abroad.[8]

Appendix Three contains approximately 1,500 examples of woodcuts and decorative blocks used by the Cromberger press at Seville, but nevertheless does not represent all the blocks used in Cromberger editions.[9] This enormous stock indicates how copiously many of the Cromberger products were illustrated.[10] The family must have invested a good deal of capital in commissioning, buying, or hiring these blocks, for they were expensive items and their use would have raised the cost of the books considerably. A large proportion of the outlay in printing a book at this period was the purchase of paper, and a publisher naturally tied up more of his money in an edition if it was illustrated. To give just one example, the Aesop which Jacobo Cromberger printed in about 1510 devoted some 33 per cent of the paper to woodcuts, thus requiring 50 per cent more investment in paper than would have been the case with an unillustrated edition. To this should be added the extra labour required to set up an illustrated book. The return on a printer's capital was frequently very slow and uncertain, and so he would have tried to keep his investment to a minimum; yet Jacobo must have thought that illustrations were necessary in this book if it were to sell at all.[11] Another not inconsiderable expense associated with this large stock of woodcuts would have been one of storage space, and the continual acquisition of this material might have contributed to Jacobo's need to change the location of his printing-office frequently in the early years of the press, as well as his grandson's decision to remodel his house near the end of his career.[12]

The obligation which the Crombergers seem to have felt to invest this capital in illustrations can be explained by the potential market for their products. Illustrations made in multiple copies from a block were considered in the fifteenth century to be a cheap alternative to the art of the illuminator and were therefore often relegated to more popular works. In Italy very few incunable editions of

[8] A. W. Pollard, *Fine Books* (London, 1912), p. 163.

[9] The expense of making photographs of so many woodcuts is prohibitive. I have therefore generally reproduced blocks only from those books I have been permitted to photocopy. Nevertheless, the great majority of blocks used by the Crombergers are reproduced in this Appendix.

[10] The stocks of printers in other European countries could be even greater than this. The Plantin–Moretus press which was founded in the 1550s possessed at least 15,000 wood-blocks and 3,000 copperplates (see Leon Voet, *The Golden Compasses: A History and Evaluation of the Printing and Publication Activities of the Officina Plantiniana at Antwerp*, 2 vols. [Amsterdam, 1969–72], i. 426).

[11] Steinberg, *Five Hundred Years*, p. 202, maintains that in England at this time illustrated books were twice as expensive as unillustrated ones.

[12] Woodcut blocks would have been about seven-eighths of an inch deep so that they could be used in the same forme as the type with which they were printed. They would therefore have been bulky to store.

learned texts were accompanied by woodcuts; these are, rather, to be found in little books of popular devotion, morality, or education as well as the famous illustrated editions of fables and *novelle*. It is true that some Italian editions of Petrarch, Dante, and Boccaccio as well as of Italian translations of Livy, Herodotus, Terence, and Ovid contained woodcuts, but the careless presswork of these books, most of which were printed at Venice, indicates that they were intended for a readership which, although able to afford something grander than the common chap-book, had little claim to scholarly learning.[13] Woodcut illustrations appear to have been regarded in a similar light in early sixteenth-century Spain, as can be seen from editions printed there by Brocar. When he was working at Logroño, he produced at least one finely illustrated book, but he later moved his main centre of activity to Alcalá some three years after Cisneros had founded the University there. Finding himself now the printer of fine editions by and for humanist scholars, he abandoned the woodcuts as unfitting additions to his texts, although he was willing enough to use them again at his branch office in Toledo, a city which was a centre for popular printing.[14] The finest products of printers with scholarly pretensions like Aldus Manutius at Venice or Brocar at Alcalá were for the most part printed without illustrations. It is significant that when Jacobo Cromberger chose to imitate Aldus in the edition of Lucan's *Pharsalia* which he printed at Seville in 1528, he did not illustrate the book; unfortunately, the title-pages of the two surviving copies of the edition are missing, but it is probable that even this preserve of the woodcut in Spain would have contained nothing but letterpress in strict accordance with Cromberger's Venetian model.

The large proportion of the Crombergers' editions which were illustrated is therefore another indication of the popular nature of much of their production, although they may have been unusually fond of illustrations even for printers of a range of popular titles.[15] When it is remembered that the cheapest and most popular of their editions, the ephemeral chap-books, have probably disappeared in larger numbers than their more expensive products, it is reasonable to conclude that their press had a particular bias towards the lower end of the market. This may well hold good for the Seville presses as a whole, as can be seen from Domínguez Guzmán's analysis of the surviving evidence of Seville printing in the first half of the sixteenth century. Some 80 per cent of this is made up of literature of entertainment, works of popular devotion, and popular history.[16]

[13] A. W. Pollard, *Italian Book Illustrations Chiefly in the Fifteenth Century*, Portfolio Monographs on Artistic Subjects, 12 (London, 1894), pp. 14–15.

[14] Norton, *Printing in Spain*, p. 48; Jesusa Vega González, *La imprenta en Toledo: estampas del Renacimiento* (Toledo, 1983), pp. 35–8, 145–8.

[15] Compare with the relative lack of illustrations in the books printed at Toledo by Hagenbach's successor (see Norton, *Printing in Spain*, p. 52).

[16] Domínguez Guzmán, *El libro sevillano*, pp. 233–71.

An examination of the illustrations reproduced in Appendix Three immediately reveals two points about the Crombergers' blocks. First, they were numerous and of a diverse nature in size, in style, and in the technique used by the wood-engraver, while, although changes in style can be to some extent explained by the chronology of acquisition of the material, the older blocks continued to be used in an eclectic manner alongside the newer ones. Second, this material had a remarkably long life. Blocks which had been used by Polono and Ungut in the fifteenth century appear repeatedly in the sixteenth. Jacobo Cromberger had obtained them either from Ungut's estate when he married Comincia de Blanquis or from Polono. Examples of such blocks are the set of foreign woodcuts treating religious subjects which will be discussed below (WC:479, 519, 621, 870, 871, etc.), the hagiographical set (WC:625, 872, etc.), and those which make up most of WC:506–616 illustrating the life of Christ.[17] Material did not come into Cromberger's possession exclusively from the partnership of Ungut and Polono, but also from the other important foreign printers of incunables at Seville, the 'Compañeros alemanes', although it is just possible that some of this material had been acquired by Ungut when he and Pegnitzer were called to Granada to print for Archbishop Talavera in that city in 1496, and that it subsequently passed to Cromberger as part of Ungut's estate. Examples of Jacobo's use of material which had originally belonged to the 'Compañeros' are WC:876, an illustration of Seville cathedral, and 1000, part of an Aesop set.[18] It is possible that, as had occurred with the types used by Polono and Ungut and later passed on to Cromberger, some of this material had been brought from Naples and had originally formed part of Moravus' stock. I have been unable, however, to substantiate this suggestion. Once these blocks had passed to Jacobo, he used them until his death, and they then formed part of the stock which he had handed over to his son, Juan, in 1525. When Juan died in 1540 this same material was inherited by Brígida Maldonado and subsequently passed to Jácome Cromberger. In the 1550s Jácome was still using blocks which had been cut over fifty years earlier.

There is one case of a metal block which had even more distant origins: this printed the large dotted illustration of St Jerome and the lion (WC:348) which Juan Cromberger used in his 1537 Seville edition of the saint's letters. This block had been seen earlier at Seville in Varela's edition of the same work printed five

[17] That woodcuts from these sets had been used by Polono and Ungut is confirmed by the reproductions in Aloys Ruppel, *Stanislaus Polonus: polski drukarz i wydawca wczesnej doby w Hiszpanii*, tr. and enlarged by Tadeusz Zapiór (Cracow, 1970), and Vindel, *El arte tipográfico*, v. 272–6.

[18] For these blocks used by the 'Compañeros' see José L. Portillo Muñoz, *La ilustración gráfica de los incunables sevillanos (1470–1500)* (Seville, 1982), figs. 13, 20. The use of WC:1000 by the 'Compañeros' suggests that they originally owned the complete Aesop set WC:878–1066, but no edition of Aesop printed by them has survived.

years earlier, but its origin can be traced back to Cologne *c.*1470.[19] The Cromberger press does not, however, appear ever to have used copperplates, which became popular among printers later in the sixteenth century and are even to be found in the occasional incunable.[20]

The long life of the blocks used by the Crombergers is proof of the care with which Jacobo and Juan handled their material and provides further evidence of the conservative nature of their printing.[21] Constant recourse to the same illustrations over a long period would not have been thought at all strange by readers because, as we have seen, the Crombergers' editions were frequently reprints of recognized favourites, many of which were medieval in origin; it would have been fitting that such works be illustrated in the time-honoured fashion and that some of the designs used should even long predate the printing age.[22] Nor are the Crombergers alone in using the same material over a long period; this seems to have been normal practice in sixteenth-century Spain, where there is evidence of the same blocks being used for the best part of a century.[23]

There is a further conclusion which can be drawn from the longevity of the blocks. Jacobo Cromberger had inherited fifteenth-century material and had continued to use it in the sixteenth, at the same time adding to his stock. This practice was continued by his son at what appears to have been the height of the press's prosperity. Yet it is noticeable that Jácome was content merely to use inherited material even when it was very worn. Not only does this use of inferior material indicate a lack of artistic judgement on the printer's part, and a dearth of skilled workmen able to select blocks in better condition; it is also yet another indication of the economic difficulties experienced by the Seville presses in the mid century.

We have said above that Spanish books of the late fifteenth and early sixteenth centuries are easily distinguished from those of other European countries and that the 'Spanish style' of the woodcuts which emerged at the end of the fifteenth century contributed to that distinctiveness. An examination of Appendix Three should make this apparent, but some general remarks may be made here about the woodcuts used by the Crombergers and we can attempt to divide them into separate categories.

They were designed to blend with the heavy gothic types and consequently used thick raised lines. There is little delicate hatching, although some simple shading

[19] Lyell, *Book Illustration*, p. 168. Varela and Cromberger were not averse to lending each other material (e.g. WC:84 and 375); see also App. I for the description of Nebrija's *Introductiones Latinae* possibly printed at Seville by Varela with Cromberger's material before the end of 1513.

[20] Millares Carlo, *Introducción*, p. 129.

[21] On the difficulty of assessing the average life of a wood-block, see Hind, *An Introduction*, i. 25.

[22] For example, the Virgin and Child in Glory of WC:350 and 874.

[23] Norton, 'Typographical Evidence', p. 99.

is attempted. The boldness of line does not produce a subtle effect, though it must have contributed to the blocks' durability. The cutting frequently gives the impression of crudity or ingenuousness, an impression created by a combination of features: unimaginative composition (WC:467), the woodcutter's inability to reproduce realistic perspective (WC:26), the incorrect proportion between figures (WC:400 or 416), and a lack of the fine cutting which is vital if an artist is faithfully to reproduce facial expression (WC:342). These features give the cuts the appearance of naïve or 'medieval' art in contrast to the more 'modern' Renaissance compositions of imported blocks (WC:30–32, 34, 36, and 341). In this the Seville blocks do not differ markedly from the local painting of at least the first half of the sixteenth century, which was generally both old-fashioned and provincial in comparison to that of Northern Europe or Italy. Indeed, many years later the composition and perspective of even a local artist of the stature of Francisco Pacheco, mentor and father-in-law of Velázquez, was similarly defective.

The woodcuts were usually printed in black ink only and there is no evidence that the Crombergers ever used more than two colours (red and black) in their illustrations. However, printing in a variety of inks was not unknown in Spain at this time, as is witnessed by Jorge Coci's edition of Pedro de la Vega's *Flos sanctorum* (Zaragoza, *c.*1521) which used six colours for the title-page.[24]

Woodcut illustrations, as opposed to the ornamental material which was normally used for title-page borders or decoration within the text, can be divided into six groups, although not all the blocks can be neatly assigned to one of these categories:

(*a*) 'title-page blocks': that is, individual blocks normally appearing on title-pages or at the beginning of different sections of a work;

(*b*) 'text blocks' (large sets): that is, sets of numerous, smaller illustrative blocks normally used within the text;

(*c*) 'text blocks' (smaller sets): that is, sets of illustrative blocks, each set containing far fewer cuts than in (*b*);

(*d*) 'factotum' or composite blocks;

(*e*) printers' marks;

(*f*) diagrams.

(a) Title-page Blocks

The individual blocks used on title-pages sometimes cover the whole page and include a xylographic title (e.g. WC: 626), but more frequently are combined with heavy border pieces and gothic type for the title.[25]

[24] Lyell, *Book Illustration*, has a facsimile of this coloured woodcut as a frontispiece.

[25] Jacobo Cromberger sometimes removed the xylographic title from 15th-c. blocks and continued to use them with movable type (e.g. WC:65/349, 97, and 400).

Many title-pages employ heraldic designs which were popular in Spain from the latter years of the fifteenth century onwards. For example, Jacobo Cromberger's 1520 edition of the *Leyes del cuaderno nuevo de las rentas de las alcabalas y franquezas* is typical of the books of law which sport woodcuts of the royal arms (WC:62, 73, 873, and 1067) on the title-page. Imposing cuts of this kind are not limited to official publications. The various editions of Guevara's *Libro áureo de Marco Aurelio con el Relox de príncipes* were similarly decorated with the arms of Spain, as befitted a serious work written by a royal chronicler; but a book did not need to be grand to be given such a monumental title-page: a fine coat of arms of Spain was used by Ramón de Petras for his Toledo edition of 1525 of nothing more distinguished than the Castilian translation of Roberto de Nola's cookery book, the *Libro de cocina*.

A somewhat different use of the heraldic design was to identify the author's patron or the person to whom the work was dedicated. Nebrija's edition of Persius' *Satires* printed by Jacobo Cromberger in 1504 and dedicated to the Archbishop of Seville, Juan de Zúñiga, bears the latter's arms on the title-page (WC:52); the same is true of Pedro Núñez Delgado's edition of the *Homelie diversorum doctorum in Evangelia* (Seville, Jacobo Cromberger, 1516?) dedicated to a fellow canon of Seville cathedral, Jerónimo Pinelo, and containing on the title-page a woodcut of the latter's arms (WC:81).

Nor was the heraldic title-page limited to books designed for Spanish readers. Jacobo Cromberger's editions printed specifically for the Portuguese market follow the same fashion. The title-pages of his Évora baptismal and breviary, both printed in 1528 for the Évora bookseller, António Lermet, and dedicated to the Cardinal Infante Dom Afonso, contain the arms of Portugal crowned by a cardinal's hat (WC:351). Similarly, the title-pages of the 1521 and 1539 editions of the *Ordenações manuelinas*, printed respectively by Jacobo and Juan Cromberger exclusively for the Portuguese market, contain a monumental woodcut coat of arms of Portugal (WC:326), which would have been commissioned especially for the first prestige edition of this work to come from the Cromberger press.

The heraldic design was, then, typical of many early Spanish books, but another widespread pattern, used particularly on the title-pages of the romances of chivalry and chronicles of which the Crombergers were major producers in the first half of the sixteenth century, was the large woodcut of a mounted knight dominating the title-page and leaving little space for the letterpress of the title. The most famous of these blocks were those used in the Amadís cycle of romances which were so frequently reprinted at Seville. Examples of these woodcuts are WC:53–56, and 74. So popular and so closely associated with the Cromberger press were the romances of chivalry that the Crombergers' designs were widely imitated in

Spain.[26] WC:53, for example, was copied by Juan de Villaquirán and Pedro de Castro in their edition of the first four books of *Amadís de Gaula* (Medina del Campo, 1545), by Sebastián Martínez in his edition of the *Crónica del rey don Alfonso onceno* (Valladolid, 1551), and by Pierres de la Floresta in his *Don Florisel de Niquea* (Zaragoza, 1568). But the fame of the Crombergers' editions of the romances of chivalry was not limited to Spain. Printers in Italy also imitated the Seville woodcuts. Jacopo Giunti and Antonio de Salamanca used a close copy of WC:54 in their edition of *Esplandián* printed at Rome in 1525, and Giovanni Antonio de' Nicolini da Sabbio's edition of the first four books of *Amadís* (Venice, 1533) had a title-page illustration which was an imitation of the Crombergers' WC:56.[27]

Other cuts used on the title-pages of Cromberger editions similarly gave some indication of the nature of the books' content. Some did this in a conventional manner: the title-pages of medical works were illustrated with a block of St Cosmas and St Damian (WC:97), patron saints of surgeons, which had been inherited from Ungut and Polono; Juan Infante's *Forma libellandi* (Seville, Juan Cromberger, 1538) was illustrated by a woodcut depicting a dispute before a judge printed from a block which had also come from Polono's stock (WC:400); the translation of St Gregory's *Moralia* was announced by a large woodcut of the saint wearing the Papal crown (WC:44); and a series of manuals for confessors employed a graphic woodcut of Satan fighting for a sinner's soul while the latter, kneeling before a bishop, confessed his misdeeds (WC:344).[28]

On the other hand, some blocks indicated the nature of the work by depicting a scene from it—such as the two versions of the Last Supper (WC:35 and 342) used on the title-pages of the many editions of Ludolph of Saxony's *Vita Christi cartujano*—or by a composition which, if it did not portray any identifiable scene in the work, nevertheless presented the major characters in a dramatic manner. This is the case with WC:99 used by Polono for his 1501 Seville edition of *La Celestina* and then employed by the Crombergers for their editions; in the 1535 and subsequent edition of the same work WC:376 replaced the worn and broken WC:99. As had happened with their romances of chivalry, the Crombergers became so closely associated with Rojas's work that the woodcuts they used in their editions were soon imitated in Italy: the illustrations found in *La Celestina* published at Rome by Antonio de Salamanca (*c.*1520) and probably printed with his own material are closely modelled on the Cromberger blocks.[29]

[26] Haebler maintained that 'no name is more intimately connected with the famous romances . . . as that of the Krombergers' (see his *The Early Printers*, p. 58).

[27] Mortimer, *Havard College . . . Italian 16th Century Books*, i. 24–5, 245–6.

[28] This last design was a common one, later being used at Lisbon and elsewhere in Spain (see Anselmo, *Bibliografia das obras impressas em Portugal no século XVI*, p. 180, and Lyell, *Early Book Illustration*, p. 290).

[29] Norton, *Printing in Spain*, pp. 152–4.

Some title-pages, however, contain woodcuts which appear to have been chosen because they were decorative or were merely the most appropriate ones which the printer had at his disposal. The first edition of Hernán Cortés's second letter from Mexico (Seville, Jacobo Cromberger, 1522) was decorated with a fifteenth-century block of an enthroned monarch (WC:94), which was suitably majestic for a missive sent to a king but gives little idea of the nature of the work itself.

These individual woodcuts, which were often large, were normally used on title-pages; very occasionally they were printed only within the book. This happened with the beautiful Crucifixion (WC:36) which was often placed in a prominent position in the preliminaries of editions of the *Vita Christi cartujano* printed by the Crombergers, but never on the title-page. It may be that this same block, which was later used by Antonio de Espinosa in Mexico, served to print the *pliegos* of devotional pictures which are recorded in their thousands in the inventories of the Crombergers' shop.

(b) Text Blocks (Large Sets)

The second category of woodcuts is made up of sets of numerous, normally smaller blocks used to print illustrations to accompany the text of books, although they are, on occasion, found alone on the title-pages of quarto or octavo editions.[30] These sets fall into three main groups found in (*i*) romances of chivalry and chronicles, (*ii*) works of popular devotion, and (*iii*) books of exemplary fables. Although these are the sorts of books which one would expect to be illustrated, the appearance of such large numbers of woodcuts in them is another indication of the Crombergers' investment in material for use in editions intended for the popular market, an investment which appears to have been profitable, for these three kinds of book account for much of their production.

(i) Romances of Chivalry and Chronicles

Despite Bland's opinion that 'the Spanish genius does not seem to have lent itself to the illustration of romance', it is surely unreasonable to ignore these woodcuts altogether, as he does, even if it is difficult to claim great artistry for them.[31] A fairer assessment of the illustration of the romances is made by Sir Henry Thomas when he claims that they are 'neat and ornamental' and 'of a strength and heaviness which harmonizes admirably with the thick Gothic type favoured at this time in Spain'.[32]

[30] For example, WC:57, which belongs to the set WC:481–505, etc., is used on the title-page of the 1545 Cromberger edition of the *Cruz de Cristo*; WC:434, which belongs to one of the sets WC:402–465, etc., is used on the title-page of the 1520 Cromberger edition of *La poncella de Francia*.

[31] David Farrant Bland, *A History of Book Illustration, the Illuminated Manuscript and the Printed Book* (London, 1958), p. 193.

[32] Sir Henry Thomas, 'Antonio [Martínez] de Salamanca, Printer of *La Celestina*, Rome *c.* 1525', *The Library*, 5th ser., 8 (1953), 45–50 (pp. 48–9).

Two, or possibly three, sets of these blocks were used by the Crombergers and, in all probability, cut for them.[33] They are partially catalogued as WC:402–465 and WC:698–701, 722. Unlike WC:124–128 and their more modern versions WC:391–393, 397, and 399, which were cut to illustrate specific scenes of *La Celestina*, these blocks could be used in many different romances of chivalry and chronicles, appearing without incongruity more than once in each work. The reason for the practicability of these cuts has less to do with their flexibility than with the formulaic nature of the works they were designed to illustrate; romances and chronicles were made up of a permutation of a limited number of stock elements which could be appropriately illustrated by an equally limited range of woodcuts. The sets contain, for example, scenes of single combat (WC:412, 432, 698, 701), and of battle (WC:404, 405, 414, 465), an attack upon a city (WC:423, 700), the knight errant's ceaseless journeys (WC:413, 431, 439, 450, 464), the knight and his lady (WC:420, 429, 430), as well as royal councils and life at court (WC:403, 406, 410, 419, 425). The woodcut blocks help the modern reader to appreciate how stereotyped were these books and how conservative the taste of the readership, or audience, for which they were written.[34] As with the famous title-page cuts of the Cromberger editions of the romances of chivalry, these smaller woodcuts were also imitated by Italian presses.

(ii) Works of Popular Devotion

The Crombergers used at least four sets designed to accompany editions of devotional works. Examples are catalogued as WC:481–623 and 628–688. These blocks are used in many editions of the vernacular books of piety which enjoyed such enormous popularity in the fifteenth and sixteenth centuries; even among the ephemeral *pliegos sueltos* which the Crombergers printed in large numbers such devotional items figure prominently.[35] Examples of the sort of books illustrated with these woodcuts are Padilla's *Retablo de la vida de Cristo* or Montesino's *Epístolas y evangelios*. The cuts used in these works depict the best-known episodes in the life of Christ and the Virgin, the Passion being particularly prominent.

The Crombergers also possessed sets of small religious vignettes (e.g. WC:371–374 and 703–708) which were occasionally used in conjunction with the above-mentioned sets and sometimes had a more decorative than illustrative function, appearing

[33] Thomas, 'Antonio [Martínez] de Salamanca', p. 49.

[34] Occasionally, however, a block from this set was cut to illustrate a particular episode of a particular book. One block (not reproduced in App. III) which appears in the 1528 Cromberger edition of *Tristán de Leonís*, and possibly also that of 1511, was designed specifically to illustrate the scene where Tristán and the Queen, who are sleeping together, are visited by an angel which comes to prophesy Tristán's death. As this block also appears in the 1510 edition of *Oliveros de Castilla* it argues for an even earlier edition, now lost, of *Tristán*.

[35] See above, p. 36.

in liturgical works, such as missals and processionals, where they represent an attempt by the printer to provide some sort of a substitute for the illuminated miniatures of the medieval manuscript books of which the printed editions are the successors. They are also firmly in the tradition of the Parisian books of hours which were exported throughout Europe.

The set of larger blocks depicting the life of Christ (most of WC:506–620) are probably of Seville origin. Many of them had been used there in the previous century by Polono and Ungut in their 1497 edition of Gulielmus Parisiensis' *Postilla super epistolas et evangelia* (Haebler 309). The set of smaller cuts, examples of which are WC:481–486 and 488–505, provides a good example of how printing allowed designs to travel internationally and, consequently, of how cautiously woodcuts must be treated as evidence for the identification of unsigned books of this period. Whereas the woodcuts from the Cromberger romances of chivalry were imitated abroad, this set of devotional cuts illustrates either how closely local woodcutters copied foreign designs or how printers imported blocks from foreign woodcutters who were producing multiple copies of such sets. The Alsatian printer, Wynkyn de Worde, Caxton's assistant and successor at Westminster from 1491 until 1535, was a contemporary of both Jacobo and Juan Cromberger. As they printed similar sorts of popular books at the same time, it comes as no surprise to discover that both presses should have possessed sets of woodcuts of the life of Christ.[36] What is striking is that only very close examination enables us to distinguish one set which Wynkyn de Worde used on several occasions from the smaller Cromberger set referred to above. Twelve examples used by de Worde are to be found in his 1521 and 1527 editions of St Austin of Abingdon's *Mirror of the Church*, and all these are almost identical with the blocks used by Cromberger. More examples from this set appeared in a series of tracts printed by Robert Copland, who hired or borrowed them from de Worde, the one dated tract of this series being the rhymed *Passion of Our Lord*, which Copland produced in 1522. Hodnett supposes that these blocks were of Flemish origin, and it is not impossible that this was also the source of the set used at Seville.[37]

[36] On de Worde's output see Bennett, *English Books and Readers 1475–1557*, pp. 185–91, and Blake, *Caxton*, p. 81.

[37] Edward Hodnett, *English Woodcuts 1480–1535*, 2nd edn., with additions and corrections (Oxford, 1973), pp. 30–1. Hodnett reproduces as his numbers 'W. de Worde 648, 662, and 664' cuts which are almost identical with Cromberger WC:483, 57, and 573 respectively. The cuts used in the two editions of the *Mirror of the Church* are the equivalent of the Cromberger WC:57, 483, 486, 491, 492, 497, 570, 573, 631, 661, 685, and 686. A Flemish connection for both Cromberger and de Worde is not unlikely. It has already been seen that Jacobo's wife or her first husband had relations in Bruges; Caxton was closely associated with the same city where he had been Governor of the English Nation of merchant adventurers and had begun his own printing career. In Spain sets based on these Flemish woodcut designs were not limited to Seville. In the edition of Pereç's *La vida de la sacratísima virgen* which Eguía printed at Toledo in 1526, small cuts represent the life of Christ and the Virgin. One of these, at least, is of the same design as Cromberger WC:482, but is not as close to the Cromberger cut as the set used in England

(iii) Books of Exemplary Fables

The Cromberger press used at least three sets, each made up of numerous blocks, which had been cut specially to illustrate the books of fables of Aesop and Bidpai. These works were among the most frequently illustrated incunables in Europe;[38] editions of them coming from the Cromberger press were so copiously illustrated that Lyell refers to them as 'picture books'.[39] Once again, the investment in woodcuts is some indication of the popularity of these collections of Spanish translations of exemplary tales; many were doubtless read to pieces and it is likely that whole editions have disappeared without trace. Yet evidence for the existence of several editions survives. Aesop was printed at least four times by the Crombergers (*c.*1510, 1521, 1526, and 1533), and they must have produced further editions which have been lost, as is suggested by references in the inventory of the stock of Jacobo Cromberger on his death in 1528, where we find an entry for 'twenty-five Hizopets', and that drawn up on the death of Juan in 1540, where there are references to 'seven hundred and twenty Ysopetes', '23 Isopetes', and '60 Isopetes'. The other best-selling collection of fables, often known as *Calila y Dimna* and printed by the Crombergers in a Spanish translation of Johannes de Capua's Latin adaptation of the tales of Bidpai, was doubtless frequently issued by the press although only two editions survive (1534 and 1546). It is also listed in the inventory of Juan Cromberger's stock, where an entry reads '39 Exemplarios', the collection being known in Spain as the *Ejemplario contra los engaños y peligros del mundo*.[40]

It is significant that the Crombergers owned two sets of blocks designed to illustrate editions of Aesop. The first (WC:878–1066) were of page width and were used in the *c.*1510 edition. A new set of smaller blocks based on the same designs was subsequently cut and appeared in the editions of 1526 and 1533 (WC:129–325). This suggests either that the first set had been used so frequently by the press that it had worn out and had to be replaced, or that the smaller set was commissioned in order to reduce the amount of paper used in the editions and therefore make the books more economical. Both hypotheses point to the popularity of Aesop, and the history of their design shows how woodcut patterns travelled throughout Europe. The blocks used by the Crombergers in their editions of 1526 and 1533 were closely imitated, directly or indirectly, by Sebastián Trugillo in his edition of Aesop printed at Seville in 1562. The Cromberger sets appear in turn to have copied the illustrations used by Johannes Hurus in the first Spanish-

is (see Lyell, *Book Illustration*, p. 228). There are many other examples of designs travelling long distances in Spain (see Norton, *Printing in Spain*, p. 57).

[38] Febvre and Martin, *L'Apparition*, p. 140.

[39] Lyell, *Early Book Illustration*, p. 172.

[40] Document dated 7 June 1529 [*sic*] (APS, Oficio 4, Libro [2] of 1529, unfoliated); and document dated 20 Sept. 1540 (APS, Oficio 17, Libro 2 of 1540, unfoliated).

language edition of the fables printed in Spain (Zaragoza, 1489) (Haebler 6), but they could also have imitated a set deriving from Hurus', like the one which appears in Joffre's edition (Valencia, 1520) which is very close to the Crombergers' set. The Zaragoza printer, who imported blocks from Germany for his copiously illustrated editions, seems to have copied his Aesop woodcuts from Sorg's edition printed at Augsburg in 1486, which had, itself, copied the original blocks used by Zainer in an edition of the same work printed at Ulm in 1477.[41] As might be expected, some of the Cromberger cuts reversed the designs of Hurus' set.

The blocks employed by the Crombergers to illustrate their editions of the tales of Bidpai are a similarly numerous group (WC:709–713 and 727–833). The design of at least some of these blocks was inspired either by the woodcuts appearing in Saint-Grille d'Alexandrie, *Miroir de la sagesse* (no place of printing, *c.*1470) or by the model for the cuts in that book, but the Cromberger designs may have come once again via Zaragoza, where Paulus Hurus printed the first Spanish-language edition of Bidpai in 1493 (Haebler 340). It is evident that printing made such pictures common property, but whatever the ultimate genealogy of the Crombergers' set, their style certainly suggests that they were cut locally in Seville.

(c) Text Blocks (Small Sets)

The third category of woodcuts consists of sets made up of just a handful of blocks which were used either to accompany the text of Cromberger editions or to decorate their title-pages. Examples of such sets are the page-width blocks cut specially to illustrate key scenes of *La Celestina* (WC:124–128), the curiously inappropriate block used only in the *c.* 1513–15 edition of that work (WC:695), and the newer versions of the same scenes (WC:391–393, 397, and 399).[42] At least three sets of woodcuts were used to illustrate the various months of the year in editions of Andrés de Li's *Repertorio de los tiempos,* of which two groups are reproduced as WC: 834–852. The appearance of completely different sets in two of the three surviving editions of this work (those of 1510 and 1529) suggests that the earlier sets wore out, an indication of the heavy use to which they were put and evidence that the Crombergers printed many editions of this popular almanac which are now lost.

More artistically interesting are two sets of undoubtedly foreign origin which were used to illustrate devotional works. The first, of which WC:479, 519, 621, 870, and 871 are examples, are small cuts with wide borders which contain white decoration on a black background in a thoroughly Italian manner. The style of

[41] Norton, *Printing in Spain*, pp. 69–70; Lyell, *Early Book Illustration*, p. 32; Hind, *An Introduction*, ii. 741.

[42] WC:695 shows one of Calisto's servants about to be hanged; the text of *La Celestina* makes great play of their being beheaded.

cutting and decoration suggests a Venetian or Florentine provenance.[43] These woodcuts are not completely uniform in design and might derive from more than one set, but all probably come from the material of Ungut and Polono, who used some of them in their edition of Gorricio's *Contemplaciones sobre el rosario de Nuestra Señora* (Seville, 1495) (Haebler 301). These blocks may have been brought from Italy by the early printers when they were called to Seville by Queen Isabella, or they could have been imported at a later date.

The second foreign set is a very fine group of woodcuts which contrast markedly with those cut locally. Examples are WC:30–32, 34, 332, and 341. Two fortunately bear the date '1530', the year in which Juan Cromberger first used them in his edition of Part One of Ludolph of Saxony's *Vita Christi cartujano*. The style of these illustrations is German or Flemish and owes much to Dürer. The acquisition of new blocks at this date is evidence that Juan was not content merely to reuse inherited material, but was willing and financially able to buy artistically superior Renaissance blocks—with their realistically portrayed figures and sophisticated use of perspective—which did much to enhance the beauty of his more ambitious editions and make them appear to us less old-fashioned than books illustrated in the traditional Spanish gothic style. It would not be surprising if he had commissioned or, at least, bought these blocks specially for his new edition of Ludolph of Saxony; it was the first one he had issued since his father's death, and the *Vita Christi cartujano*, although much reprinted, had always been something of a prestige publication. The Castilian translation and its first edition had been ordered by Ferdinand and Isabella, although they did not finance the printing. In Portugal Queen Eleanor had commissioned an edition of the Portuguese translation, and this was printed at considerable expense by Valentim Fernandes and Nicolau de Saxónia at Lisbon in 1495 (Haebler 373).[44]

(d) 'Factotum' or Composite Blocks

The cost of illustrating books was high not only because woodcuts occupied space on paper, which was expensive—something that was unavoidable in any illustrated book—but also because of the cost involved in having the blocks cut in the first place. If a printer wished to use woodcuts which were appropriate to a particular book rather than just using whatever blocks he happened to have in his workshop, and if he printed a wide range of titles, he would obviously require a large stock of blocks. One solution to this problem was to use a small number of reasonably appropriate blocks and repeat them several times in the same book. Frequently,

[43] Compare with the woodcuts from these two cities reproduced in Max Sander, *Le Livre à figures italien depuis 1467 jusqu'à 1530: essai de sa bibliographie et de son histoire*, 5 vols. (Milan, 1942) and a supplement by Carlo Enrico Rava (Milan, 1969).

[44] Norton, *Printing in Spain*, p. 9; Bataillon, *Erasmo y España*, pp. 44–5; Anselmo, *Origens*, pp. 277–8, 442–8. The first Spanish-language edition had been printed at Alcalá by Polono in 1502–3.

however, printers would use whatever cuts they happened to have to hand whether appropriate or not. This could have unfortunate results, as is witnessed by the case of a New Testament illustrated with woodcuts designed for an edition of Ovid's *Metamorphoses*.[45] There was, however, another solution. Just as movable types were superior to the early block books with their immutable xylographic text, so the same principle was adopted for illustrations in the development of 'factotum', or composite, blocks. These were far more versatile than the traditional single unalterable blocks and, although considerably less elegant, provided variety with a minimum of expense. Factotum blocks were composed of individual pieces which could be combined to make up a complete picture. The pieces normally represented people, trees, or buildings, and were permutated to suit the requirements of different scenes in a book.

Factotum blocks had first been used by the Strasburg printer Johann Grüninger in his 1496 edition of Terence. The idea seems to have spread rapidly, for it is found in France—where Antoine Vérard published a translation of Terence at Paris in about 1500 using the same method of illustration—in the Low Countries, and in England.[46] In Spain such blocks became the stock-in-trade of most printers; some, at least, of the sets used by the Cromberger press had been inherited from Polono, who was the first printer in Spain to employ this method of illustration.[47]

The Crombergers used these blocks mainly in their cheaper and more popular quarto books. Their editions of *La Celestina* were always illustrated in this manner, which, for several reasons, was particularly appropriate to Rojas's book. First, *La Celestina* was enormously popular and went through numerous editions in the sixteenth century; such a work, which clearly appealed to a modest readership as well as to a more elevated one, would be a candidate for illustrations of some sort.[48] Second, the history of Spanish editions of *La Celestina* in the early sixteenth century is one of an increasingly economic use of paper; printers evidently intended to make the book cheaper to produce, and probably also to buy, and therefore preferred an inexpensive form of illustration. Last, it is no coincidence that the sort of blocks which had been developed to illustrate Terence should have been copied in *La Celestina*, which, based as it was on the tradition of the Italian humanistic comedies in Latin which had taken Plautus and Terence as their models, is, arguably at least, a drama for which a graphic *dramatis personae* at the beginning of each act is particularly apt.[49]

[45] Steinberg, *Five Hundred Years*, p. 158.

[46] Pollard, *Fine Books*, p. 150.

[47] Norton, *Printing in Spain*, p. 152.

[48] For the history of 16th-c. editions of *La Celestina* see Clara Louisa Penney, *The Book Called Celestina in the Library of the Hispanic Society of America* (New York, 1954); Clive Griffin, 'Four Rare Editions of *La Celestina*', *Modern Language Review*, 75 (1980), 561–74; and Ivana Gallo and Emma Scoles, 'Edizioni antiche della *Celestina* sconosciute o non localizzate dalla tradizione bibliografica', *Cultura neo-latina*, 43 (1983), 103–19.

[49] The edition of *La Celestina* printed by Marchio Sessa at Venice in 1531 used the same woodcuts

Examples of factotum blocks from at least three sets are WC:14–17, 100–123, 377–390, 394–396, 398, 435, 436, 443–446, 692–694, 702 and 1074. Two of these sets were used for *La Celestina*, one until 1528 and the other from 1535 onwards. The replacement of the early set (WC:100–123) by the more modern one (WC:377–390, 394–396, and 398) suggests that the former had worn out. A study of these woodcuts provides evidence, such as we have already seen with other sets, of the heavy use to which they were put, and therefore of the demand for editions of *La Celestina, pliegos sueltos*, and similar publications in which these factotum blocks appeared.

(e) Printers' Marks

The Crombergers had no distinctive device such as Aldus' anchor and dolphin, Gryphius' lion, or Plantin's compasses, and they appear to have used only two separate printers' marks (OM:80 and WC:696). Nevertheless, they did employ a series of monograms with the letters 'IC', which are frequently to be found within a circle surmounted by a cross. One is tempted to wonder whether the number of names which Jacobo and Juan could choose for the son they hoped would follow them in the press was somewhat limited by the investment already made in cutting these monograms. These little devices either form a permanent part of a frame or border-piece, or else they were cut on a small block which could be inserted into a hole in these frames or border-pieces. Examples of the monograms are OM:1, 71, 79, 84, 126, 207, 230, 234, 267, 314, 350, 357, 358, and in WC:58. Not all books of this period carried such marks, but on occasion they were required in the terms of a royal privilege to a printer, so that copies could be identified as being the product of his press. This was the case when in 1511 the king demanded that Brocar add his printer's mark to a series of editions on which a privilege had been granted.[50]

(f) Diagrams

Diagrams are to be found mainly in treatises on philosophy or navigation, and the blocks from which they were printed were made more versatile by being designed to hold type (e.g. WC:27 and 28). Printing contributed to the advancement in knowledge of surgery and anatomy, particularly through the use

he had previously employed for Plautus. In the preface to the 1553 edition of *La Celestina* printed at Venice by Gabriel Giolito de' Ferrari and his brothers, the editor, Alonso de Ulloa, makes clear his assumption of the similarity between the Spanish work and that of the classical comedians when he complains (fol. 2^{v}) that in all previous editions of *La Celestina*, 'despite its being a comic drama . . . it has been printed like a prose narrative, there being no typographical break from the beginning of each act until its end, and without the characters' names being printed in the margin as one stops speaking and another begins. I think that this is an important oversight in a work like *La Celestina* . . . because we have all seen the comedies of Terence and Plautus printed as they should be: in other words, as each character begins his speech, his name is printed in the margin'.

[50] Beltrán de Heredia, *Cartulario*, ii. 380–1.

of woodcuts; large diagrams, again containing type, were used in some of the Crombergers' editions of medical works. These woodcuts range from charts indicating how a diagnosis of illness could be reached by an examination of urine (WC:715) to a cut of 'astrological man' (WC:716), a view of female anatomy (WC:718), keys to blood-letting (WC:717, 854, and 855), and a guide to the treatment of wounds and injuries (WC:719). Some of these are the largest woodcuts to be found in Cromberger editions and are used on folding pages the size of a full sheet (i.e. two folio pages). They are copies of woodcuts used in Spanish and Italian incunable editions of medical works. Some liturgical editions also carry small diagrams which enable the reader to calculate the date of holy days (e.g. WC:352 and 353).

To the category of diagrams could be added the one map encountered in a Cromberger edition (WC:343).[51] Seville was the centre for fitting out and victualling the fleets engaged in trade with the Indies; maps and charts were therefore required in large numbers in the city. So careful were the authorities to ensure that pilots sailing on the Atlantic routes were well trained and well informed that the Casa de Contratación founded the Universidad de Mareantes at Seville, and this institution must have required a steady supply of charts. Nevertheless, there is no evidence to show that the Crombergers tried to exploit this local market. It is of course possible that the press did produce maps, and that the hard use for which they were destined has prevented any of them from surviving. It is more probable, however, that charts were normally drawn by hand because the sort of blocks used by the Seville presses were neither large enough nor cut with sufficient precision to print adequately detailed maps. There is no reference to any printing of this sort in the two surviving inventories of the Crombergers' shop. Juan did own two navigational charts, which probably related to his own mercantile interests, but these could not have been printed maps because their dimensions were nine by over four feet.[52] Indeed, it appears that map-printing was not carried out extensively at Seville or even elsewhere in Spain.[53] In the 1540s Lázaro Nuremberger, Sebastian Cabot, and one 'Gabriel Mutgel' (Kaspar Nützel?) signed a contract in Seville in which they agreed to have a mappemond printed which Cabot had drawn of the new discoveries in America combined with the traditional geography of Ptolemy. But they did not give the order to Lázaro's

[51] It has been claimed that another map, now lost, was included in Cromberger's edition of Cortés's second letter (see Hernán Cortés, *Letters from Mexico*, ed. and tr. A. R. Pagden [New York, 1971], p. lx).

[52] Gestoso, p. 83. Some maps were, however, printed from wood-blocks, the first being Leonard Holl's edition of Ptolemy's *Geography*, printed at Ulm in 1482, but copperplates had been used earlier than this. Woodcuts were still being employed for maps until the mid-sixteenth century, after which the copperplate, which gave a superior result, displaced them (see John Jackson and W. A. Chatto, *A Treatise on Wood Engraving*, 2nd edn. [London, 1861], pp. 199–204, 417–19).

[53] The maps discussed and reproduced by Francisco Vindel in his *Mapas de América en los libros españoles de los siglos XVI al XVIII (1503–1798)* (Madrid, 1955), are few in number and unimpressive in quality for the period 1503–60.

relations, the Crombergers, as would have been logical if their press had enjoyed a reputation for this sort of printing. Rather, they decided that Cabot's original should be sent to Nuremburg, the centre of navigational studies in Northern Europe, where a suitable block or plate was made and copies of the map run off.[54]

Most of the woodcuts found in editions printed by the Crombergers can be subsumed into these six categories. Their stock was very mixed, but the majority of their blocks appear to have been cut locally and their style frequently makes a Seville book distinguishable from the products of other Spanish cities, especially those in the east of the country such as Barcelona and Valencia, where the appearance of woodcuts was markedly different. Woodcuts therefore provide useful clues for the attribution of a book to a particular city or, in some cases, to an individual printer. They can even be used to establish the chronology of editions, as Norton was able to show in his illuminating study of the early Seville editions of *La Celestina*.[55]

Nevertheless, a note of caution should be sounded here, for woodcuts are a reliable aid to attribution only where corroborative evidence is available. Many printers were itinerant in sixteenth-century Spain and both used and acquired their material in different places. As has been seen, many were foreigners who looked to their homelands for supplies or who brought their equipment from abroad. They were businessmen or artisans who naturally tended to settle in cities on the main trade routes, thus maximizing the outlets for their wares; this also meant that they acquired new material which had travelled along these routes with merchants and had been used elsewhere. Individual blocks could and often did travel considerable distances: a woodcut royal coat of arms used by Varela at Toledo passed to Juan de Villaquirán in that city, yet later appeared back in Varela's hands at Seville;[56] the blocks employed in Jorge Coci's Livy, printed at Zaragoza in 1520, which were acquired by the Zaragozan printer Juan Millán and used in his edition of Pulgar's *Crónica de los Reyes Católicos* of 1567, had

[54] Hazañas, i. 162–3. Hazañas dates this contract to 11 Mar. 1541 and provides an inaccurate reference in APS; I have been unable to trace this document as even the year he gives is incorrect. Cabot, when bequeathing his financial interest in the mappemond to his wife's niece in 1548, by which time it must have been printed, talks of the contract having been signed 'about three years ago' (APS, Oficio 19, Libro 1 of 1548, fol. 894^{v}). Lázaro Nuremberger probably had contacts with map-printers in Germany. As early as 1518 he was writing from Lisbon to a fellow factor of the Hirschvogel company, Michael Behaim, who was in Germany. Michael was the nephew of the famous Nuremberg cosmographer and navigator Martin Behaim, who had settled in Portugal and advised King John II on his maritime explorations. Martin's maps were printed in Germany (see Otte, 'Jakob und Hans Cromberger', pp. 133, 156). In his will drawn up in Seville in 1564 Lázaro mentioned the publication of the mappemond: 'Item, I declare that I have not yet settled up with Gaspar Nizel, the German from Nuremberg in Germany, concerning some navigational charts called "papamundi" [*sic*] which we sent to Nuremberg to be made and which were, indeed, produced there' (see Werner, 'Zur Geschichte Tetzelscher Hammerwerke', pp. 219–20).

[55] Norton, *Printing in Spain*, pp. 141–56.

[56] Norton, 'Typographical Evidence', p. 101.

come to Aragon from Germany.[57] We have already seen that the metal block of St Jerome used by Varela and Juan Cromberger in the 1530s had been made in Cologne some fifty years earlier; it had probably already passed through the hands of at least one other Spanish printer before reaching Seville, for Joan Joffre is said to have employed it at Valencia in 1520.[58] Some of the Crombergers' own blocks crossed the Atlantic and were used in Mexico.[59] Woodcuts were bought, lent, hired, pledged, and inherited. In Spain, at least, they were frequently used over a period of many years, as one might expect in a craft which was constantly undercapitalized. The result is that many blocks had long and chequered careers.

The problem becomes even more labyrinthine when one examines the diffusion of the designs, rather than simply of the blocks themselves. Designs could be closely imitated by woodcutters working in different countries, and not only to deceive a potential purchaser or those attempting to enforce the provisions of the frequently ineffectual printing privileges.[60] It was easier for a woodcutter to copy a design than to invent one, and it should not be forgotten that he would have been considered a mere craftsman just as painters were in sixteenth-century Spain.[61] During the early history of the craft, at least, the wood-engraver belonged to the same class and guild as the carpenter. Great woodcut designers like Dürer and Holbein generally left the actual cutting of blocks to artisans, who were not expected to use their imagination but simply to carry out a commission. So when asked to copy a series of cuts from a previous edition, the engraver would have done just that. For this reason many cuts borrow designs used by other printers but reverse them, for the woodcutter would copy the design from an impression in a book, or even glue the impression torn out of the book on to his block as a guide. His new cut would therefore print a reverse image of the original.[62] This is probably what happened to WC:357, where the letters 'I.N.R.I.' print backwards. An additional reason for the diffusion of designs was that wood-engravers themselves were either often itinerant or sent their wares to various cities for sale, a fact which may account for the similarity of woodcuts used by various printers in different centres, for the engraver would have a stock set of patterns, which he would constantly repeat.[63]

[57] Lyell, *Book Illustration*, p. 128.

[58] Hind, *An Introduction*, i. 194.

[59] Juan Pablos used WC:334 and 338 in Mexico in 1549, while Antonio de Espinosa used WC:36 in 1565. Many of the Crombergers' characteristic border-pieces, or close copies of them, travelled with Pablos when he emigrated to the New World.

[60] On the piracy of Cromberger books and imitation of their material by pirates, see Griffin, 'More on the Spanish *Scala celi*'; for virtually identical blocks being used in different cities simultaneously, compare my App. III with Vega González, *La imprenta en Toledo*, pp. 131–211.

[61] Even as late as the 1630s in Spain most artists were considered to be mere craftsmen like carpenters or glaziers (see Jonathan Brown and J. H. Elliott, *A Palace for a King: The Buen Retiro and the Court of Philip IV* [New Haven and London, 1980], p. 92).

[62] Pollard, 'The Transference', p. 368.

[63] Hind, *An Introduction*, i. 93.

Let one example of this diffusion within Spain suffice. A block which the Crombergers first used in 1506 and which thenceforth frequently appeared in their books depicts a scribe sitting at his desk in a vaulted room, facing to the right and writing with his left hand (WC:41). The design was not original. It probably derives directly or indirectly from a block used at Barcelona by Rosembach in his 1493 edition of San Pedro's *Cárcel de amor* (Haebler 606) and is a reversed copy of it. But it could equally well have come from a family of copies, examples of which are seen in a Burgos edition of the *Cárcel* (Basilea, 1496) (Haebler 604) and a Toledan *Proverbios de Séneca* (Hagenbach, 1500) (Haebler 619). It is clear that books, and therefore designs, travelled quickly and widely in Spain.

In addition to this confusing diffusion, so many editions have disappeared that there are major gaps in our knowledge of individual blocks' history. Woodcuts can, then, only normally be used for the attribution and dating of unsigned and undated books when supported by other evidence. However, when cuts are combined with the other material used by a particular printer they can effectively contribute to this process of identification.

Ornamental Material

There can be no watertight division between ornamental material and woodcuts, for some blocks were used both for illustration and to make up a decorative border (e.g. WC:40, OM:351, and OM:352), but, in general, ornamental material was used to frame title-pages, to accompany illustrations within the text, and, occasionally, to decorate a page. Most of this material was printed from wooden blocks, but some metal ones are recorded in Spain; the Crombergers owned at least those metal ornaments which are referred to in Appendix One as type ornaments (e.g. OM:5, 58, 75 and 251 *bis*).[64] Type ornaments were a relatively late development.

Not only are the heavy borders made up of heterogeneous pieces typical of the decoration of title-pages in early Spanish printing, but the very combination of such pieces distinguishes one printer's stock from that of another. The various stages at which pieces were severed from their original frames or cut down in size are also useful as an aid to the dating of unsigned books.[65] The fragments cut from previously four-sided frames ('compartments') are numerous in Cromberger

[64] Norton, *Printing in Spain*, p. 67.

[65] An example of how accurately the cutting down of a border fragment can be dated is OM:173. This was complete in the *Libro del conde Partinuplés*, which was signed by Jacobo Cromberger on 2 Dec. 1519, but in his *Los misterios de Jerusalén*, dated 2 Jan. 1520, it had been truncated. Similarly, OM:62 was cut down in Mar. or Apr. 1527. However, the dating of unsigned works on the basis of such evidence can be misleading because printers sometimes owned more than one copy of a design, a fact which only comes to light when both copies are used together. While one copy is truncated, the other may well remain intact.

editions, and most borders of surviving books are decorated with such material. We are forced to conclude that the printers either acquired collections of such oddments or that they readily cut up complete compartments. If the latter is true, the infrequency with which intact compartments are found is an indication of the large number of editions that have disappeared. On occasion, it has proved possible to piece together such fragments and give them a provisional date. An example of such a set, although not one which ever formed a compartment made of one piece, is OM:54, 56, 97 and 194, which were undoubtedly commissioned to accompany WC:326 on the title-page of the 1521 edition of the *Ordenações manuelinas*. The pieces are of the same design and all contain King Manuel's characteristic armillary sphere. This provides a date *post quem* for all subsequent undated books which use these pieces. Very rarely, as in OM:316, a border piece actually bears the date of cutting. The design of some border pieces indicates the sort of book for which they were originally made: OM:96 and 204, for example, would have been commissioned for books printed for the Franciscan Order; OM:263 for books designed for the Portuguese market.

The Crombergers appear to have realized that, as it was their practice to use a mixture of pieces for borders, it would be more practical to have single pieces cut rather than to commission whole compartments which would in any case later be dismembered. They therefore developed self-contained pieces which were designed to print a single side of a title-page border (e.g. OM:181 and 182).

A tentative chronology of styles of ornamental material used by the Crombergers can be drawn up as an aid to dating. The style of border pieces in which a white pattern stands out against a black background, exemplified by OM:122, 133, 149, and 216, has all the hallmarks of an Italian origin; it therefore comes as no surprise to encounter at least OM:149 and 216 in books printed by the 'Compañeros alemanes', who may well have brought them from Italy.[66] The designs which contain small human or animal figures within twisted foliage (e.g. OM:12, 14, 16, 18, 107, 112, 217, 222, 269, and 335) are early blocks, many of them having passed to Jacobo Cromberger from Polono and Ungut. Equally early may be the 'gothic' compartments used for folio editions (OM:98 and its virtual negative OM:150). As the sixteenth century progressed and artistic tastes changed, so Renaissance plateresque designs were introduced into the press's stocks.[67] For example, the medallion motif of OM:23, 28, 29, 174, 175, 220 and 221 is markedly modern in comparison with some of the inherited fifteenth-century material. Likewise, classically inspired mythical figures (OM:26 and 79), combatants attired in Roman garb (OM:102 and 125), and the architectural

[66] These two pieces are used in the 'Tres Compañeros'' edition of the *Cuaderno de las leyes nuevas de la Hermandad* printed at Seville in 1498. OM:216, at least, is very similar to the style of borders being used at Milan in the late 15th c. (see Sander, *Le Livre à figures*, v, No. 868).

[67] Bohigas, *El libro español*, pp. 171–4.

designs of OM:1, 2, 24, 25, 37, and 38, are all tell-tale signs of the influence of the artistic and architectural developments which took place during the reign of Charles V.[68] The ornate frames, OM:71, 79 and 267, which were clearly conceived in one piece and could not easily be dismembered, were also a later development.

As has already been noted in the examination of the press's woodcuts, Seville blocks were not designed in isolation from the fashions set in other centres of printing. For example, the use of vignettes to make up borders in religious works, such as WC:362–374 or the small blocks which could be inserted into the bottom part of the frame (OM:319), is probably derived from the decoration of the books of hours which were imported from France in such large numbers. Like woodcuts, ornamental blocks themselves not only passed from printer to printer but also travelled from city to city.[69]

We have seen that woodcuts used simultaneously at Seville and Westminster are barely distinguishable, that the illustrations of the Crombergers' romances of chivalry were copied in Italy, and that the Crombergers, in their turn, copied illustrations from earlier editions of the fables of Aesop and Bidpai. This diffusion of designs was naturally not limited to woodcut illustrations but was as frequent in ornamental material. This is only to be expected with cast ornaments, but is also evident in ornamental material impressed from wood-blocks. For example, considerable confusion is caused in the history of Seville printing by the fact that Varela owned a set of border pieces which were almost identical with the Crombergers' OM:192, 193, 247, and 248, and that both presses used their own version of this border in unsigned editions of Osuna's *Abecedario espiritual.* Likewise, the Crombergers' Renaissance architectural border, made up of OM:27 (or 36), 28, 29, and 30 (or 39), appears in almost indistinguishable versions in works printed by Eguía at Logroño in 1529, by Villaquirán at Valladolid in 1536, and by an unknown Valencian printer in 1537. A block similar to the top section of this frame also appears in a 1547 Toledo edition printed by Fernando de Santa Catalina.[70] The design of the imposing compartment used by the Crombergers in their folio editions of liturgical works, which was still in one piece in the 1525 Cordoban missal but was later dismembered (OM:319), has a similarly complex history and was probably imitated at Seville from foreign editions. The first work in which I have encountered the design is Luc'Antonio Giunti's Valencian missal printed at Venice in 1509. It subsequently appears in the Ávila missal printed at Salamanca by Juan de Porras in 1510 (Haebler, 436[5], incorrectly dated to 1500)

68 For parallel developments in French book-illustration see Martin and Chartier, *Le Livre conquérant,* pp. 221–2.

69 Norton, *Printing in Spain*, pp. 67, 100.

70 Vindel, *Manual gráfico-descriptivo*, v, Nos. 1608, 1685, and 1687; and Armando Gusmão, *Tipografia espanhola*, Livros impressos no século XVI existentes na Biblioteca Pública e Arquivo Distrital de Évora, 2 (Évora, 1966), p. 196.

and Juan de Celaya's *In tertium volumen sententiarum* (Valencia, 1527). This, incidentally, says something about the influence of imported editions of service books in Spain.

But the most Byzantine of histories of the transference of a design of ornamental material is that made for the London printer, Edward Whitchurch. This decorated compartment was used by Whitchurch in 1548–9 on the title-pages of both volumes of his edition of Erasmus' paraphrase of the New Testament but, quite extraordinarily, in 1554 the old branch office of the Cromberger press in Mexico City used a close imitation in its edition of the *Dialecta resolutio cum textu Aristotelis*. The woodcutter was careful to change the English royal coat of arms and that of Catherine Parr for the arms of Spain and the device of the Augustinians, but nevertheless retained the English printer's initials 'EW' at the bottom of the frame.[71]

Ornamental Initials

Ornamental initials were first used in the Mainz psalter of 1457.[72] Manuscripts had contained illuminated initials, which printers attempted to imitate with decorated capitals impressed from blocks of wood or soft metal. Many early books, including some produced by Jacobo Cromberger, do not employ this form of decoration, merely leaving a blank space often containing a small printed letter which indicated to an illuminator what initial he was to paint. Nevertheless, woodcut initials were common in fifteenth-century printed books and became the norm in the following century. They arrived in Spain early, no doubt owing to the influence of German immigrant printers. Both metal and wood-blocks were used there in the fifteenth century, although it is often impossible to identify the material of the block from which an initial was printed.[73] With the exception of alphabets of simple gothic letters without borders (OI:2, 6, 25, 30, and 58), which were probably cast in metal, the Crombergers' initials were doubtless printed from wood-blocks.

An index of initials, ranging in size from the small plain letters measuring about

[71] Lucy Eugenia Osborne, 'The Whitchurch Compartment in London and Mexico', *The Library*, 4th ser., 8 (1927), 303–11.

[72] Charles Sayle, 'Initial Letters in Early Printed Books', *Transactions of the Bibliographical Society*, 7 (1902–4), 15–47 (p. 16). See also Victor Scholderer, *Johann Gutenberg, The Inventor of Printing*, 2nd edn. (London, 1970), pl. IV.

[73] A. W. Pollard, 'Some Pictorial and Heraldic Initials', in *Old Picture Books with Other Essays on Bookish Subjects* (London, 1902), pp. 124–45 (p. 131), refers to a set of initials cast in soft metal which were employed by Centenera at 15th-c. Zamora. When the same initials were used simultaneously in different towns, it is an indication that they were printed from cast metal blocks. However, caution must be exercised because woodcutters were so skilled that they were capable of producing virtually identical copies of a single design. Indeed, some of the Cromberger alphabets contained multiple, and indistinguishable, copies of the same letters.

5 x 5 mm. to impressive historiated initials of 53 x 50 mm., forms Appendix Four. This index, when used in conjunction with those for types and ornamental material, is designed to enable unsigned books printed by the Crombergers to be attributed to them with certainty. The appearance of ornamental initials can also be of use in dating undated products of the press. For example, the one letter of OI:31 (possibly an alternative form from OI:26) was intact in 1534; by the following year it had been broken (as reproduced in Appendix Four), but continued to appear in a damaged state for many years, thus providing a convenient guide to dating. The history of these alphabets also enables us to trace the dispersal of the press's material to other printers after the disappearance of Jácome Cromberger, and is therefore a useful tool for the historian of the Seville presses.

Ornamental initials were an expensive item in a printer's outlay;[74] it is some indication of the Crombergers' prosperity and of their predilection for these initials that they should have owned such a large number of alphabets. Appendix Four records some 130 sets. This is a remarkable figure. Plantin was working at a period when ornamental initials were at the height of their popularity, and ran a prosperous and prolific printing-shop noted for its initials, but he owned or used only 58 recorded sets.[75]

Some, but not all, of Ungut's and Polono's initials passed to Jacobo (OI:5, 10, 13, and 22), and the design of some of his other alphabets suggests that they had been brought to Seville from Italy or were closely modelled on Italian originals. The frequent imports of Italian service books to Spain may account for this influence. The handsome roman alphabet, OI:14, bears a remarkable resemblance to a set used at Venice by Ratdolt in 1484. The numerous sets designed for liturgical use similarly suggest a direct or indirect Venetian influence and, in turn, provide further evidence of the Crombergers' willingness to invest large sums of money in order to win contracts for this sort of printing.[76] On the other hand, several of the alphabets of white gothic letters set against a black flowered background are similar to examples encountered in books printed at fifteenth-century Lyons.[77]

The durability of the recorded blocks is surprising when the heavy use to which they were put is considered. Set OI:5, for example, was used by Polono and Ungut in the late fifteenth century; it then became one of the most frequently seen

[74] Febvre and Martin, *L'Apparition*, p. 166.

[75] Stephen Harvard, *Ornamental Initials: The Woodcut Initials of Christopher Plantin: A Complete Catalogue* (New York, 1974), p. ix.

[76] C. Castellani, *Early Venetian Printing* (Venice and London, 1895), especially pp. 66–7 (Ratdolt, 1482), p. 72 (Ratdolt, 1484), and p. 153 (Arrivabene, 1499). The Crombergers' 'liturgical' sets are OI:4, 11, 15, 17, 18, 20, 21, 28, 39, 61, 63, 69, 70, 72–76, 82, 86–90, 99, 103, 107, 109, 110, 114, 118, 120, 121, 125, 127, 129, and 130. Other Spanish printers such as Brocar had very similar 'liturgical' sets.

[77] A. Claudin, *Histoire de l'imprimerie en France au XVe et au XVIe siècle*, 4 vols. (Paris, 1900–14), iv. *passim*. Compare especially Cromberger OI:7 with Claudin, *Histoire de l'imprimerie*, iv. 323.

sets in the books of Jacobo and Juan Cromberger, and it was still in use in the 1550s in the last surviving editions printed by Jácome. Indeed, so closely was it identified with the Cromberger press that the scribe who drew up the inventory of Juan's stock of books in 1540 delicately impressed the first letter of the document from a block belonging to this alphabet, which he must have found lying among the deceased printer's material.[78]

Jacobo Cromberger was not content merely to reuse this inherited material. In 1506 he introduced his distinctive alphabet of 'ribbon initials' (OI:1) which, with OI:5, became the press's staple sets, appearing in almost every one of the family's editions including some of Jácome's last known products. It was so successful that Varela even had a close copy made for himself.[79]

As we would expect, an analysis of the Crombergers' ornamental initials corroborates the findings of the study of the press's financial history and of the examination of its other printing material. Innovations were made by Jacobo; Juan continued in his father's footsteps; and under Jácome decadence set in. The use made of the initials by the last named was shoddy. Not only were they badly worn, but they were used in a haphazard fashion so that the elegant uniformity of the earlier products of the press was lost. Thus, for example, while Jacobo's edition of Valera's *Crónica de España abreviada* of 1517 had contained four carefully chosen alphabets, that of Jácome printed in 1553 or 1554 used ill-assorted letters from twenty-two sets. Jacobo employed predominantly gothic alphabets, which suited the fashions of his day (although it must be admitted that his OI:5 was particularly inappropriate to his experimental italic edition of the *Pharsalia* of 1528). The style of ornamental initials began to change elsewhere in Europe in the 1520s and 1530s, but Spanish books remained stubbornly gothic in appearance, and therefore Jacobo's initials continued to be used under Juan. By the middle of the century, even in Seville such letters were beginning to look old-fashioned and a printer like Montesdoca introduced woodcut initials of a decidedly modern stamp to accompany his books printed in roman types in the city at this time.[80] Yet, with the exception of one set (OI:133) probably acquired from another printer when already very worn, Jácome did not invest in new alphabets; characteristically, he continued to use what was left of the sets he had inherited from his father and grandfather.

A close study of the woodcuts and ornamental blocks used by a single press not only provides help in the recognition and dating of its products. It is also of interest to the art historian, and gives the student of the early presses information

[78] Document dated 20 Sept. 1540 (APS, Oficio 17, Libro 2 of 1540, unfoliated).
[79] Norton, *Printing in Spain*, p. 149 and pl. 11.
[80] Klaus Wagner, *Martín de Montesdoca*, pp. 155–67, especially his alphabets I-VII.

about the rapid diffusion of printed books, the relationship between various printers, and the history of a particular office.[81] For the bibliographer and the student of literature, such a study bears witness to the popularity of certain works and also provides worrying evidence of the large number of sixteenth-century editions which have been entirely lost. A scrutiny of the Crombergers' woodcuts suggests, for example, that many editions of popular works like Aesop's *Fables* or Andrés de Li's *Repertorio de los tiempos* have disappeared without trace. While the fact that dismembered compartments are seldom seen in their original state indicates a heavy loss of unidentifiable titles, the existence of unsuspected editions of identifiable works of importance can reasonably be deduced from evidence provided exclusively by woodcut material.[82]

[81] See above, p. 109.

[82] See App. I for Petrarch, *Los triunfos* (Seville, Cromberger press, before 27/1/1527?), and pp. 187, 193, nn. 18, 34 above. For evidence of the usefulness of woodcuts to this end see Anthony J. Farrell, 'Version H of "The Seven Sages": A Descriptive Bibliography of Spanish Editions', *La corónica*, 9 (1980), 56–66 (pp. 58–9), and the notes to my description of Piccolomini's *Historia de dos amantes* (1512) in App. I.

CONCLUSION

THIS book shows how a combination of archival research with a close study of an individual press's surviving books and its printing material can provide not only a history of that press, but also a picture of the tastes of the book-buying public, and therefore an insight into the intellectual and spiritual atmosphere of the time at which the printing-office was operating. Although Spanish archives do not provide the sort of evidence which has been found for the extraordinarily well-documented history of a printer like Plantin, Spain is nevertheless fortunate in having numerous legal archives which can provide the raw material for a history of the early presses. In his pioneering *Printing in Spain* Norton drew upon published documents from these collections for his account of Spanish printing during the first two decades of the century. Nevertheless, a large amount of relevant archival material still remains unpublished. Indeed, only the Barcelona archives have been systematically exploited by historians researching into the early presses, and Barcelona is not a particularly important site for printing in the sixteenth century.[1]

Norton's other major work, his *Descriptive Catalogue*, provides a detailed list of editions issued from all Iberian presses during the first twenty years of the century. It is unlikely that any single scholar will be able to continue Norton's bibliographical research on such a broad front for the period after 1520. Yet students of printing, of Spanish literature, and, more generally, of Spanish culture will continue to build their houses upon sand unless a fuller and more accurate list of Spanish publications (and, indeed, of books printed in Spanish and Latin, and sent to Spain from abroad) than those now available is drawn up. One way of providing such a bibliography is to study the output of individual presses. An examination of the surviving books from those presses must once again be combined with a scrutiny of archival material if the researcher is to obtain evidence for some, at least, of the editions which have disappeared. Only when a series of monographs on the presses, complete with lists of the editions produced, has been written will a full picture of sixteenth-century printing in Spain emerge. It is my opinion that only through team-work can this be achieved efficiently and within a reasonable time.[2] However, the present book is an example of what one person can do, and it provides a model which future researchers may find useful, whether working individually or in groups.

[1] Madurell and Rubió, *Documentos*.

[2] It is to be hoped that Jośe Simón Díaz's current (1985) ambitious project to co-ordinate the work of bibliographers listing items printed in Spain between 1500 and 1600 will prove the validity of this opinion.

For a variety of reasons the Cromberger press is a particularly interesting case to study. For half a century it dominated printing in Castile's leading centre of book-production; it was one of the most important presses in the whole of Spain; many of its books have survived, some of which were extremely influential first editions; and a large body of documents dealing with the family's business interests can be found. Juan Cromberger is also considered a leading figure in the history of printing because he established the first press in the New World, although it has been shown above that the importance of his Mexican press was greatly exaggerated by previous writers. A study of the Crombergers reveals how closely their wider commercial interests were linked to their printing activities, and this tells us something about the nature of the industry in sixteenth-century Spain: a printer who did not enjoy substantial patronage had to manage his affairs with great care if he were to survive. Diversification was an urgent necessity. Thus Jacobo did not limit himself to printing, but began to publish and sell his own editions early in his career. He did not restrict his stock to editions from his own presses, but soon entered into agreements with other printers and publishers in Spain to exchange books. His own output contained both sizeable volumes which produced a long-term income and chap-books which could be rapidly sold and reprinted. Even an analysis of the titles he issued and the arrangements under which they were printed indicates that he was wary of becoming too specialized. The Crombergers' commercial interests indicate this same emphasis on diversification, whether it was in printing, publishing, and bookselling in Seville, the acquisition of property in the city and of land outside it, or in general commerce with the New World and investment in Mexican silver-mines. This diversification had many advantages: it lessened the risk of bankruptcy caused by the failure of costly editions or a slump in demand for books; it doubtless meant that profits from one activity could be used to finance another—something which was particularly important for printers, who had to bear heavy overheads and yet bridge the gap between their investment in an edition and the eventual income from that investment; and it enabled the Crombergers to build up a wide network of outlets for their wares.

Printing was a notoriously risky occupation, as the hard-headed capitalist, Andrés Ruiz, observed in 1575 on the eve of entering the business with other members of his family: 'As for printing, it is a new enterprise and I have never had a high opinion of it. Please God that it does not ruin us.'[3] But all enterprise in the 'boom-town' atmosphere of Seville was hazardous. After a long period of prosperity the Crombergers eventually succumbed, whether because of the failure of the press or of their other activities we cannot be sure, but this history of their rise and decline was typical of that of many another merchant dynasty in the volatile atmosphere of sixteenth-century Seville.

[3] Lapeyre, *Une Famille*, p. 572.

BIBLIOGRAPHY

The many bibliographies, printed library catalogues and sale catalogues consulted are not included unless they are referred to in the printed part of this book or its appendices; the place of publication of journals is given only where there is a possibility of confusion or where the journals are somewhat obscure.

Manuscript Sources

London

British Library

MS Add. 21447

MS Egerton 1129 [referred to as 'Méndez MS' in App. I]

Madrid

Archivo Histórico Nacional

Inquisición, Libro 323

Inquisición, Legajos 2072(1), 4442(1)

Mexico City

Archivo General de la Nación

Inquisición, vols. 5, 16, 43

Mercedes, Libros 1, 2, 4

Nuremberg

Staatsarchiv Nürnberg

Nürnberg Bürgerbüchern, Rep. 52b

Seville

Archivo de la Catedral

Actas Capitulares for 1553–4, 1555–6

Archivo General de Indias

Contratación, Legajos 63B, 204, 215, 229, 236, 239, 689A & B, 926, 1059, 1079, 1451, 1788, 2146, 2439, 2464, 2490, 2724, 4674, 4675A, 5536, 5537, 5538, 5767

Indiferente General, Legajo 420

Justicia, Legajos 720, 1004, 1005

Archivo de Protocolos

Approximately 350 Libros, mainly from the period 1500–60 (from Oficios 1, 3, 4, 5, 6, 7, 9, 10, 11, 15, 16, 17, 18, 19, 20, 21, 23), were examined with profit. Many more were consulted in vain

Printed Primary Sources

ABIZANDA Y BROTO, MANUEL, *Documentos para la historia artística y literaria de Aragón procedentes del Archivo de Protocolos de Zaragoza*, 3 vols. (Zaragoza, 1915–32)

BELTRÁN DE HEREDIA, VICENTE, OP, *Cartulario de la Universidad de Salamanca*, 6 vols. (Salamanca, 1970–2)

Cartas de Indias, published by the Ministerio de Fomento (Madrid, 1877)

COLÓN, HERNANDO, *Catalogue of the Library of Ferdinand Columbus; Reproduced in Facsimile from the Unique Manuscript in the Columbine Library of Seville by Archer M. Huntington* (New York, 1905) [normally referred to as '*Registrum B*']

Documentos para la historia de la tipografía americana, ed. with an introduction by Emilio Valtón (Mexico City, 1936)

FERNÁNDEZ DEL CASTILLO, FRANCISCO, *Libros y libreros en el siglo XVI*, 2nd edn. (Mexico City, 1982)

GESTOSO Y PÉREZ, JOSÉ, *Noticias inéditas de impresores sevillanos* (Seville, 1924)

GREENLEAF, RICHARD E., *Zumárraga and his Family Letters to Vizcaya 1536–1548: A Collection of Documents in Relation to the Founding of a Hospice in his Birthplace* (Washington, 1979)

HAZAÑAS Y LA RÚA, JOAQUÍN, *La imprenta en Sevilla: noticias inéditas de sus impresores desde la introducción del arte tipográfico en esta ciudad hasta el siglo XIX*, 2 vols. (Seville, 1945–9)

KROELL, ANNE, 'Le Voyage de Lazarus Nürnberger en Inde (1517–1518)', *Bulletin des études portugaises et brésiliennes*, 41 (1980), 59–87

Lazarus Nurenberger: Being a Photograph of the Manuscript Account of his Voyage to India in 1517 and 1518 Preserved in the Central Library of the Slovakian Academy of Sciences in Bratislava, Czechoslovakia, ed. with an introduction by Ronald Bishop Smith (Lisbon, 1974)

MADURELL MARIMÓN, JOSÉ MARÍA, *Documentos para la historia de la imprenta y librería en Barcelona (1474–1533)*, with notes by Jorge Rubió y Balaguer (Barcelona, 1955)

MILLARES CARLO, AGUSTÍN, 'Algunos documentos referentes a tipógrafos de México en el siglo XVI', in id., *Investigaciones biobibliográficas iberoamericanas: época colonial* (Mexico City, 1950), pp. 111–42

—— and MANTECÓN, JOSÉ IGNACIO, *Índice y extractos de los protocolos del Archivo de Notarías de México D.F.*, 2 vols. (Mexico City, 1945–6)

REUSCH, FR. HEINRICH, *Die Indices Librorum Prohibitorum des sechzehnten Jahrhunderts*, Bibliothek des Litterarischen Vereins in Stuttgart, 176 (Tübingen, 1886)

SÁINZ DE ZÚÑIGA, C. Mª AJO G. Y, *Historia de las universidades hispánicas: orígenes y desarrollo desde su aparición a nuestros días*, 2 vols. (Madrid, 1957; Ávila, 1958)

TERNAUX-COMPANS, H., *Voyages, relations et mémoires originaux pour servir à l'histoire de la découverte de l'Amérique publiés pour la première fois en français*, 20 vols. (Paris, 1837–41)

Secondary Sources

ADAMS, H. M., *Catalogue of Books Printed on the Continent of Europe, 1501–1600, in Cambridge Libraries*, 2 vols. (Cambridge, 1967)

AGUILAR PIÑAL, FRANCISCO, 'Diego Alejandro Gálvez y el origen de la imprenta sevillana', *Archivo hispalense*, 2nd ser., 50–1, Nos. 153–8 (1969), 107–16

—— 'Adiciones a la tipografía hispalense del siglo XVI', *ibid.*, 193–201

AITON, ARTHUR SCOTT, *Antonio de Mendoza: First Viceroy of New Spain* (New York, 1927)

ALCIATI, ANDREA, *Los emblemas* (Lyons, 1549)

AMEZÚA Y MAYO, AGUSTÍN G. DE, *Cómo se hacía un libro en nuestro Siglo de Oro* (Madrid, 1946)

ANDRÉS, MELQUÍADES, *La teología española en el siglo XVI*, 2 vols. (Madrid, 1977)

ANSELMO, ANTÓNIO JOAQUIM, *Bibliografia das obras impressas em Portugal no século XVI* (Lisbon, 1926)

ANSELMO, ARTUR, *Origens da imprensa em Portugal* (Lisbon, 1981)

ANTONIO, NICOLÁS, *Bibliotheca Hispana nova sive Hispanorum scriptorum qui ab anno MD. ad MDCLXXXIV. floruere notitia*, 2nd edn., 2 vols. (Madrid, 1788)

ARAUJO, EDUARDO F., *Primeros impresores e impresos en Nueva España* (Mexico City, 1979)

ARMSTRONG, ELIZABETH, *Robert Estienne, Royal Printer: An Historical Study of the Elder Stephanus* (Cambridge, 1954)

ASENSIO, EUGENIO, 'El erasmismo y las corrientes espirituales afines', *Revista de filología española*, 36 (1952), 31–99

AVIS, FREDERICK C., 'Miscellaneous Costs in Sixteenth Century English Printing', *Gutenberg-Jahrbuch* (1976), 306–10

BALSAMO, LUIGI and TINTO, ALBERTO, *Origini del corsivo nella tipografia italiana del 1500* (Milan, 1967)

BARRANTES, VICENTE, 'Apuntes para un catálogo de impresores desde la introducción del arte en España hasta el año 1600', *Revista contemporánea* [Madrid], 26 (1880), 385–407, and 27 (1880), 43–72, 421–36

BATAILLON, MARCEL, *Erasmo y España: estudios sobre la historia espiritual del siglo XVI*, tr. Antonio Alatorre, 2nd Spanish-language edn. (Mexico, 1966)

BEARDSLEY, THEODORE S., JR., *Hispano-Classical Translations Printed Between 1482–1699* (Pittsburgh, 1970)

—— 'Spanish Printers and the Classics: 1482–1599', *Hispanic Review*, 47 (1979), 25–35

BEAULIEUX, CHARLES, *Catalogue de la Réserve XVIᵉ siècle (1501–1540) de la Bibliothèque de l'Université de Paris*, 2 vols. (Paris, 1910–23)

BENNASSAR, BARTOLOMÉ, *Valladolid au siècle d'or: une ville de Castille et sa campagne au XVIe siècle*, Civilisations et Sociétés, 4 (Paris and The Hague, 1967)

—— *La España del Siglo de Oro*, tr. Pablo Bordonava (Barcelona, 1983)

BENNETT, H. S., *English Books and Readers 1475–1557* (Cambridge, 1969)

BERGER, PHILIPPE, 'La Lecture à Valence de 1474 à 1504', *Mélanges de la Casa de Velázquez*, 11 (1975), 99–118

—— 'Contribution à l'étude du déclin du valencien comme langue littéraire au seizième siècle', *Mélanges de la Casa de Velázquez*, 12 (1976), 171–94

—— 'La Lecture à Valence de 1474 à 1560 (évolution des comportements en fonction des milieux sociaux)', in Manuel C. Díaz y Díaz *et al.*, *Livre et lecture en Espagne*, pp. 97–110

BERMÚDEZ PLATA, CRISTÓBAL, *Catálogo de pasajeros a Indias*, 3 vols. (Seville, 1940–6)

—— 'Las impresiones de las Bulas de la Santa Cruzada para las Indias', *Archivo hispalense*, 2nd ser., 6, No. 15 (1946), 9–40

—— 'Relaciones entre Jacobo Cromberger y Hernán Cortés, con noticias de imprentas sevillanas', *Anuario de estudios americanos* [Seville], 4 (1947), 665–88

BEZZEL, IRMGARD, 'Bartholomäus May (ca.1515–1576) aus Bern, ein Sammler spanischer Drucke', *Iberoromania*, 1 (1969), 235–98

Blackwell's Rare Books, Catalogue A 81, *Fyfield Miscellany*, 5 (Oxford, 1985)

BLAKE, N. F., *Caxton and His World* (London, 1969)

BLAND, DAVID FARRANT, *A History of Book Illustration, the Illuminated Manuscript and the Printed Book* (London, 1958)

BLISS, P. D., *A History of Wood-engraving* (London, 1964)

BMC, see *Catalogue of Books Printed in the XVth Century*

BOEHMER, EDWARD, *Bibliotheca Wiffeniana. Spanish Reformers of Two Centuries from 1520. Their Lives and Writings According to the Late Benjamin B. Wiffen's Plan and with the Use of his Materials*, 3 vols. (Strasburg and London, 1874–1904)

BOHIGAS, PEDRO, *El libro español: ensayo histórico* (Barcelona, 1962)

BONNANT, GEORGES, 'Note sur quelques ouvrages en langue espagnole imprimés à Genève par Jean Crespin (1557–1560)', *Bibliothèque d'humanisme et Renaissance*, 24 (1962), 50–7

BRAUDEL, FERNAND, *The Mediterranean and the Mediterranean World in the Age of Philip II*, tr. Siân Reynolds, 2nd English edn., 2 vols. (London, 1972)

BRIQUET, C. M., *Les Filigranes: dictionnaire historique des marques du papier dès leur apparition vers 1282 jusqu'en 1600*, Jubilee edn., ed. Allan Stevenson, 4 vols. (Amsterdam, 1968)

BROWN, JONATHAN, and ELLIOTT, J. H., *A Palace for a King: The Buen Retiro and the Court of Philip IV* (New Haven and London, 1980)

BRUN, ROBERT, *Le Livre français* (Paris, 1948)

—— *Le Livre français illustré de la Renaissance: étude suivie du catalogue des principaux livres à figures du XVIe siècle* (Paris, 1969)

BUJANDA, J. M. DE, 'La Censure littéraire en Espagne au XVIe siècle', *Canadian Journal of History*, 7 (1972), 1–15
BURGER, KONRAD, *Die Drucker und Verleger in Spanien und Portugal von 1501–1536, mit chronologischer Folge ihrer Druck- und Verlagswerke* (Leipzig, 1913)
BUTSCH, A. F., *Die Bücher-Ornamentik der Renaissance* (Leipzig, 1878–81)

Carlos V y su época: exposición bibliográfica y documental (Barcelona, 1958)
CARPENTER, KENNETH E. (ed.), *Books and Society in History: Papers of the Association of College and Research Libraries, Rare Books and Manuscripts Preconference, 24–28 June, 1980, Boston, Massachusetts* (New York and London, 1983)
CARREÑO, ALBERTO MARÍA, 'La primera biblioteca del continente americano', *Divulgación histórica* [Mexico City], 4 (1943), 428–31, 488–92
—— '¿El primer impreso en América?', *Boletín de la Biblioteca Nacional de México*, 5 (1954), 5–33
CARTER, HARRY, *A View of Early Typography up to about 1600* (Oxford, 1969)
CARTER, JOHN, and MUIR, PERCY H. (eds.), *Printing and the Mind of Man: A Descriptive Catalogue Illustrating the Impact of Print on the Evolution of Western Civilization During Five Centuries*, 2nd, rev. and enlarged edn. (Munich, 1983)
CARVER, ALEXANDER B., 'Esteban Martín, the First Printer in the Western Hemisphere: An Examination of Documents and Opinion', *Library Quarterly*, 39 (1969), 344–52
CASTAÑEDA, CARLOS E., 'The Beginning of Printing in America', *Hispanic American Historical Review*, 20 (1940), 671–85
CASTAÑEDA, VICENTE, and HUARTE, AMALIO (eds.), *Colección de pliegos sueltos, agora de nuevo sacados* (Madrid, 1929)
CASTELLANI, C., *Early Venetian Printing* (Venice and London, 1895)
Catálogo colectivo de obras impresas en los siglos XVI al XVIII existentes en las bibliotecas españolas, edición provisional: siglo XVI, 15 vols. (Madrid, 1972–84) [only vols. 1–13 seen]
Catálogo de los fondos americanos del Archivo de Protocolos de Sevilla, Colección de documentos inéditos para la historia de Hispano-América, 10, 11, 14, etc., 5 vols. (Madrid, Barcelona, Buenos Aires, 1930-7)
Catálogo dos livros do século XVI existentes na Biblioteca da Faculdade de Medicina de Coimbra (s.l., s.d.)
Catálogo dos reservados da Biblioteca Geral da Universidade de Coimbra (Coimbra, 1970)
Catalogue of Books Printed in the XVth Century now in the British Museum, 12 vols. (London, 1908–71) [normally referred to as '*BMC*']
CAZALLA, JUAN DE, *Lumbre del alma*, ed. with an introduction by J. Martínez de Bujanda (Madrid, 1974)
CHAMBERS, MARLENE, 'Early Printing and Book Illustration in Spain', *The Library Chronicle*, 39 (1973), 3–17
CHAUNU, PIERRE, 'Les Romans de chevalerie et la conquête du Nouveau Monde', *Annales: économies, sociétés, civilisations*, 10 (1955), 216–28
—— *Séville et l'Amérique aux XVIe et XVIIe siècles* (Paris, 1977)

—— and CHAUNU, HUGUETTE, *Séville et l'Atlantique 1504–1650*, 8 vols. (Paris, 1955–9)

CHEVALIER, MAXIME, *Lectura y lectores en la España de los siglos XVI y XVII* (Madrid, 1976)

CIPOLLA, CARLO M., *Literacy and Development in the West* (Harmondsworth, 1969)

CLAIR, COLIN, *A History of European Printing* (New York and London, 1976)

CLAUDIN, A., *Histoire de l'imprimerie en France au XVe et au XVIe siècle*, 4 vols. (Paris, 1900–14)

COLLANTES DE TERÁN, FRANCISCO, 'Un taller alemán de imprenta en Sevilla en el siglo XV', *Gutenberg-Jahrbuch* (1931), 145–65

COLLONGE, C., 'Les Allemands dans l'Espagne du XVIe siècle', *Les Langues néo-latines*, 59 (1965), 1–22

CORTÉS, HERNÁN, *Letters from Mexico*, ed. and tr. A. R. Pagden (New York, 1971)

CRUICKSHANK, D. W., 'Some Aspects of Spanish Book-Production in the Golden Age', *The Library*, 5th ser., 31 (1976), 1–19

—— ' "Literature" and the Book Trade in Golden-Age Spain', *Modern Language Review*, 73 (1978), 799–824

CUESTA GUTIÉRREZ, LUISA, 'La imprenta y el libro en la América hispana colonial', *Gutenberg-Jahrbuch* (1957), 160–7

DAVIS, NATALIE ZEMON, 'The Protestant Printing Workers of Lyons in 1551', in G. Berthoud *et al.*, *Aspects de la propagande religieuse* (Geneva, 1957), pp. 247–57

—— 'Strikes and Salvation at Lyons', *Archiv für Reformationsgeschichte*, 56 (1965), 48–64

—— 'A Trade Union in Sixteenth-century France', *Economic History Review*, 2nd ser., 19 (1966), 48–69

DESLANDES, VENÂNCIO, *Documentos para a historia da typographia portugueza nos seculos XVI e XVII*, 2nd edn. (Lisbon, 1888)

DEVOTO, DANIEL, 'Amadis de Galia', *Bulletin hispanique*, 74 (1972), 406–35

DEYERMOND, A. D., *The Middle Ages*, A Literary History of Spain [1] (London, 1971)

DÍAZ Y DÍAZ, MANUEL C., *et al.*, *Livre et lecture en Espagne et en France sous l'Ancien Régime: colloque de la Casa de Velázquez* (Paris, 1981)

DOMÍNGUEZ GUZMÁN, AURORA, *El libro sevillano durante la primera mitad del siglo XVI* (Seville, 1975)

—— 'Veinte años de impresiones sevillanas (1551–1570)', *Cuadernos bibliográficos* [Madrid], 37 (1978), 1–57

DOMÍNGUEZ ORTIZ, ANTONIO, *Orto y ocaso de Sevilla*, 3rd edn. (Seville, 1981)

DONNELLY, PAUL J., 'A Study of Spanish Translations of Erasmus's *Colloquia* (1525–1536), with Special Reference to the Translations of Alonso de Virués: Together with a Critical Edition of the Versions of *Uxor Mempsigamos, Funus*, and Part of *Convivium Religiosum*' (D.Phil. thesis, Oxford, 1979), 2 vols.

DROZ, E., 'Notes sur les impressions genevoises transportées par Hernández', *Bibliothèque d'humanisme et Renaissance*, 22 (1960), 119–32

EBERT, FRIEDRICH ADOLF, *A General Bibliographical Dictionary, from the German*, 4 vols. (Oxford, 1837)

EISENBERG, DANIEL, 'A Seville 1549 Edition of the Sergas de Esplandián', *Journal of Hispanic Philology*, 1 (1976–7), 61–3

—— *Castilian Romances of Chivalry in the Sixteenth Century: A Bibliography*, Grant and Cutler Research Bibliographies and Checklists, 23 (London, 1979)

—— *Romances of Chivalry in the Spanish Golden Age* (Newark, 1982)

—— '"Una uña de la gran bestia"', *Journal of Hispanic Philology*, 8 (1984), 93–8

EISENSTEIN, ELIZABETH L., *The Printing Press as an Agent of Change: Communications and Cultural Transformations in Early-Modern Europe*, 2 vols. (Cambridge, 1979)

ELLIOTT, J. H., *Imperial Spain 1469–1716*, 2nd edn. (Harmondsworth, 1970)

—— *The Old World and the New 1492–1650* (Cambridge, 1970)

ERASMUS, DESIDERIUS, *Opus Epistolarum Des. Erasmi Roterodami*, ed. P. S. Allen *et al.*, 12 vols. (Oxford, 1906–58)

—— *Tratado del Niño Jesús . . . (Seville, 1516): ahora fielmente reimpreso en facsímile con un estudio preliminar de Eugenio Asensio* (Madrid, 1969)

ESCALANTE, HILDAMAR, 'Juan Pablos, primer impresor en América', *Revista nacional de cultura* [Caracas], 38 (1943), 76–84

ESCUDERO Y PEROSSO, FRANCISCO, *Tipografía hispalense: anales bibliográficos de la ciudad de Sevilla desde el establecimiento de la imprenta hasta fines del siglo XVIII* (Madrid, 1894)

ESPEJO, CRISTÓBAL, 'La carestía de la vida en el siglo XVI', *Revista de archivos, bibliotecas y museos*, 41 (1920), 36–54, 170–204, 329–54

FALKENSTEIN, KARL, *Beschreibung der Königlichen Öffentlichen Bibliothek zu Dresden* (Dresden, 1839)

FARRELL, ANTHONY J., 'Version H of "The Seven Sages": A Descriptive Bibliography of Spanish Editions', *La corónica*, 9 (1980), 56–66

FEBVRE, LUCIEN, and MARTIN, HENRI-JEAN, *L'Apparition du livre* (Paris, 1958)

FERNÁNDEZ, LUIS, SJ, 'La real imprenta del Monasterio de Nuestra Señora del Prado (1481–1835)', in Benito Colombás *et al.*, *Studia Hieronymiana: VI Centenario de la Orden de San Jerónimo*, 2 vols. (Madrid, 1973), ii. 137–214

FRANCO SILVA, ALFONSO, *Regesto documental sobre la esclavitud sevillana (1453–1513)* (Seville, 1979)

—— *La esclavitud en Sevilla y su tierra a fines de la Edad Media* (Seville, 1979)

GALLARDO, BARTOLOMÉ JOSÉ, *Ensayo de una biblioteca española de libros raros y curiosos formado con los apuntamientos de B. J. Gallardo, coordinados y aumentados por M. R. Zarco del Valle y J. Sancho Rayón*, 4 vols. (Madrid, 1863–89)

GALLO, IVANA, and SCOLES, EMMA, 'Edizioni antiche della *Celestina* sconosciute o non localizzate dalla tradizione bibliografica', *Cultura neo-latina*, 43 (1983), 103–19

GARCÍA CHICO, E., 'Documentos referentes a la imprenta en Medina del Campo', *Castilla: boletín del Seminario de Estudios de Literatura y Filología de la Universidad de Valladolid, Facultad de Historia*, 2 (1941–3), 233–98

García de Enterría, María Cruz (ed.), *Pliegos poéticos españoles de la Biblioteca Nacional de Viena*, Joyas bibliográficas, serie conmemorativa, 18, 2 vols. (Madrid, 1975)

—— (ed.), *Pliegos poéticos españoles en bibliotecas de Portugal*, Joyas bibliográficas, serie conmemorativa, 24, Homenaje a Edward M. Wilson, 2 vols. (Madrid, 1982)

García Icazbalceta, Joaquín, *Don Fray Juan de Zumárraga primer obispo y arzobispo de México*, Colección de escritores mexicanos, 41–4, 2nd edn., rev. Rafael Aguayo Spencer and Antonio Castro Leal, 4 vols. (Mexico City, 1947)

—— *Bibliografía mexicana del siglo XVI: catálogo razonado de libros impresos en México de 1539 a 1600 con biografías de autores y otras ilustraciones, precedido por una noticia acerca de la introducción de la imprenta en México*, 2nd edn., rev. Agustín Millares Carlo (Mexico City, 1954)

García de Santa María, Gonzalo, *El Catón en latín y en romance (1493/94)*, facsimile, ed. A. Pérez Gómez, Incunables poéticos castellanos, 9 (Cieza, 1964)

Gaselee, Stephen, *The Early Spanish Printing Press: A Lecture Delivered Before the Anglo-Spanish Society of Great Britain and the Spanish-speaking Countries on Wednesday, December 19th, 1923* (London, 1924)

Gaskell, Philip, *A New Introduction to Bibliography* (Oxford, 1972)

Gestoso y Pérez, José, *Ensayo de un diccionario de los artífices que florecieron en Sevilla desde el siglo XIII al XVIII inclusive*, 3 vols. (Seville, 1899–1909)

Gibbs, Jack, 'The Status of the Cromberger Editions of Antonio de Guevara's *Libro aureo de Marco Aurelio* and *Relox de principes*', *Bulletin of Hispanic Studies*, 54 (1977), 199–201.

Gil Fernández, Luis, *Panorama social del humanismo español (1500–1800)* (Madrid, 1981)

Gillet, Joseph E. (ed.), *Propalladia and Other Works of Bartolomé de Torres Naharro*, 4 vols. (Bryn Mawr and Philadelphia, 1943–61)

Gilmont, Jean-François, 'Printers by the Rules', *The Library*, 6th ser., 2 (1980), 129–55

—— *Jean Crespin: un éditeur réformé du XVI*[e] *siècle*, Travaux d'humanisme et Renaissance, 186 (Geneva, 1981)

Goldschmidt, E. Ph., *Medieval Texts and Their First Appearance in Print*, Bibliographical Society Publication for 1940 (London, 1943)

Gomes da Silva, Nuno J. Espinosa, 'Algumas notas sobre a edição das *Ordenações Manuelinas* de 1521', *Scientia Juridica* [Braga], 30 (1980), 5–16

Goñi Gaztambide, José, 'El impresor Miguel de Eguía procesado por la Inquisición (c. 1495–1546)', *Hispania sacra*, 1 (1948), 35–88

Granata Catalogue (Almería, 1980)

Grantel Santander, Luis S., and Grantel, Mercedes S., *El libro médico español renacentista* (Salamanca, 1980)

Greenleaf, Richard E., *The Mexican Inquisition of the Sixteenth Century* (Albuquerque, 1969)

Grice-Hutchinson, Marjorie, *Early Economic Thought in Spain 1177–1740* (London, 1978)

GRIFFIN, CLIVE, 'Four Rare Editions of *La Celestina*', *Modern Language Review*, 75 (1980), 561–74

—— 'More on the Spanish *Scala celi*', *The Library*, 6th ser., 5 (1983), 385–91

—— 'The Crombergers of Seville and the First Italic Book Printed in Spain', in Jean-François Gilmont (ed.), *Palaestra typographica: aspects de la production du livre humaniste et religieux au XVIe siècle* (Aubel, 1984), pp. 57–96

—— 'Giacomo Caviceo's *Libro del Peregrino*: The Fate of an Italian Wanderer in Spain', in Anna Laura Lepschy, John Took, and Dennis E. Rhodes (eds.), *Book Production and Letters in the Western European Renaissance: Essays in Honour of Conor Fahy*, Publications of the Modern Humanities Research Association, 12 (London, 1986), pp. 132–46

GUICCIARDINI, FRANCESCO, *Viaje a España de Francesco Guicciardini Embajador de Florencia ante el Rey Católico*, tr. José María Alonso Gamo (Valencia, 1952)

GUILLÉN, CLAUDIO, 'Un padrón de conversos sevillanos (1510)', *Bulletin hispanique*, 45 (1963), 49–98

GUSMÃO, ARMANDO, *Tipografia espanhola*, Livros impressos no século XVI existentes na Biblioteca Pública e Arquivo Distrital de Évora, 2 (Évora, 1966)

GUTIÉRREZ, CONSTANTINO, SJ, *Españoles en Trento* (Valladolid, 1951)

HAEBLER, KONRAD, *The Early Printers of Spain and Portugal*, Bibliographical Society's Monographs, 4 (London, 1897)

——*Typographie ibérique du quinzième siècle: reproduction en facsimile de tous les caractères typographiques employés en Espagne et en Portugal jusqu'à l'année 1500* (The Hague and Leipzig, 1901)

——*Bibliografía ibérica del siglo XV: enumeración de todos los libros impresos en España y Portugal hasta el año de 1500*, 2 vols. (The Hague and Leipzig, 1904–17)

——'Typefounding and Commerce in Type during the Early Years of Printing', *Ars Typographica* [New York], 3 (1926), 3–35

HAMILTON, E. J., *American Treasure and the Price Revolution in Spain 1501–1650* (Cambridge, Mass., 1934)

HARING, CLARENCE HENRY, *Trade and Navigation between Spain and the Indies in the Time of the Habsburgs* (Cambridge, Mass., 1918)

——*The Spanish Empire in America*, 3rd edn. (New York and London, 1963)

HARVARD, STEPHEN, *Ornamental Initials: The Woodcut Initials of Christopher Plantin: A Complete Catalogue* (New York, 1974)

HAZAÑAS Y LA RÚA, JOAQUÍN, *La imprenta en Sevilla: ensayo de una historia de la tipografía sevillana y noticias de algunos de sus impresores desde la introducción del arte tipográfico en esta ciudad hasta el año de 1800* (Seville, 1892)

——*Maese Rodrigo 1444–1509* (Seville, 1909)

HELLINGA, WYTZE, and HELLINGA, LOTTE, *The Fifteenth-century Printing Types of the Low Countries*, 2 vols. (Amsterdam, 1966)

HEREDIA HERRERA, ANTONIA, 'La "Carta de Diligencias" de bienes de difuntos', *Archivo hispalense*, 2nd ser., 57, No. 174 (1974), 39–48

HIND, ARTHUR M., *An Introduction to a History of Woodcut with a Detailed Survey of Work Done in the Fifteenth Century*, repr. of 1935 edn., 2 vols. (New York, 1963)

HIRSCH, RUDOLPH, *Printing, Selling and Reading 1450–1550* (Wiesbaden, 1967)

HODNETT, EDWARD, *English Woodcuts 1480–1535*, 2nd edn., with additions and corrections (Oxford, 1973)

HOFFMAN, PAUL E., 'The Archivo de Protocolos de Sevilla', *Itinerario: Leyden Centre for the History of European Expansion*, 5 (1981), 39–45

HOUSTON, RAB, 'Literacy and Society in the West, 1500–1850', *Social History*, 8 (1983), 269–93

HUARTE, A., 'La "E" cromberguiana', *Revista de bibliografía nacional* [Madrid], 4 (1943), 135–46

HUERGA, ÁLVARO, *Agustín de Esbarroya: Purificador de la conciencia* (Madrid, 1973)

ICAZA, FRANCISCO A. DE, *Diccionario autobiográfico de conquistadores y pobladores de Nueva España*, 2 vols. (Madrid, 1923)

—— 'Miscelánea histórica', *Revista mexicana de estudios históricos*, 2 (1928), 5–112

IFE, BARRY, *Reading and Fiction in Golden-Age Spain: A Platonist Critique and Some Picaresque Replies* (Cambridge, 1985)

Index Aureliensis: Catalogus librorum sedecimo saeculo impressorum, 7 vols. to date (Baden-Baden, 1965–) [the catalogue has reached 'Ch']

IRIGOIN, JEAN, 'L'Introduction du papier italien en Espagne', *Papiergeschichte: Zeitschrift der Forschungsstelle Papiergeschichte in Mainz*, 10 (1960), 29–32

JACKSON, JOHN, and CHATTO, W. A., *A Treatise on Wood Engraving*, 2nd edn. (London, 1861)

JENNETT, SEÁN, *Pioneers in Printing* (London, 1958)

JENNINGS, O., *Early Woodcut Initials* (London, 1908)

JOHNSON, A. F., 'The Classification of Gothic Types', *The Library*, 4th ser., 9 (1929), 357–80

—— *Type Designs: Their History and Development* (London, 1934)

—— 'The Supply of Types in the Sixteenth Century', *The Library*, 4th ser., 24 (1943), 47–65

KAŠPAR, OLDŘICH, *Soupis španělských tisků bývalé zámecké knihovny v Roudnici nad Labem nyní deponovaných ve Státní knihovně ČSR v Praze: Registro de los impresos españoles de la antigua biblioteca del castillo de Roudnice nad Labem actualmente depositada en la Biblioteca Estatal de la República Socialista Checa en Praga* (Prague, 1983)

KELLENBENZ, HERMANN, 'Os mercadores alemães por volta de 1530', *Revista portuguesa de história*, 9 (1960), 125–40

—— 'Die Beziehungen Nürnbergs zur Iberischen Halbinsel: besonders im 15. und in der ersten Hälfte des 16. Jahrhunderts', in Gerhard Hirschmann *et al.*, *Beiträge zur Wirtschaftsgeschichte Nürnbergs*, 2 vols. (Nuremberg, 1967), i. 456–93

KINDER, A. GORDON, *Casiodoro de Reina: Spanish Reformer of the Sixteenth Century* (London, 1975)

KINGDON, ROBERT M., 'The Business Activities of Printers Henri and François Estienne', in G. Berthoud *et al.*, *Aspects de la propagande religieuse* (Geneva, 1957), pp. 258–75
—— 'Patronage, Piety and Printing in Sixteenth-century Europe', in D. H. Pinkney and T. Ropp (eds.), *A Festschrift for Frederick B. Artz* (Durham, NC, 1964), pp. 19–36
KNAPP, W. I., *The Earliest Decree on Printing or Thierry Martin of Spain (Seville, December 25 1477)* (New Haven, 1881)
[KRAUS, H. P.] *Fifty Years: An Anniversary Catalogue of 120 Outstanding Books Printed Before 1700 and Sold by H. P. Kraus from 1932 to 1982 with a Bibliography of all H. P. Kraus Rare Book Catalogues Issued During Those Years* (New York, 1982)

LADERO QUESADA, MIGUEL ÁNGEL, *La ciudad medieval (1248–1492)*, Historia de Sevilla, 2, 2nd edn. (Seville, 1980)
—— and QUINTANILLA RASO, MARÍA CONCEPCIÓN, 'Bibliotecas de la alta nobleza castellana en el siglo XV', in Manuel C. Díaz y Díaz *et al.*, *Livre et lecture*, pp. 47–59
LAPEYRE, HENRI, *Une Famille de marchands les Ruiz: contribution à l'étude du commerce entre la France et l'Espagne au temps de Philippe II* (Paris, 1955)
LAWRANCE, JEREMY, 'Nuño de Guzmán: Life and Works' (D.Phil. thesis, Oxford, 1983)
—— 'The Spread of Lay Literacy in Late Medieval Castile', *Bulletin of Hispanic Studies*, 62 (1985), 79–94
LE FLEM, JEAN-PAUL, 'Los aspectos económicos de la España moderna', in *La frustración de un imperio (1476–1714)* (Barcelona, 1982), vol. v of Manuel Tuñón de Lara (ed.), *Historia de España* (Barcelona, 1982–), pp. 11–133
LENKEY, SUSAN V., 'Printers' Wives in the Age of Humanism', *Gutenberg-Jahrbuch* (1975), 331–7
LEÓN, NICOLÁS, 'Noticia de una obra en tarasco', *Anales del Museo Michoacano* [Morelia], 1 (1888), 62–4
LEONARD, IRVING A., *Romances of Chivalry in the Spanish Indies with Some 'Registros' of Shipments of Books to the Spanish Colonies* (Berkeley, 1933)
—— *Books of the Brave: Being an Account of Books and of Men in the Spanish Conquest and Settlement of the Sixteenth-century New World* (Cambridge, Mass., 1949)
LÓPEZ ESTRADA, FRANCISCO, 'Sobre la imprenta en Sevilla en el siglo XVI', *Archivo hispalense*, 2nd ser., 18, No. 57 (1953), 37–48
LOWRY, MARTIN, *The World of Aldus Manutius: Business and Scholarship in Renaissance Venice* (Oxford, 1979)
—— 'The Social World of Nicholas Jenson and John of Cologne', *La Bibliofilia*, 83 (1981), 193–218
LYELL, JAMES P. R., *Early Book Illustration in Spain* (London, 1926)
LYNCH, JOHN, *Spain under the Habsburgs*, 2nd edn., 2 vols. (Oxford, 1981)

MCKERROW, R. B., *An Introduction to Bibliography for Literary Students* (Oxford, 1928)
MCMURTRIE, DOUGLAS C., 'The First Typefounding in Mexico', *The Library*, 4th ser., 8 (1927), 119–22
MADURELL MARIMÓN, JOSÉ MARÍA, 'Licencias reales para la impresión y venta de libros 1519–1705', *Revista de archivos, bibliotecas y museos*, 72 (1964–5), 111–248

—— 'Antiguos fundidores de letras en Barcelona', *Gutenberg-Jahrbuch* (1970), 289–97
—— *Claudi Bornat* (Barcelona, 1973)
MAGGS BROS., *Books Printed in Spain and Spanish Books Printed in Other Countries*, Catalogue No. 495 (London, 1927)
MANUEL II (ex-King of Portugal), *Early Portuguese Books 1489–1600 in the Library of His Majesty the King of Portugal Described by H.M. King Manuel in Three Volumes: Livros antigos portuguezes 1489–1600 da Bibliotheca de Sua Majestade Fidelíssima descriptos por S. M. El-Rei D. Manuel em três volumes*, 3 vols. (Cambridge, 1929–35) [referred to as 'King Manuel' in App. I]
MARINIS, TAMARO DE, 'Nota per Mattia Moravo', *Gutenberg-Jahrbuch* (1930), 115–18
MARTIN, HENRI-JEAN, and CHARTIER, ROGER (eds.), *Le Livre conquérant: du Moyen Âge au milieu du XVIIe siècle* (Paris, 1983), vol. i of eid. (eds.), *Histoire de l'édition française* (Paris, 1983–)
MATEU Y LLOPIS, FELIPE, 'Licencia, tasa y privilegio de impresión en la España de los siglos XVI y XVII', *Gutenberg-Jahrbuch* (1965), 197–200
MEDINA, JOSÉ TORIBIO, *El veneciano Sebastián Caboto al servicio de España y especialmente de su proyectado viaje a las Molucas por el Estrecho de Magallanes y al reconocimiento de la costa del continente hasta la gobernación de Pedrarias Dávila*, 2 vols. (Santiago de Chile, 1908)
—— *La imprenta en México 1539–1821*, 8 vols. (Santiago de Chile, 1908–12)
MEDINA, PEDRO DE, *Obras*, ed. Ángel González Palencia (Madrid, 1944)
MERCADO, PEDRO DE, *Dialogos de philosophia natural y moral*, 2nd edn. (Granada, 1574)
MILLARES CARLO, AGUSTÍN, *Introducción a la historia del libro y de las bibliotecas* (Mexico City, 1971)
MOLL, JAIME, 'Problemas bibliográficos del libro del Siglo de Oro', *Boletín de la Real Academia Española*, 59 (1979), 49–107
MONTOTO, SANTIAGO, *Impresos sevillanos* (Madrid, 1948)
MORALES PADRÓN, FRANCISCO, *La ciudad del quinientos*, Historia de Sevilla, 3 (Seville, 1977)
MORGADO, ALONSO, *Historia de Sevilla* (Seville, 1587)
MORTIMER, RUTH, *Harvard College Library, Department of Printing and Graphic Arts, Catalogue of Books and Manuscripts: Part II: Italian 16th Century Books*, 2 vols. (Cambridge, Mass., 1974)
MOXON, JOSEPH, *Mechanick Exercises on the Whole Art of Printing [1683–84]*, ed. Herbert Davis and Harry Carter, 2nd edn. (New York, 1962)
MURO, LUIS, *La expedición Legazpi-Urdaneta a las Filipinas (1557–1564)* (Mexico City, 1975)

NAVAGERO, ANDREA, *Il viaggio fatto in Spagna et in Francia* (Venice, 1563)
NORTON, F. J., *Italian Printers 1501–1520: An Annotated List with an Introduction*, Cambridge Bibliographical Society Monographs, 3 (London, 1958)
—— *Printing in Spain 1501–1520* (Cambridge, 1966)
—— 'Some Spanish Printed Bifolia of the Early Sixteenth Century: Francisco de Peñalver's

Carta de la gota and its Antecedents', in Dennis E. Rhodes (ed.), *Essays in Honour of Victor Scholderer* (Mainz, 1970), pp. 271–5

—— 'Typographical Evidence as an Aid to the Identification and Dating of Unsigned Spanish Books of the Sixteenth Century', *Iberoromania*, 2 (1970), 96–103

—— *A Descriptive Catalogue of Printing in Spain and Portugal 1501–1520* (Cambridge, 1978)

—— and WILSON, EDWARD M., *Two Spanish Verse Chapbooks: Romançe de Amadis (c.1515–19), Juyzio hallado y trobado (c.1510): A Facsimile Edition with Bibliographical and Textual Studies* (Cambridge, 1969)

NOVALÍN, JOSÉ LUIS G., *El inquisidor general Fernando de Valdés (1483–1568)*, 2 vols. (Oviedo, 1968–71)

NÚÑEZ DELGADO, PEDRO, *Epigrammata*, ed. Cristóbal Núñez (s.l., 1537)

ODRIOZOLA, ANTONIO, 'Algunos problemas bibliográficos que plantean las obras de Nebrija, primera serie', *Bibliografía hispánica*, 4 (1945), 213–41

—— 'La caracola del bibliófilo nebrisense: extracto seco de bibliografía de Nebrija en los siglos XV y XVI', *Revista de bibliografía nacional* [Madrid], 7 (1946), 3–114

—— 'Los libros impresos por Juan Parix en Segovia y Toulouse y los atribuibles a Turner y Parix en esta última ciudad (1472–1478) (una investigación sobre protoincunables)', in J. M. Ruiz Asencio *et al.*, *Homenaje a D. Agustín Millares Carlo*, 2 vols. (Las Palmas, 1975), i. 281–308 and 18 plates

—— 'Los protoincunables (1472–1479) impresos por Juan Parix en Segovia (España) y Toulouse (Francia) con tres figuras', *Gutenberg-Jahrbuch* (1976), 130–7

ORTIZ, LUIS, *Memorial del contador don Luis Ortiz a Felipe II*, transcribed by Jaime Fernández Laville (Madrid, 1970)

OSBORNE, LUCY EUGENIA, 'The Whitchurch Compartment in London and Mexico', *The Library*, 4th ser., 8 (1927), 303–11

OTTE, ENRIQUE, 'Jakob und Hans Cromberger und Lazarus Nürnberger, die Begründer des deutschen Amerikahandels', *Mitteilungen des Vereins für Geschichte der Stadt Nürnberg*, 52 (1963–4), 129–62

PAGDEN, ANTHONY R. D., 'The "School of Salamanca" and the "Affair of the Indies"', *History of Universities* [Avebury], 1 (1981), 71–112

PALAU Y DULCET, A., *Manual del librero hispano-americano*, 2nd edn., 28 vols. (Barcelona and Oxford, 1948–77)

PARKER, GEOFFREY, *Philip II* (London, 1979)

PEETERS-FONTAINAS, J. F., 'L'Officine espagnole de Martin Nutius à Anvers', *De Gulden Passer*, 35 (1957), 1–106

PEIXOTO, JORGE, 'Aspectos económicos do livro em Portugal no século XVI', *Gutenberg-Jahrbuch* (1965), 142–9

PÉLIGRY, CHRISTIAN, 'Les Difficultés de l'édition castillane au XVIIe siècle à travers un document de l'époque', *Mélanges de la Casa de Velázquez*, 13 (1977), 257–84

—— 'Les Éditeurs lyonnais et le marché espagnol aux XVIe et XVIIe siècles', in Manuel C. Díaz y Díaz *et al.*, *Livre et lecture*, pp. 85–93

PENNEY, CLARA LOUISA, *The Book Called Celestina in the Library of the Hispanic Society of America* (New York, 1954)
—— *Printed Books 1468–1700 in the Hispanic Society of America: A Listing* (New York, 1965)
PÉREZ GÓMEZ, ANTONIO, *Glosas a las Coplas de Jorge Manrique: noticias bibliográficas* (Cieza, 1963)
PÉREZ PASTOR, CRISTÓBAL, *La imprenta en Medina del Campo* (Madrid, 1895)
PERRY, MARY ELIZABETH, *Crime and Society in Early Modern Seville* (Hanover, NH, and London, 1980)
PETTAS, WILLIAM A., *The Giunti of Florence: Merchant Publishers of the Sixteenth Century* (San Francisco, 1980)
PIKE, RUTH, *Enterprise and Adventure: The Genoese in Seville and the Opening of the New World* (Ithaca, 1966)
—— *Aristocrats and Traders: Sevillian Society in the Sixteenth Century* (Ithaca and London, 1972)
PLACE, EDWIN B. (ed.), *Amadís de Gaula*, 4 vols. (Madrid, 1959–69)
PLATTER, THOMAS, *Autobiographie*, tr. Marie Helmer, Cahiers des Annales, 22 (Paris, 1964)
Pliegos poéticos españoles en la Universidad de Praga, with an introduction by R. Menéndez Pidal, Joyas bibliográficas, serie conmemorativa, 7 and 8, 2 vols. (Madrid, 1960)
Pliegos poéticos góticos de la Biblioteca Nacional [of Madrid], with an introduction by José Antonio García Noblejas, Joyas bibliográficas, serie conmemorativa, 1, 2, 5, 6, 9, and 10, 6 vols. (Madrid, 1957–61)
POLLAK, MICHAEL, 'The Durability of Fifteenth-century Type', *Library Quarterly*, 40 (1970), 371–90
—— 'The Performance of the Wooden Printing Press', *Library Quarterly*, 42 (1972), 218–64
POLLARD, A. W., *Italian Book Illustrations Chiefly in the Fifteenth Century*, Portfolio Monographs on Artistic Subjects, 12 (London, 1894)
—— 'The Transference of Woodcuts in the Fifteenth and Sixteenth Centuries', *Bibliographica: Papers on Books, Their History and Art* [London], 2 (1896), 343–68
—— *Old Picture Books with Other Essays on Bookish Subjects* (London, 1902)
—— *Fine Books* (London, 1912)
—— *Early Illustrated Books: A History of the Decoration and Illustration of Books in the 15th and 16th Centuries* (London, 1917)
PORTILLO MUÑOZ, JOSÉ L., *La ilustración gráfica de los incunables sevillanos (1470–1500)* (Seville, 1982)
PUVILL (Librería), Catalogues Nos. 162 (Barcelona, 1978), 175 (Barcelona, 1983)

QUARITCH, BERNARD, *A Catalogue of Important and Rare Books on Botany, Agriculture, Forestry, Fruit-Culture, Gardens and Gardening, Herbals, Early Medicine and Surgery etc.*, Catalogue No. 443 (London, 1931)

—— *Hispanic Books*, Catalogue No. 1048 (London, 1985)

QUINTANILLA RASO, MARÍA CONCEPCIÓN, 'La biblioteca del marqués de Priego (1518)', in Miguel Ángel Ladero Quesada (ed.), *En la España medieval: estudios dedicados al profesor D. Julio González González* (Madrid, 1980), pp. 347–83.

REDONDO, AGUSTÍN, 'Les Éditions des *Ordenações manuelinas* faites par les imprimeurs sévillans de la première moitié du XVIe siècle Jacobo et Juan Cromberger', *Mélanges de la Casa de Velázquez*, 2 (1966), 103–25

RODRÍGUEZ-MOÑINO, ANTONIO, *Los pliegos poéticos de la colección del Marqués de Morbecq (siglo XVI)* (Madrid, 1962)

—— *Construcción crítica y realidad histórica en la poesía española de los siglos XVI y XVII*, 2nd edn. (Madrid, 1968)

—— *Diccionario bibliográfico de pliegos sueltos poéticos (siglo XVI)* (Madrid, 1970)

ROGERS, DAVID, 'Printing in 1478: The Background to the First Press at Oxford', *Journal of the Printing Historical Society*, 13 (1978–9), 67–77

ROMERO DE LECEA, CARLOS, 'Hernando de Talavera y el tránsito en España "Del manuscrito al impreso"', in Benito Colombás, *et al.*, *Studia Hieronymiana: VI Centenario de la Orden de San Jerónimo*, 2 vols. (Madrid, 1973), i. 315–77

—— *La imprenta y los pliegos poéticos*, Colección de estudios y ensayos Joyas bibliográficas (Madrid, 1974)

—— *et al.*, *Historia de la imprenta hispana* (Madrid, 1982)

[ROSENWALD, LESSING J.] *The Lessing J. Rosenwald Collection: A Catalog of Gifts of Lessing J. Rosenwald to the Library of Congress, 1943–1975* (Washington, 1977)

RUBIÓ Y BALAGUER, JORGE, 'Integración de los impresores alemanes en la vida social y económica de Cataluña y Valencia en los siglos XV—XVI', *Spanische Forschungen der Görresgesellschaft*, 1st ser., 20 (1962), 103–22

RUBIO MAÑÉ, J. IGNACIO, *Introducción al estudio de los virreyes de Nueva España, 1535–1746*, 4 vols. (Mexico City, 1955–63)

—— 'La expedición de Miguel López de Legazpi a Filipinas', *Boletín del Archivo de la Nación* [Mexico City], 2nd ser., 5 (1964), 427–798

—— 'Más documentos relativos a la expedición de Miguel López de Legazpi a Filipinas', *Boletín del Archivo de la Nación*, 2nd ser., 11 (1970), 82–156, 453–556

RUPPEL, ALOYS, *Stanislaus Polonus: polski drukarz i wydawca wczesnej doby w Hiszpanii*, tr. and enlarged by Tadeusz Zapiór (Cracow, 1970)

RUSSELL, P. E., 'Towards an Interpretation of Rodrigo de Reinosa's "poesía negra"', in R. O. Jones (ed.), *Studies in Spanish Literature of the Golden Age Presented to Edward M. Wilson* (London, 1973), pp. 225–45

SALVÁ Y MALLÉN, PEDRO, *Catálogo de la Biblioteca de Salvá enriquecido con la descripción de otras muchas obras*, 2 vols. (Valencia, 1872)

SAN PEDRO, DIEGO DE, *Obras completas*, ed. Keith Whinnom and (vol. iii only) Dorothy S. Severin, 3 vols. (Madrid, 1971–9)

Sánchez Herrero, José, 'Centros de enseñanza y estudiantes de Sevilla durante los siglos XIII al XV', in Miguel Ángel Ladero Quesada (ed.), *Estudios dedicados al profesor D. Ángel Ferrari Núñez,* En la España medieval, 4, 2 vols. (Madrid, 1984), ii. 875–98

Sander, Max, *Le Livre à figures italien depuis 1467 jusqu'à 1530: essai de sa bibliographie et de son histoire,* 5 vols. (Milan, 1942) and a supplement by Carlo Enrico Rava (Milan, 1969)

Santander Rodríguez, Teresa, *Hipócrates en España: siglo XVI* (Madrid, 1971)

Saunders, A. C. de C. M., *A Social History of Black Slaves and Freedmen in Portugal 1441–1555* (Cambridge, 1982)

Sayle, Charles, 'Initial Letters in Early Printed Books', *Transactions of the Bibliographical Society,* 7 (1902–4), 15–47

Scammell, G. V., 'Shipowning in the Economy and Politics of Early Modern England', *The Historical Journal,* 15 (1972), 385–407

Schäfer, Ernesto, 'La Casa de la Contratación de las Indias de Sevilla, durante los siglos XVI y XVII', *Archivo hispalense,* 2nd ser., 5, No. 13 (1945), 149–62

—— 'La Universidad de los Mareantes de Sevilla', *Archivo hispalense,* 2nd ser., 5, No. 14 (1945), 271–85

Schiff, Mario, *La Bibliothèque du Marquis de Santillane,* Bibliothèque de l'École des Hautes Études: Sciences historiques et philologiques, 153 (Paris, 1905)

Scholderer, Victor, *Johann Gutenberg, The Inventor of Printing,* 2nd edn. (London, 1970)

Selig, Karl Ludwig, 'A German Collection of Spanish Books', *Bibliothèque d'humanisme et Renaissance,* 19 (1957), 51–79 [referred to as 'Selig, Munich' in App. I]

Serís, Homero, *Nuevo ensayo de una biblioteca española de libros raros y curiosos, formado en presencia de los ejemplares de The Hispanic Society of America en Nueva York y de la Ticknor Collection en la Biblioteca Pública de Boston,* vol. i (A-B) (New York, 1964)

Serrano y Sanz, M., 'Noticias biográficas de Fernando de Rojas, autor de *La Celestina,* y del impresor Juan de Lucena', *Revista de archivos, bibliotecas y museos,* 6 (1902), 245–99

Sierra Corella, Antonio, *La censura de libros y papeles en España y los índices y catálogos españoles de los prohibidos y expurgados* (Madrid, 1947)

Simón Díaz, José, *Bibliografía de la literatura hispánica,* 12 vols. to date (1950–) [referred to as '*BLH*' in App. I]

—— *El libro español antiguo: análisis de su estructura,* Teatro del Siglo de Oro, bibliografías y catálogos, 1 (Kassel, 1983)

Stark, Lewis Morgrage, 'Bibliographical Notes: Americana', *Papers of the Bibliographical Society of America,* 34 (1940), 87

Steinberg, S. H., *Five Hundred Years of Printing,* 3rd edn., rev. James Moran (Harmondsworth, 1974)

Stols, Alexandre A. M., *Antonio de Espinosa, el segundo impresor mexicano,* 2nd edn. (Mexico City, 1964)

Strauss, Gerald, *Nuremberg in the Sixteenth Century* (New York, 1966)

TELLECHEA IDÍGORAS, J. IGNACIO, 'Biblias publicadas fuera de España, secuestradas por la Inquisición de Sevilla en 1552', *Bulletin hispanique*, 64 (1962), 236–47

TENORIO, NICOLÁS, 'Algunas noticias de Meinardo Ungut y Lanzalao Polono', *Revista de archivos, bibliotecas y museos*, 5 (1901), 633–8

THOMAS, SIR HENRY, 'The Output of Spanish Books in the Sixteenth Century', *The Library*, 4th ser., 1 (1920–1), 69–94

—— *Short-Title Catalogue of Books Printed in Spain and of Spanish Books Printed Elsewhere in Europe Before 1601 now in the British Museum* (London, 1921) [referred to as 'Thomas' in App. I]

—— 'The English Translations of Guevara's Works', in Lucio Gil Fagoaga and Gerhard Moldenhauer (eds.), *Estudios eruditos in memoriam de Adolfo Bonilla y San Martín*, 2 vols. (Madrid, 1927–30), ii. 565–82

—— 'Antonio [Martínez] de Salamanca, Printer of *La Celestina*, Rome *c.*1525', *The Library*, 5th ser., 8 (1953), 45–50

—— (ed.) *The Crafty Farmer: A Spanish Folk-Tale Entitled: How a Crafty Farmer with the Advice of his Wife Deceived Some Merchants*, tr. with an introduction by Henry Thomas (London, 1938)

THOMPSON, LAWRENCE S., *Essays in Hispanic Bibliography* (s.l., 1970)

TORRE REVELLO, JOSÉ, *Los orígenes de la imprenta en la América española* (Madrid, 1927)

—— *Orígenes de la imprenta en España y su desarrollo en América española* (Buenos Aires, 1940)

UPDIKE, DANIEL BERKELEY, *Printing Types: Their History, Forms, and Use*, 2nd edn., 2 vols. (Cambridge, Mass., 1937)

VALLS I SUBIRÀ, ORIOL, *The History of Paper in Spain*, tr. Sarah Nicholson, 3 vols. (Madrid, 1978–82)

VALTÓN, EMILIO, *Impresos mexicanos del siglo XVI* (Mexico City, 1935)

VÁZQUEZ DE PRADA, VALENTÍN, 'La Inquisición y los libros sospechosos en la época de Valdés-Salas (1547–1566)', in J. Ignacio Ruiz de la Peña *et al.*, *Simposio "Valdés-Salas", Oviedo, 8–11 diciembre 1968* (Oviedo, 1970), pp. 147–55

—— *Los siglos XVI y XVII* (Madrid, 1978), vol. iii of id. (ed.), *Historia económica y social de España* (Madrid, 1973–)

VEGA GONZÁLEZ, JESUSA, *La imprenta en Toledo: estampas del Renacimiento* (Toledo, 1983)

VERMASEREN, B. A., 'De Spaanse Uitgaven op Godsdienstig Gebied van M. Nutius en J. Steelsius: een voorlopige oriëntatie', *De Gulden Passer*, 50 (1972), 26–99

VERVLIET, HENDRIK D. L., *Sixteenth-century Printing Types of the Low Countries* (Amsterdam, 1968)

—— 'Printing in Fifteenth Century Antwerp', in Simon van Venlo, *Little Book of the Office or Service of the Mass: Boexken van der Officien ofte Dienst der Missen*, ed. Ludo Simons, 2 vols. (Antwerp, 1982), ii. 43–9

VILAR, PIERRE, *A History of Gold and Money 1450–1920*, tr. Judith White, 2nd English edn. (London, 1984)

VINDEL, FRANCISCO, *Manual gráfico-descriptivo del bibliófilo hispano-americano (1475-1850)*, 12 vols. (Madrid, 1930–4)

—— *Escudos y marcas de impresores y libreros en España durante los siglos XV a XIX (1485–1850)*, 2 vols. (Barcelona, 1942; Madrid, 1950)

—— *El arte tipográfico en España durante el siglo XV*, 9 vols. and an appendix (Madrid, 1945–54)

—— *El primer libro impreso en América fue para el rezo del Santo Rosario (Méjico, 1532–34)* (Madrid, 1953), and the *Apéndice* to that work (Madrid, 1954)

—— *Réplica en 'Carta abierta' de Francisco Vindel al dictamen emitido por la Academia Mexicana de la Historia correspondiente de la Real de Madrid sobre su descubrimiento e investigación de 'El primer libro. . .'* (Madrid, 1954)

—— *Mapas de América en los libros españoles de los siglos XVI al XVIII (1503–1798)* (Madrid, 1955)

—— *En papel de fabricación azteca fue impreso el primer libro en América (apuntes que comprueban la falta de veracidad en un dictamen de la Academia Mexicana de la Historia)* (Madrid, 1956)

VINDEL, PEDRO, *Repertorio de libros raros, curiosos y antiguos que se hallan de venta en esta casa* (Madrid, 1926)

VOET, LEON, 'Production and Sales Figures of the Plantin Press in 1566', in S. van der Woude (ed.), *Studia Bibliographica in Honorem Herman De La Fontaine Verwey* (Amsterdam, 1967), pp. 418–36

—— *The Golden Compasses: A History and Evaluation of the Printing and Publication Activities of the Officina Plantiniana at Antwerp*, 2 vols. (Amsterdam, 1969–72)

WAGNER, C. P. (ed.), *El Libro del cavallero Zifar* (Ann Arbor, 1929)

WAGNER, HENRY R., *Nueva bibliografía mexicana del siglo XVI: suplemento a las bibliografías de Don Joaquín García Icazbalceta, Don José Toribio Medina y Don Nicolás León* (Mexico City, 1940)

—— 'Early Silver Mining in New Spain', *Revista de historia de América* [Mexico City], 14 (1942), 49–71

WAGNER, KLAUS, 'Apuntes para el coste de vida en Sevilla: agosto 1544—febrero 1545', *Archivo hispalense*, 2nd ser., 55, No. 170 (1972), 119–30

—— La *Suma de geografía* del bachiller Martín Fernández de Enciso: privilegio, honorarios del autor, tirada', *Estudios y notas* [Madrid], 24 (1972), 67–9

—— 'Los impresores sevillanos Estacio y Simón Carpintero', *Archivo hispalense*, 2nd ser., 58, No. 178 (1975), 135–42

—— '¿Qué costaron los estudios universitarios en Salamanca a principios del siglo XVI?', *Archivo hispalense*, 2nd ser., 59, No. 180 (1976), 149–51

—— 'El negocio de las *Siete Partidas*', *La Bibliofilia*, 78 (1976), 67–82

—— 'La biblioteca del Dr. Francisco de Vargas, compañero de Egidio y Constantino', *Bulletin hispanique*, 78 (1976), 313–24

—— *Regesto de documentos del Archivo de Protocolos de Sevilla referentes a judíos y moros* (Seville, 1978)

—— *El doctor Constantino Ponce de la Fuente: el hombre y su biblioteca* (Seville, 1980)

—— 'Erasmo en el comercio librero sevillano', in Gerhard Schmidt and Manfred Tietz (eds.), *Stimmen der Romania: Festschrift für W. Theodor Elwert zum 70. Geburtstag* (Wiesbaden, 1980), pp. 451–6

—— 'La reforma protestante en los fondos bibliográficos de la Biblioteca Colombina', *Revista española de teología,* 41 (1981), 393–463

—— *Martín de Montesdoca y su prensa: contribución al estudio de la imprenta y de la bibliografía sevillanas de siglo XVI* (Seville, 1982)

—— 'La Giralda como marca de impresor (a propósito del *Missale Hispalense* de 1558)', in T. Falcón Márquez (ed.), *Homenaje al profesor Dr Hernández Díaz* (Seville, 1982), i. 269–73

—— 'Una impresión mal conocida de Juan Cromberger. A propósito del único ejemplar del *Exercitatorio de la vida espiritual,* Sevilla, 1534', *Cuadernos bibliográficos* [Madrid], 44 (1982), 1–8

WERNER, THEODOR GUSTAV, 'Zur Geschichte Tetzelscher Hammerwerke bei Nürnberg und des Kupperhüttenwerks Hans Tetzels auf Kuba', *Mitteilungen des Vereins für Geschichte der Stadt Nürnberg,* 55 (1967–8), 214-25

WHINNOM, KEITH, 'The Problem of the "Best-Seller" in Spanish Golden-Age Literature', *Bulletin of Hispanic Studies,* 52 (1980), 189–98

WILSON, EDWARD M., and CRUICKSHANK, DON W., *Samuel Pepys's Spanish Plays* (London, 1980)

WITTEN, LAURENCE, 'The Earliest Books Printed in Spain', *Papers of the Bibliographical Society of America,* 53 (1959), 91–113

WOODFIELD, DENIS B., *Surreptitious Printing in England 1550–1640* (New York, 1973)

YCIAR, JUAN DE, *Recopilacion subtilissima intitulada orthographia practica* (Zaragoza, 1548)

ZAVALA, SILVIO, 'La amalgama en la minería de Nueva España', *Historia mexicana,* 11 (1961–2), 416–21

INDEX OF EDITIONS PRINTED BY THE CROMBERGER PRESS AT SEVILLE

The numbers in round brackets refer to the numbering of the bibliographical descriptions in Appendix One; this index does not contain references to all the authors whose works are included in collections like the *cancioneros*, or *pliegos sueltos*.

GENERAL INDEX